Algorithms and Computation in Mathematics • Volume 2

Editors

Manuel Bronstein   Arjeh M. Cohen
Henri Cohen   David Eisenbud
Bernd Sturmfels

Springer
Berlin
Heidelberg
New York
Hong Kong
London
Milan
Paris
Tokyo

Wolmer V. Vasconcelos

Computational Methods in Commutative Algebra and Algebraic Geometry

With Chapters by David Eisenbud,
Daniel R. Grayson, Jürgen Herzog
and Michael Stillman

With 11 Figures

 Springer

Wolmer V. Vasconcelos

Rutgers University
Department of Mathematics
Hill Center for the
Mathematical Sciences
New Brunswick, NJ 08903, USA

e-mail:
vasconce@rings.rutgers.edu

David Eisenbud

MSRI Mathematical Sciences
Research Institute
1000 Centennial Drive
Berkeley, CA 94720, USA

e-mail:
de@msri.org

Daniel R. Grayson

University of Illinois
at Urbana-Champaign
Department of Mathematics
1409 W. Green St.
Urbana, IL 61801, USA

e-mail:
dan@math.uiuc.edu

Jürgen Herzog

Universität Essen
FB 6 Mathematik und Informatik
Universitätsstrasse 2
41141 Essen, Germany

e-mail:
Juergen.Herzog@uni-essen.de

Michael Stillman

Cornell University
Department of Mathematics
Ithaca, NY 14853, USA

e-mail:
mike@math.cornell.edu

1st ed. 1998. Corr. 2nd printing 1999, 3rd printing 2004

Mathematics Subject Classification (2000): 13-XX, 13-01, 13D02, 13H10, 13P10, (secondary:13D40, 13H15, 13F20); 14-XX, 14-02, 14-04, 14B15; 16-XX,16-04, 116E10; 18-XX, 18G10, 18G20

Library of Congress Control Number: 2004104811

ISSN 1431-1550

ISBN 3-540-21311-2   Springer-Verlag Berlin Heidelberg New York

Springer-Verlag is a part of Springer Science+Business Media

springeronline.com

© Springer-Verlag Berlin Heidelberg 1998
Printed in Germany

Typeset by the author using a Springer LATEX macro package
Cover design: *design & production* GmbH, Heidelberg

Printed on acid-free paper   46/3142db - 5 4 3 2 1 0-

This is for Aurea and in fond memory of Zindinha

Preface

The interplay between computation and many areas of algebra is a natural phenomenon in view of the algorithmic character of the latter. The existence of inexpensive but powerful computational resources has enhanced these links by the opening up of many new areas of investigation in algebra. At the same time it made available the theoretical tools of this area of mathematics to help deal with problems of interest in physics, engineering and other disciplines. We aim here to discuss how certain devices that permit the rapid processing of polynomials and matrices make it possible to examine parts of two areas of algebra – commutative algebra and algebraic geometry – where those data structures play critical roles.

Among the main tasks in computational algebra are the constructions of decompositions and closures of objects in the ring of polynomials. Among the former are finding primary decompositions and modules of syzygies, and among the latter, the computation of integral closures and of rings of invariants. As a rule, they are assisted by any *a priori* knowledge available. Another frequent task is to certify that a given object has a certain property.

This book is an attempt to deal with these issues, despite the pace of development in the field (ring?). The material was drawn mostly from the published literature, both classical and recent, including conference proceedings on computer algebra that tended to focus on algebraic structures. There are now several sources where some of this material can be also found. Basic textbooks dealing with Gröbner basis theory and the role of Buchberger's algorithm are [AL90], [BKW93] and [CLO92], each emphasizing different aspects of the theory. Although dedicated to commutative algebra proper, [Ei95] must be added to this list as it contains big chunks of Gröbner basis stuff; it is also highly recommended for its foundational and motivational aspects. Other sources of methods are the manuals and accompanying tutorial notes of computer algebra systems.

An important aspect of the text are several unsolved problems scattered throughout. In a manner of speaking, most of the proposed methods are to be considered preliminary, in need of improvement.

A great deal of the impetus for the writing of this book came from sets of lectures given at the Encuentro Latinoamericano de Algebra y Geometria Algebraica,

Guanajuato, Mexico, and at the Workshop on Commutative Algebra, International Centre for Theoretical Physics, Trieste, Italy, both in the summer of 1992, and at the Summer School for Commutative Algebra, Barcelona, Spain, in the summer of 1996. We are grateful to the organizers of the meetings for the opportunities these events provided. In many ways this is a much expanded account of two earlier reports [Vas93], [Vas91] and particularly Chapter 10 of [Vas94], from which it inherited its interest in blowup algebras.

It was the author's original intention to write a text at the level that required [AM69] for its main prerequisite. Although it remains adequate for lots of the material here, it is necessary to bring in dosages of applied homological algebra at various critical junctions. Appendix A is an overview, mostly with full proofs, of the commutative algebra required as background in addition to a textbook at the level of [AM69]. The reader may be better served by looking up her/his own favorite reference.

We are thankful to many friends and colleagues from whom we learned the stuff. That the understanding of the subject shown here is not very thorough is only partly their fault. The number of colleagues with whom we engaged in useful conversations and discussions is too large to be all listed here but Alberto Corso, Jeff Dawson, Craig Huneke, Teo Mora, Ezra Miller, Lorenzo Robbiano, Aron Simis, Bernd Sturmfels, Moss Sweedler, Bernd Ulrich and Rafael Villarreal deserve special mention for they influenced the organization and choice of topics. We also benefited from comments by readers and anonymous reviewers of earlier drafts.

We are particularly grateful to David Eisenbud and Jürgen Herzog for permitting the inclusion in this text of their illuminating discussions of computation of cohomology and Hilbert functions, and to Daniel Grayson and Michael Stillman, the developers of *Macaulay 2*, who along with David Eisenbud wrote an introduction to their software, which we hope will be useful to other readers. (Current versions of *Macaulay 2*, for several platforms, can be found at its homepage (http://www.math.uiuc.edu/~Macaulay2)).

The author is keenly aware of some unbalance in this book, with topics ranging from freshman calculus to local cohomology, without enough detail of each and jumping over many topics in between! There is also a degree of obsession in highlighting simple, but explicit formulas, that are felt to represent models of efficiencies. It is only hopeful that the given references point to missing scholarly discussion of the topics.

The National Science Foundation also has our heartfelt thanks for the support of the research that went into this text.

It is a certain event that errors will be found in the text. To atone for this, we will maintain in the author's homepage (http://www.math.rutgers.edu/~vasconce) an updated list of corrections and complements.

August 12, 1997
New Brunswick, New Jersey   WOLMER V. VASCONCELOS

Note on present printing: Several typos and errors that were pointed out to the author are corrected. Publication data on the original bibliography is updated, and a listing of related references is upended.

Contents

Introduction

Lua, gosto de te ver assim,
desmistificada, desmetaforizada, satélite.
Manuel Bandeira

Large scale computations in algebraic geometry and commutative algebra are circumscribed by an intrinsic intractability, due to the worst case non-elementary complexity of several of the fundamental algorithms which are involved. It is a panorama in which the processing of every medium sized problem is expected to lie on the brink of combinatorial explosion. That many large computations do go through, particularly when the data has an interesting embedded mathematical structure, is surprising. It is a shifting terrain which we must learn to navigate.

This state of affairs demands a continued refinement of techniques to ensure that the cost of computation be borne largely by theoretical means. To inform a computation of mathematical knowledge constitutes a most challenging problem. To endow the algorithms with this kind of intelligence remains a formidable task.

Our aim is to describe theoretical and practical issues in some basic constructions in algebraic geometry and commutative algebra, and how in turn they may affect effective implementation by symbolic computation programs. Gröbner basis and factorization algorithms play fundamental roles here. They form the tools that mediate between the theoretical steps of the several constructions.

There will not be a great deal of overlap with the details and methods of the constructions discussed in two classical papers, [Her26] and [Sei74]. These authors were very thorough in their analyses of the constructions, and of the cost of carrying them out. They were based however on elimination theory, which tended to view each and every problem under a worst case scenario. As an approach, it leads to the overuse of generic changes of variables, which in turn may create huge polynomials, when one wants to carry out most computations in the natural variables of the problem. Another serious difficulty with these methods is that they lack programmability. The approach here is in scope less comprehensive, but in counterpoint more dynamic, viewing each problem through its particular features. This lack of universality is an obvious drawback, which in turn makes the challenge of finding the solution pathways much more interesting.

We should emphasize here our usage of the word *method*: It is meant as describing algorithmic pathways to various constructions but without necessarily the specification of the nifty algorithms required. The constructions that we are going to deal with are those for radicals of ideals, with steps towards a facilitation of full primary decomposition, the integral closure of a ring, general properties of morphisms, primality and flatness testing, the setting up of ideal transform computations, and the computation of cohomology. These are among the items in a wish list of constructions one would like to do in a direct and efficient manner. They lie between more primitive operations closer to the fundamental algorithms (factorization algorithms, Hilbert functions, syzygies, bits of elimination theory, etc.) and the finer properties of the algebraic objects. They provide approaches to elements of the computation of rings of invariants of connected groups and the solution of systems of equations with finitely many zeros.

There are many related problems where no significant headway has been made. One that comes to mind is that of deciding whether an affine domain is factorial. Already the case of a hypersurface ring, $A = k[x_1, \ldots, x_n]/(f)$, presents great difficulties. General theoretical approaches may even be lacking!

As a matter of general strategy, the constructions are mediated through the funnel of homological algebra, and can be characterized as successive layers of syzygy computations, complemented by factorization methods. This has an advantage that often one can supply explicit formulas for the objects to be computed, which might lead to predictions. There is a great deal of emphasis on generating data that may be used for symbolic/numeric equation solvers.

A method that pervades many of the techniques we will discuss is that of the Hilbert function $H_A(\mathbf{t})$ of a graded ring A. Its computation and usages is a bright spot on a dark landscape. It becomes available – via the truly nice theorem of Macaulay and the efficiency in the processing of monomial ideals – once a Gröbner basis has been determined, or even more generally, when the initial ideal of the Gröbner basis has been computed, or by indirection from relationships given by exact sequences with other modules. The thrust in the study of an algebra A becomes how to uncover some of its geometric and algebraic properties that are reflected on its Hilbert function $H_A(\mathbf{t})$. An extensive examination by Jürgen Herzog details general, basic and subtle points of the theory of Hilbert functions. It is a guide to our use of this tool.

The reader will not find here counts for the complexity of the proposed constructions. In fact, this is an area undergoing a period of intense activity that calls for a text by an expert. There are other reasons for the omission, the main one being the near impossibility of estimating the complexity of Gröbner basis or Wu–Ritt computations of ideals whose generators have many relationships. Nevertheless, close attention was paid to the known optimizations of these computations. The reader will also not find many explicit examples here: The sizes of the output precluded a generous sprinkling. Nevertheless, a couple of large examples with surprisingly brief outputs are discussed. A few illustrative examples are included.

The following lists the main topics we will discuss:

- Gröbner Basis Techniques
- Integral and Flat Morphisms
- Cohen–Macaulay Algebras
- Primary Decomposition
- Primality Testing
- Nullstellensätze
- Computing in Artin Algebras
- Algebraic Equation Solvers
- Hybrid Solvers
- Integral Closure of Rings, Ideals and Morphisms
- Ideal Transforms
- Rings of Invariants
- Computation of Cohomology

This is obviously a very meager catalog, short of a satisfying wish list. Some of the topics missing include a full examination of the computation of rings of invariants, the theory of resultants, and automated theorem proving, each of which is deserving a full report on itself. Fortunately, a recent book by Sturmfels [Stu93] addresses many aspects of computational invariant theory. We will limit ourselves to a general discussion, computation issues in special cases and the description of a new algorithm for reductive groups ([DK97]).

We will now outline the contents of the book. Each chapter being highlighted by the treatment of one of the topics above, we limit ourselves to a brief discussion on the choices and the relationships they bear to one another.

The first chapter is a fast paced review of orderings in rings of polynomials and Gröbner basis techniques. They stress the connection between the Buchberger algorithm and the computation of modules of syzygies. It ends with comments about the implementation of these methods in several current symbolic computation systems.

Chapter 2 assembles a collection of various operations in commutative algebra, particularly those that impact on the homomorphisms between commutative rings. The constructions of ideal quotient and ring of endomorphisms are singled out for special treatments as they are ubiquitous in the text. There are discussions on carrying out of Noether normalizations and the testing of flatness and of Cohen–Macaulayness.

Chapter 3 deals with a myriad of issues related to primary decomposition of ideals. The task is viewed as taking place in two steps. The first consists of methods to decompose an ideal into an intersection of ideals of different codimensions. We appeal repeatedly to homological algebra, particularly the theory of local duality, to extract these equidimensional components. It is, in some sense, rather naive, since it takes place in the context of the Artin–Rees lemma but does not seek to predict upper bounds on the indices involved in the lemma. Furthermore, it requires of the program the ability to compute the basic derived functors.

The second task, finding the primary decompositions of equidimensional ideals, is much harder. We consider several reductions to the zero-dimensional case, but steer away from methods that involve factorization of polynomials over function fields of positive transcendence degree over the base field. The actual discussion of zero-dimensional decomposition is pushed forward to the next chapters.

The next two chapters are dedicated to various constructions to determine the radical of an ideal.

Chapter 4, dealing with decomposition in Artin algebras, has for goal finding idempotents, or one of its surrogates, nontrivial zero divisors. They both serve the purpose of effectively breaking up the algebra into smaller factors. We also seek to factor in methods grounded on Berlekamp's trick. The other aim of this chapter is to provide an interface between symbolic and numerical solvers. Basically we seek the means to prepare sets of equations in a manner that may be useful to root finders.

The other topic of this chapter has a practical character: To generate a radical, zero–dimensional ideal by a set of elements whose cardinality is exactly the number of variables. Numerical equation solvers usually work out of this setting. In principle, at the cost of forming generic linear combinations, this is always possible. This road is not pursued since it is much cheaper to attempt to achieve the same goal on a closely related ideal in a ring with one more indeterminate. We present one such trick inspired by K-theory.

In Chapter 5, we exploit systematically various relationships between an ideal and its attached Jacobian ideals. It is here that major efficiencies are to be had. To be efficient, it requires the ability to find regular sequences of appropriate lengths repeatedly. We describe some current approaches to this fundamental issue of computational algebra. Among the applications, we examine the question of extracting the isolated zeros of a system of polynomial equations.

Three main aspects of the integral closure are treated in Chapter 6: the integral closure of an affine ring, of an ideal and of a morphism. The first of these has a long history and has seen many different approaches. The path we chose passes through the theory of the canonical module (in the theory of local duality) together with Jacobian methods, in order to benefit from the results developed in Chapter 5. The integral closure of ideals has many uses in the theoretical literature and presents tantalizing computational opportunities.

Chapter 7 is an account on the elements that may help to set up ideal transforms. The topic warrants attention because it is an element in the computation of rings of invariants. It has the drawback of being not always terminating, so it requires lots of theoretical attention to details. Its more direct methods discuss the processing of subrings, particularly of rings of invariants.

Chapter 8, by David Eisenbud, details his beautiful approach to the computation of the cohomology of projective schemes.

The last chapter has a more theoretical flavor and consists in an examination of various complexities of an algebra A over a field k, in other words, measures of comparison to a polynomial ring over k. Here we bring to the fore the reduction number of a graded algebra A, and study its relationship to the several degrees defined for A: Castelnuovo regularity, arithmetic and geometric degrees and a novel family of cohomological degrees. Some light is also shone on how to look at the exponent in the Nullstellensatz.

In Appendix A, we collect several cornerstones and some of the basic navigational tools of Commutative Algebra, with proofs assembled from many sources. It is to be used when the reader is without her/his favorite handbook.

Another contribution that enriches the text considerably is Appendix B, where Jürgen Herzog gives an account, tailored to our needs, of the theory of Hilbert functions.

Appendix C, by David Eisenbud, Daniel Grayson and Michael Stillman, is a hands-on introduction to *Macaulay 2*. Carefully it tells the reader to use M2 to begin exploring algebras, modules and many of the geometric structures they support.

1

Fundamental Algorithms

The settings for computations that we will consider are rings of polynomials

$$R = R_n(k) = k[x_1, \ldots, x_n],$$

where k is a finite field or a finite extension of $\mathbb{Q}$, or in a few cases, rings where coding can be done as efficiently as with those basic fields. The problems themselves are concerned with affine rings over k – and are therefore adequately modeled by an ideal of some $R_n(k)$ – or by a subring of $R_n(k)$:

- R/I, $I \subset R$
- $k[f_1, \ldots, f_m]$, $f_i \in R$.

In several circumstances, to ascertain properties of individual elements of the rings, it is convenient to have them represented as matrices over simpler rings. The processing itself is based on division algorithms, Buchberger's algorithm in the case of R/I and SAGBI bases for subrings of R.

This chapter gives a bird's eye view of the Gröbner basis methods that undergird large sectors of computational algebra. They are often used in conjunction with more classical tools, such as the theory of resultants and factorization algorithms. There is also a modern theory of elimination that addresses the issue of sparsity very closely, but it will not be examined here. In Chapter 7, we consider methods tailored for the direct processing in subrings of rings of polynomials. The topics treated are:

- Polynomial rings and their orderings
- Division algorithms
- Buchberger algorithm
- Computation of syzygies
- Computation of Hilbert functions

The treatment here is only intended to sketch out basic concepts and algorithms and point out its capabilities, focusing instead on the interface between the algorithms and algebra itself. It will become clear that this interfacing takes place over an open set not just a thin layer of activities.

There are many excellent sources that deal with these topics and greater detail and depth: [AL90], [BKW93], [CLO92] and [Ei95] are our favorites. The entry gate to most of the applications is the ability of several computer algebra systems to compute syzygies. They are mentioned briefly at the end of the chapter with instructions on how some can be obtained.

1.1 Gröbner Basics

Division algorithms are key tools for processing in rings of polynomials. The most straightforward of these is probably *pseudo–division*. It consists in a minor modification of ordinary long division of polynomials in one variable with coefficients in a field. For more general coefficients, it works as follows: Let $f(x)$ and $g(x)$ be elements of $R[x]$,

$$f(x) = a_r x^r + \cdots + a_0, \quad a_r \neq 0.$$

If $\deg g(x) = s \geq r$, then there are polynomials $q(x), p(x)$ such that

$$a_r^{s-r+1} g(x) = p(x)f(x) + q(x), \quad \deg q(x) < \deg f(x). \tag{1.1}$$

Despite its simplicity, it has many theoretical applications in commutative algebra: see the treatment of the *Nullstellensatz* in [Kap74] or Theorem 5.37. It is also the working horse in the algorithm of Ritt–Wu ([GM91], [Rit50], [Wu78]).

Our path however will take us in a different direction, the theory of *Gröbner bases*[1]. This division method has had a profound impact on computer algebra. The discussion will be focused on its uses, very little being said about costs and actual implementations.

Several recent textbooks have examined these devices minutely and with a great deal of depth. There is also a growing and rich literature on various extensions of this theory from fields to rings, in a manner of speaking. Some of these sources are found, by indirection, in the bibliography listed. A regretful omission is any examination of the geometry of Gröbner basis theory. But this is easy to remedy, as the survey of Bayer and Mumford [BM93], with its rich vistas and eloquence, has appeared.

Polynomial Rings, Monomials, Orderings and Weight Vectors

Let k be a field, and let R be the polynomial ring $k[x_1, \ldots, x_n]$. Suppose I is an ideal of R given by a set $\{f_1, \ldots, f_m\}$ of generators. The study of the ring R/I is helped by the knowledge of canonical bases for the k–vector space R/I. The purpose of division algorithms in R is to provide us with such bases. We now examine the elements for such algorithms.

Let us fix a ring of polynomials $R = k[x_1, \ldots, x_n]$ over a field k. Denote by $\mathbb{M}$ the set of all monomials

---

[1] The terminology *Gröbner basics* is the felicitous expression by Bernd Sturmfels to denote the myriad issues associated with the computation of these sets.

$$\mathbf{x}^\alpha = x_1^{\alpha_1} \cdots x_n^{\alpha_n} \qquad (1.2)$$

(including 1). $\mathbb{M}$ is a multiplicative monoid isomorphic to the additive monoid $\mathbb{N}^n$.

Given an element $f \in R$, it is written

$$f = \sum_{\alpha \in \mathbb{N}^n} c_\alpha \mathbf{x}^\alpha \qquad (1.3)$$

in a manner that facilitates the processing under multiplication or division. This is usually achieved by picking orders on $\mathbb{M}$ that are compatible with multiplication.

Definition 1.1. *An* admissible partial order *T is a partial order $>_T$ on $\mathbb{M}$ with the property*

- *$m >_T 1$ for any non constant monomial m;*
- *If $m_1 >_T m_2$ and $m_3 \in \mathbb{M}$ then $m_1 \cdot m_3 >_T m_2 \cdot m_3$.*

If $>_T$ is a total order we say that it is a term *order (or* term ordering *or even a* monomial ordering*).*

A basic example of a term ordering is the *lexicographic* order (*lex* for short):

$$m_1 = x_1^{a_1} \cdots x_n^{a_n} >_{lex} x_1^{b_1} \cdots x_n^{b_n} = m_2$$

if

$$a_1 = b_1, \cdots, a_{r-1} = b_{r-1}, a_r > b_r, \text{ for } 1 \le r \le n.$$

Product of Orderings

More general term orderings arise by combining several admissible partial orders through their lexicographic product of orderings. If $T_1, \ldots, T_s$ are such partial orders, the product order

$$T = T_1 \times_{lex} T_2 \times_{lex} \cdots \times_{lex} T_s \qquad (1.4)$$

is defined as above

$$m_1 >_T m_2 \iff m_1 =_{T_1} m_2, \ldots, m_1 =_{T_{r-1}} m_2, m_1 >_{T_r} m_2, \text{ for } 1 \le r \le s.$$

Degree Orderings

Among term orderings noteworthy are those that place emphasis on the degrees of the polynomials. They are obtained as the product of Deg, the total degree partial order, and T_2 another partial order. For instance, if T_2 is *lex*, their product is the so–called graded lexicographic ordering:

$(a_1, \ldots, a_n) < (b_1, \ldots, b_n) \iff$
first nonzero entry of $(\sum b_i - \sum a_i, b_1 - a_1, \ldots, b_n - a_n)$ is positive.

Particularly striking properties are enjoyed by the *reverse lexicographic order*, defined by changing the last requirement above to: the last nonzero entry is negative. Macaulay introduced it in his fundamental studies on Hilbert functions ([Mac27]). Bayer and Stillman [BSa87] have discovered many of its interesting properties and incorporated its efficiencies into their *Macaulay* program.

As these examples already indicate, it is necessary to consider more general partial orderings of $\mathbb{M}$ as constituent blocks for term orderings. A simple mechanism is to embed the monoid $\mathbb{N}^n$ into a real vector space V: each element $w \in V^*$ (the dual of V) induces a partial order on $\mathbb{M}$, compatible with its composition law, by

$$\mathbf{x}^{\mathbf{a}} < \mathbf{x}^{\mathbf{b}} \Leftrightarrow w(\mathbf{a}) < w(\mathbf{b}). \tag{1.5}$$

We refer to w as a *weight vector*. A result of Robbiano ([Ro85]) asserts that every compatible partial ordering of $\mathbb{M}$ is realized in this manner, and that every term ordering is the lexicographic product of at most n such orderings. The study of all possible orderings is subsumed in the theory of the Gröbner fan of Mora and Robbiano [MRo88].

Initial Ideals

Orderings are the means to pass back and forth between the monoid of all monomials and $\mathbb{N}^n$. We define the homomorphism

$$\log : \mathbb{M} \longrightarrow \mathbb{N}^n,$$

by

$$\log(x_1^{a_1} \cdots x_n^{a_n}) = (a_1, \ldots, a_n).$$

Let us fix, for our discussion, a term order which we denote simply by $>$.

If $f = \sum_{\alpha} c_{\alpha} \mathbf{x}^{\alpha} \in R$, the *support* of f is the subset of $\mathbb{N}^n$

$$\mathrm{supp}(f) = \{\, \alpha \mid c_{\alpha} \neq 0 \,\},$$

while the *Newton polytope* of f is the convex hull of $\mathrm{supp}(f)$.

If $0 \neq f \in R$, $f = C(f) \cdot M(f) + \sum a_i M_i$, where $M(f)$ is the highest monomial that occurs in the representation of f; $0 \neq C(f)$ is its leading coefficient; the product $L(f) = C(f) \cdot M(f)$ is the *leading term* or *initial term* $\mathrm{in}(f)$ of f. We define $\log(f) = \log(M(f))$.

A more common notation is to denote the *leading coefficient* of f by

$$\mathrm{lc}(f) = c_{\beta}, \ \beta = \max\{\, \alpha \in \mathrm{supp}(f) \,\},$$

$\mathit{in}(f) = \mathbf{x}^{\beta}$ is the *initial monomial* and $\mathrm{lt}(f) = c_{\beta} \cdot \mathbf{x}^{\beta}$ is the *leading term* of f:

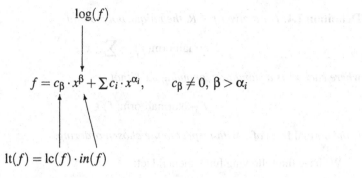

For a nonzero ideal I, define $\log(I)$ to be the union of $\log(f)$, $0 \neq f \in I$. This defines a sub–monoid of $\mathbb{N}^n$, stable under the addition of quadrants:

$$\log(I) = \log(I) + \mathbb{N}^n.$$

By the Hilbert basis theorem there are finitely many elements $g_1, \ldots, g_r$ in I so that

$$\log(I) = \bigcup (\log(g_i) + \mathbb{N}^n), \ 1 \leq i \leq r.$$

Definition 1.2. *The* initial ideal *of I for the term order $>$ is the ideal*

$$in_>(I) = (x^a, \text{ for all } a \in \log(I)). \tag{1.6}$$

The ideal $in_>(I)$ is highly dependent on the chosen ordering; once $>$ is fixed, the ideal is denoted simply by $in(I)$.

Definition 1.3. *The set $\{g_1, \ldots, g_r\}$ of elements of I is a* Gröbner basis *if*

$$in(I) = (in(g_1), \ldots, in(g_r)). \tag{1.7}$$

Normal Form Representation

We begin to use the notion of orderings for the study of arbitrary ideals. The discussion assumes the choice of a term order.

A first simple observation–but still of great significance–is that the images of the monomials

$$x^a \text{ where } a \in \mathbb{N}^n \setminus \log(I) = \Delta_I,$$

form a basis for the k–vector space R/I. They are the *standard monomials* associated to the Gröbner basis. The Gröbner basis G is said to be *reduced* if

$$\text{supp}(f - \text{lt}(f)) \subset \Delta_I, \quad \forall f \in G.$$

Definition 1.4. *For a given $f \in R$, the unique polynomial*

$$\text{NormalForm}(f) = \sum c_a x^a, \qquad (1.8)$$

where each x^a is a standard monomial, such that

$$f - \text{NormalForm}(f) \in I,$$

is the normal form *of f with respect to the chosen ordering.*

We have the following fundamental fact:

Theorem 1.5 (Macaulay Theorem). *Given an ideal I there exists a monomial ideal $in(I)$ such that the set $B = \{x^a \notin in(I)\}$ is a basis of R/I. More concretely, the mapping*

$$\boxed{\text{NormalForm}: R/I \longrightarrow R/in(I)} \qquad (1.9)$$

is an isomorphism of k-vector spaces.

1.2 Division Algorithms

The setting in this section is a ring of polynomials $R = k[x_1, \ldots, x_n]$ over a computable field. We sketch out Buchberger algorithm but leave a fuller treatment to excellent dedicated sources; see [AL90], [BKW93], [CLO92].

Syzygies

In order to inform the discussion, we recall a notion which gets involved practically in all stages of the processing of ideals.

Definition 1.6. *Let R be a ring and let I be an ideal generated by $F = \{f_1, \ldots, f_m\}$. The* syzygies *of the f_i's are the tuples $(r_1, \ldots, r_m) \in R^m$ such that*

$$\sum_{i=1}^{m} r_i f_i = 0.$$

The simplest of all syzygies are *Koszul syzygies*,

$$f \cdot g - g \cdot f = 0,$$

where f, g are elements of the ring R. A refinement is

$$A \cdot g - B \cdot f = 0,$$

if $f = A \cdot h$, $g = B \cdot h$, where h is a common divisor. For larger sets of elements this step will be further refined using the division that Gröbner bases permit.

Taken together the syzygies of the set F form a submodule of R^m, the *module of syzygies* of the f_i's. When another set of generators for I is chosen, the corresponding module of syzygies is closely related to the first in a manner originally observed by Fitting.

When the f_i's are monomials, the module of syzygies of F is generated by Koszul relations (see Theorem A.81 for a fuller explanation). For more general sets F, we are going to use this as a tool to get the syzygies of appropriate sets of generators of the ideal (F).

There are other notions of syzygies associated to the set F, of which we recall two. First, consider a homomorphism

$$\varphi : k[T_1, \dots, T_m] \mapsto R, \quad \varphi(T_i) = f_i.$$

$I = \ker(\varphi)$ is the ideal of *algebraic* syzygies of the f_i's. A more general notion, useful in the theory of blowup algebras (see Chapter 7), deals with mappings such as φ but with a different source:

$$\varphi : k[x_1, \dots, x_r, T_1, \dots, T_m] \mapsto R, \quad \varphi(T_i) = f_i,$$

where $k[x_1, \dots, x_r] \subset R$.

It is surprising that ultimately all these kinds of syzygies are going to be dealt with in the same manner.

Buchberger Algorithm

Here one seeks to divide a polynomial g by a finite collection $f_1, \dots, f_m$ of polynomials,

$$g = \sum_{i=1}^m h_i f_i + \text{remainder}, \tag{1.10}$$

in which 'remainder' has some appropriate minimizing property.

The following deceptively simple statement embodies the efficacy of Gröbner bases as the generating set of choice for an ideal of a polynomial ring.

Proposition 1.7. *Let I be an ideal of R and let $>$ be a term ordering of R. A set*

$$\{g_1, \dots, g_r\} \subset I,$$

is a Gröbner basis of I with respect to $>$ if and only if every nonzero element of I can be written as

$$f = \sum a_i g_i, \quad \text{with } \log(f) \geq \log(a_i g_i).$$

In particular, a Gröbner basis of I is a generating set for I.

The proof is contained in the very definition of the Gröbner basis. Note the close parallel with the Euclidean algorithm in the ring $k[t]$ and the elements of Gaussian elimination.

The previous proposition is a basis for solving several general questions about the ideal I, particularly the membership problem. There remains to find such bases. This results from the following analysis due to Buchberger (see [Buc85] for extensive details and related bibliography; see also [MM86] and [Ro86]).

Let I be defined by a generating set $F = \{f_1, \ldots, f_m\}$. One must have a criterion to decide whether $\log(F) = \{\log(f_1), \ldots, \log(f_m)\}$ generates $\log(I)$, and if not, a device to add new elements to the f_i's. These steps come together in the same argument.

Reduction

We begin with the observation on how to add a possible new generator to F. Let f be a nonzero element of I. If $in(f)$ is not a multiple of any of the $in(f_i)$'s, one has a new generator. However, even if $in(f)$ is already a multiple of a $in(f_i)$, f may still contribute a new generator. To see this, suppose that the leading monomial $M(f)$ of f is divisible by the leading monomial $M(f_i)$ of f_i, and pick $q \in R$ such that $\log(f - qf_i) < \log(f)$. It is also usual to effect this operation on the next largest monomial of f which does not belong to the span of the $M(f_i)$'s. On iterating we end up with an element

$$g = f - \sum_i a_i f_i,$$

with the property that $g = 0$ or $\log(g)$ is not divisible by $\log(F)$. In either case we say that f *reduces* to g relative to F.

If $g = 0$, f is ignored; otherwise adding $\log(g)$ to the submonoid of $\mathbb{N}^n$ generated by the $\log(f_i)$'s gives rise to a larger submonoid. The Hilbert basis theorem guarantees that such additions cannot go on forever.

S–resultant

The issue is how to pick appropriate elements of I. The basic step goes to the core of both the Euclidean and Gaussian algorithms. It is embodied in the notion of the (resultant) S–polynomial attached to two polynomials $f, g \in R$: If $M(f)$ and $M(g)$ are their leading monomials, set

$$S(f,g) := a_g \cdot f - (C(f)/C(g)) \cdot b_f \cdot g, \tag{1.11}$$

where $a_g \cdot M(f) = b_f \cdot M(g)$ is the least common multiple of $M(f)$ and $M(g)$. The collections of such objects have a very natural place in the theory of Taylor resolutions ([Tay66]).

The Buchberger algorithm is made up of the following result and the scheme that follows to produce the required elements.

Theorem 1.8. *A set of generators* $F = \{f_1,\ldots,f_m\}$ *of the ideal* I *is a Gröbner basis of* I *if and only if the S–polynomial* $S(f_i,f_j)$ *of each pair* (f_i,f_j) *of elements of* F *reduces to* 0 *with respect to* F.

Proof. The proof of the necessity is clear. For the converse, let f be an element of I. We may assume that f is its own normal form with respect to F. Let $f = \sum_j h_j f_j$, and consider the $\log(h_j f_j)$'s (for $h_j \neq 0$). $\log(f)$ cannot be equal to one of the $\log(h_j f_j)$, as it is already in normal form. This means that there must be some cancelling out at the top monomial occurring in the products $h_j f_j$. More precisely, suppose

$$M(h_1) \cdot M(f_1) = M(h_2) \cdot M(f_2) = \cdots = M(h_k) \cdot M(f_k)$$

are the top monomials that occur in the right hand side of the representation of f. Their cancelling out means that the vector of leading terms

$$(L(h_1),\ldots,L(h_k))$$

is a syzygy of

$$(L(f_1),\ldots,L(f_k)).$$

But it is an elementary fact that such relations are combinations of the syzygies of pairs $\{C(f_r)M(f_r), C(f_s)M(f_s)\}$. This means that we have a representation

$$f = \sum_j h'_j f_j + \sum a_{rs} S(f_r, f_s),$$

where $\log(h'_j f_j) < \log(h_1 f_1)$. An easy induction completes the proof. $\qquad\square$

Algorithm 1.9 (Buchberger Algorithm) *Let* $F = \{f_1,\ldots,f_m\}$ *be a set of generators of the ideal* I, *and let* $>_T$ *be a term order for* $\mathbb{M}$.

> $G := F$.
> $B := \{(f_1,f_2) \mid f_1, f_2 \in F, \text{ and } f_1 \neq f_2\}$.
>   *while* $B \neq \emptyset$ *do*
> $(f_1,f_2) := $ a pair in B
> $B := B \setminus \{f_1, f_2\}$
> $g := $ normal form of $S(f_1, f_2)$ with respect to G
>   *if* $g \neq 0$, *then*
> $B := B \cup \{(g,h) \mid h \in G\}$
> $G := G \cup \{g\}$

Example 1.10. Let $R = k[x,y,z]$ be a polynomial ring over a field k with the reverse lexicographic ordering and let $f_1 = y^4 - x^2 z^2$, $f_2 = x^3 - y^2 z$ and $f_3 = xy - z^2$. Applying the algorithm in this setting gives the reduced Gröbner basis for the ideal I generated by the f_i's:

$$f_1,\ f_2,\ f_3,\ xz^5 - z^6,\ yz^5 - z^6,\ y^2 z^3 - xz^4,\ y^3 z - x^2 z^2,\ x^2 z^3 - xz^4.$$

Example 1.11. Let $R = k[x_1 \ldots, x_8]$ be a polynomial ring over a field k and let G be the bipartite graph

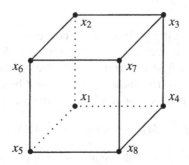

Let $k[G]$ be the k-subring of R spanned by the set of monomials $f_{ij} = x_i x_j$ so that x_i is adjacent to x_j and let $P(G)$ be the toric ideal of $k[G]$, that is, $P(G)$ is the kernel of the graded homomorphism

$$\varphi: B = k[t_{ij}] \longrightarrow k[G], \quad \text{induced by} \quad \varphi(t_{ij}) = f_{ij}.$$

$P(G)$ is a Cohen-Macaulay prime ideal of codimension 5, whose generators are determined by the edge cycles contained in the graph ([Vil95], [DG96]): to the cycle

$$\{\alpha_1, \beta_1, \ldots, \alpha_s, \beta_s\}, \quad \text{associate the binomial} \quad T_{\alpha_1} \cdots T_{\alpha_s} - T_{\beta_1} \cdots T_{\beta_s}.$$

If the terms in B are ordered by the reverse lexicographical ordering

$$t_{14} > t_{23} > t_{12} > t_{56} > t_{37} > t_{26} > t_{34} > t_{78} > t_{15} > t_{48} > t_{67} > t_{58},$$

then a reduced Gröbner basis for $P(G)$ is

$$
\begin{aligned}
&h_1 = t_{15}t_{48} - t_{14}t_{58}, \; h_5 = t_{14}t_{26}t_{78} - t_{12}t_{48}t_{67}, \; h_9 = t_{26}t_{78}t_{15} - t_{12}t_{67}t_{58} \\
&h_2 = t_{37}t_{26} - t_{23}t_{67}, \; h_6 = t_{14}t_{56}t_{37} - t_{34}t_{15}t_{67}, \; h_{10} = t_{56}t_{37}t_{48} - t_{34}t_{67}t_{58} \\
&h_3 = t_{12}t_{56} - t_{26}t_{15}, \; h_7 = t_{23}t_{78}t_{15} - t_{12}t_{37}t_{58}, \; h_{11} = t_{56}t_{78} - t_{67}t_{58} \\
&h_4 = t_{14}t_{23} - t_{12}t_{34}, \; h_8 = t_{23}t_{56}t_{48} - t_{26}t_{34}t_{58}, \; h_{12} = t_{34}t_{78} - t_{37}t_{48}.
\end{aligned}
$$

A curious feature here is that this basis is actually shorter than a 'natural' basis provided by all the edge cycles.

Consider the ideal $I = (h_1, h_2, h_3, h_4, h_{12})$. Notice that I is a complete intersection because the leading terms of the h_i's are relatively prime (see Exercise 1.14) and therefore $\{h_1, h_2, h_3, h_4, h_{12}\}$ is a Gröbner basis for I.

The theoretical cost of these computations can be staggering, doubly exponential in the number of variables, according to [MMe82]. This feature was already present in the classical analysis of the cost of computation in polynomial ideal theory by Grete Hermann [Her26]. On the other hand, the dynamic behavior of Buchberger algorithm benefits from the average cost of the computation (linear in the number of variables). Furthermore, unlike the classical methods that had to work out always from a worst case assumption, Gröbner bases algorithms are eminently programmable.

Except for a few cases, it is impossible to predict what the normal form of the S–polynomial of two elements will look like. One of the exceptions, exhibited in the examples above, is that of an ideal generated by binomials: all polynomials in the process will be binomials. But even here, the worst case complexity is not any better ([Stu91]).

Remark 1.12. The theorem of Macaulay already provides an indication of the high degrees that may occur amongst the generators of $in(I)$, even when the generators of I have relatively low degrees. Suppose I is a homogeneous ideal of $R = k[x_1, \ldots, x_n]$, of codimension $g = n - d$. Let

$$\frac{e_0}{(d-1)!} t^{d-1} + \text{ lower order terms,}$$

be the Hilbert polynomial of the ring R/I (see Appendix B). Let

$$in(I) = Q_1 \cap \cdots \cap Q_s \cap Q,$$

be a decomposition of $in(I)$ in which Q_i, $1 \leq i \leq s$, are the primary components of $in(I)$ of codimension g and Q is an ideal of codimension $> g$. Since R/I and $R/in(I)$ have the same Hilbert polynomial, we must have

$$e_0 = \sum_{i=1}^{s} \ell(R/Q_i)_{\mathfrak{p}_i} \deg R/\mathfrak{p}_i,$$

where $\mathfrak{p}_i = \sqrt{Q_i} = (x_{i_1}, \ldots, x_{i_d})$. This means that

$$\deg R/\mathfrak{p}_i = 1$$
$$\ell(R/Q_i)_{\mathfrak{p}_i} = \dim_k k[x_{i_1}, \ldots, x_{i_d}]/Q_i \cap k[x_{i_1}, \ldots, x_{i_d}].$$

The point is that there are only $\binom{n}{g}$ prime ideals $\mathfrak{p}_i$ to choose from, so that if e_0 is large, then some of the $Q_i \cap k[x_{i_1}, \ldots, x_{i_d}]$ must have generators in high degree, and therefore $in(I)$ will have generators in possibly much higher degrees.

This raises the hypothetical issue of whether there are other classes of ideals that would, in a context such as that of Macaulay theorem, play a role similar to monomial primes, which could be handled algorithmically but that existed in richer diversity.

Exercise 1.13. Let k be a field of characteristic $p > 0$ and let I be an ideal of the polynomial ring $k[x_1, \ldots, x_n]$. Denote by $I^{[p]}$ the ideal generated by the pth powers of all elements of I. Prove that if $\{f_1, \ldots, f_m\}$ is a Gröbner basis of I (for some term ordering), then $\{f_1^p, \ldots, f_m^p\}$ is a Gröbner basis of $I^{[p]}$ (for the same ordering).

Exercise 1.14. Let f, g be polynomials whose leading monomials form a regular sequence. Prove that $S(f, g)$ reduces to 0 relative to $\{f, g\}$. In particular, it follows that if an ideal I is generated by the polynomials $f_1, \ldots, f_m$ whose leading monomials for some term order form a regular sequence, then $\{f_1, \ldots, f_m\}$ is a regular sequence and a Gröbner basis of I.

1.3 Computation of Syzygies

Let $R = k[x_1, \dots, x_n]$ be the ring of polynomials over a field k and let I be an ideal given by a set $\{f_1, \dots, f_m\}$ of generators. In concrete situations, these generators carry along many mutual relationships. Furthermore, they were likely obtained in a 'natural' setting.

There are several strategies adapted for computation of the properties of the ring R/I:

- Transformation of the object into another with similar numerical data (*Hilbert Functions*).
- Comparison of objects by looking at their *syzygies*. Broadly, it is simpler to study an algebraic object **M** when it is *free*: the methods typical of linear algebra may be imported. Lacking freeness, one uses a presentation

$$\mathbf{K} \longrightarrow \mathbf{F} \longrightarrow \mathbf{M} \to 0,$$

where **F** is a free object; **M** is equally well coded by the relations **K**. Typical examples are the (linear) relations of a set of generators of a module of an R–module or the (algebraic) relations of the generators of an affine k–algebra.
- Classical and modern elimination theory.
- Factorization techniques.

Syzygies are devices to permit the study of ideals and modules by unraveling relationships between their generators. At the moment, there is no more fruitful approach to computation in commutative algebra than based on the theory of syzygies. For a general discussion, see [MM86]. It still lacks the efficiencies achieved in factorization techniques used in the study of one or two polynomials, but is vastly preferable to the classical determinantal based methods.

Gröbner Bases and Syzygies

The reader will not find here a discussion of the varied ways computation of syzygies can be performed. We refer to basic references on the subject such as [AL90], [BKW93], [CLO92].

Here we just outline how the first–order syzygies of an ideal $I = (F) = (f_1, \dots, f_n)$ may be determined in principle. The generators are going to be taken as ordered, and we write one of its syzygies as a column vector $v \in R^n$ such that $F \cdot v = 0$.

Let $G = \{g_1, \dots, g_m\}$ be a Gröbner basis of I. For each pair of polynomials $\{g_i, g_j\}$ in G the S–polynomial $S(g_i, g_j)$ has a reduction

$$S(g_i, g_j) = a_j g_i - a_i g_j = \sum_{k=1}^{m} h_k g_k, \tag{1.12}$$

by Theorem 1.8. In particular the vector

$$\begin{bmatrix} h_1 \\ \vdots \\ h_i - a_j \\ \vdots \\ h_j + a_i \\ \vdots \\ h_m \end{bmatrix} \qquad (1.13)$$

is a syzygy of G. The following elementary fact ([Sc80]) is extremely useful.

Theorem 1.15. *The syzygies of G are generated by the vectors* (1.13).

Proof. We leave its proof to the reader. □

To be more useful it is essential to extend the notion of Gröbner bases from ideals of $R = k[x_1, \ldots, x_n]$ to submodules of a free R–module P. One could, for instance, view a submodule of P as an ideal of a big polynomial ring $k[x_1, \ldots, x_n, e_1, \ldots, e_m]$, generated by linear forms in the variables e_i and the quadratic polynomials $e_i e_j$. But this approach, in general, is going to be wasteful of resources.

It is better (see [Ei95], [Mol91], [MM86]) to proceed as follows. Given a free module P, order one of its bases, $\{e_1, \ldots, e_m\}$, and define a monomial of P as an element of the form $\mathbf{x}^A e_i$, where $\mathbf{x}^A$ is a monomial of R. A *monomial submodule M* is generated by such elements. The notion of a Gröbner basis can then be applied to the submodules of P as well, and the calculus of syzygies above can be repeated for submodules of free modules.

Assembling the Syzygies

It is necessary to convert the syzygies of a Gröbner basis G into the syzygies of the basis F from which it was derived. It involves the two matrices that convert one set of generators into the other. The following observations were kindly pointed out to us by H.-G. Gräbe.

Denote by

$$F = (f_1, \ldots, f_n)$$
$$G = (g_1, \ldots, g_s).$$

the ordered given generators and the computed Gröbner basis of a module Q, respectively. This gives rise to two transition matrices A and B

$$G = F \cdot A$$
$$F = G \cdot B,$$

the first of which is obtained in the execution of the Buchberger algorithm, while the other is the representation of the elements of F by the normal form algorithm.

The syzygies of F are the column vectors $v \in R^m$ such that $F \cdot v = 0$. The module of syzygies of G is that described above, and at issue is how to convert from one module to the other.

Proposition 1.16. *If S_G is a basis for the syzygies of G then the columns of*

$$[E - A \cdot B | A \cdot S_G]$$

is a basis for the syzygies of F, with E the $n \times n$ identity matrix.

Proof. The columns of this matrix are clearly syzygies of F. Conversely, if $F \cdot v = 0$, then $B \cdot v \in S_G$. In this case

$$v = (E - A \cdot B) \cdot v + A \cdot B \cdot v$$
$$\subset \text{column space}(E - A \cdot B) + \text{column space}(A \cdot S_G),$$

which proves the assertion.   □

Gröbner Bases over Rings

If k is a ring and $R = k[x_1, \ldots, x_n]$, any term ordering $<$ on the monomials $\mathbf{x}^\alpha$ permits the development of several aspects of Gröbner bases techniques to this extended setting. We recall here some aspects to be used later but refer the reader to [AL90] for additional details.

Let us fix a term order. Let I be an ideal of R and denote by $in(I)$ the ideal of all leading terms of elements of I.

Definition 1.17. A Gröbner basis of I is a family of polynomials

$$h_\alpha = f_\alpha + \text{lower order terms} \in I,$$

whose leading terms f_α span $in(I)$.

If R is Noetherian, the ideal $in(I)$ will be finitely generated. However, if I is finitely generated but R is not Noetherian then $in(I)$ may well be not finitely generated.

Unlike the field case, we now must keep track of the full leading term instead of just the leading monomial. If $f_\alpha, \alpha \in \Sigma$ is a set of monomials that generates $in(I)$, the leading coefficients of the f_α's now play an important role. There are significant similarities and contrasts, of which we consider a few cases.

Proposition 1.18. *Suppose that the leading coefficients of the f_α's are all 1. Then R/I is a free k–module with a basis given by the standard monomials.*

Proof. Clear from the definition of $in(I)$.   □

Gröbner Bases and Localization

The next useful result describes the behavior of the initial ideal formation under localization.

Proposition 1.19. *Let k be a commutative domain and let $<$ be a term order for the monomials of $R = k[x_1,\dots,x_n]$. For an ideal $I \subset R$ let G be a Gröbner basis. If S is a multiplicative set of k then $S^{-1}G$ is a Gröbner basis of $S^{-1}I$. In other words*

$$in(S^{-1}I) = S^{-1}in(I).$$

Proof. If $h \in S^{-1}I$, there exists $s \in S$ such that $s \cdot h \in I$. This means that the contribution of the leading term of h to $in(S^{-1}I)$ already lies in $S^{-1}in(I)$. The converse is clear. $\qquad\Box$

Computation of Gröbner Bases

There are two instances when the computation over Gröbner bases over $R[\mathbf{T}]$ can be dealt with much in the same manner as the case of a field.

Suppose k is a field, $R = k[\mathbf{x}]$, $A = R[\mathbf{T}]$, and $<$ is a product term order for monomials of A so that $x_i < T_j$. If I is an ideal of $A = R[\mathbf{T}]$, and we consider associated Gröbner bases G_1 and G_2, the first when only the order on the monomials on the T_j's are taken into account and the other when all monomials are considered.

$$G_1 = \{g_\alpha(x)\mathbf{T}^\alpha + \text{lower order terms}, \ \alpha \in \Sigma\}$$
$$G_2 = \{\mathbf{x}^\beta\mathbf{T}^\gamma + \text{lower order terms}, \ (\beta,\gamma) \in \Sigma'\}.$$

It is clear that we can obtain a Gröbner basis equivalent to G_1 by arranging each polynomial in G_2 in the distributed form with respect to the T_j's.

A different kind of difficulty is that of a ring such as $R = \mathbb{Z}$. Here it is clear that Buchberger algorithm will produce a Gröbner basis provided that in the course of taking the S–resultant of pairs of polynomials, say $f = a\mathbf{T}^\alpha + \cdots$ and $g = b\mathbf{T}^\beta + \cdots$, one sets $S(f,g) = c \cdot f - d \cdot g$, where

$$c = \frac{b}{\gcd(a,b)}\mathbf{T}^\gamma \quad d = \frac{a}{\gcd(a,b)}\mathbf{T}^\delta,$$

so that $\alpha + \gamma = \beta + \delta$ and $\text{supp}(\gamma) \cap \text{supp}(\delta) = \emptyset$.

1.4 Hilbert Functions

Let $A = k[x_1,\dots,x_n]/I$ be a graded ring over the field k and denote by $H_A(t)$ its Hilbert function. The key to the computation of $H_A(t)$ is Theorem 1.5: If $<$ is a term ordering and $I' = in(I)$ is the corresponding initial ideal, then $H_A(\mathbf{t}) = H_{A'}(\mathbf{t})$, where $A' = k[x_1,\ldots,x_n]/I'$.

We indicate some of the known approaches to find the coding of Hilbert functions by Hilbert–Poincaré series of algebras defined by monomial ideals. The more delicate points of these strategies, the aspects that must be carefully assembled to obtain optimization of coding, will not be treated here. There is a detailed discussion of the issues in [Bi97].

Suppose $I = (m_1,\ldots,m_r)$, where the m_i are monomials in the indeterminates $x_1,\ldots,x_n$. A theoretical approach is via Taylor resolutions, and derives the Hilbert–Poincaré series directly from the projective resolution of R/I. But this resolution can have as many as 2^r terms, which militates against its use if r is large.

Monomials of Degree Two

We illustrate the kind of assemblage that takes place by considering a very straightforward case.

The usual path has been to "filter" the graded module R/I by other graded modules. Let us indicate this by treating one example in great detail. Suppose the monomials m_i are of degree two and square-free. The monomials model a graph G whose vertex set is $\{x_1,\ldots,x_n\}$, and whose edges are $\{x_k,x_\ell\}$ if $x_k x_\ell$ is one of the m_i; the algebra R/I is denoted $k[G]$ (see [Vil90]). Adding variables to the monomial ideal corresponds to considering graphs with isolated vertices.

Let us derive the Hilbert–Poincaré series of $k[G]$ in terms of series for graphs with fewer vertices ([Wa92]).

Proposition 1.20. *Let G be a graph, and $P_G(\mathbf{t})$ the Hilbert–Poincaré series of the associated ring $k[G]$. For any vertex $x \in V(G) = \{x_1,\ldots,x_n\}$ we have*

$$P_G(\mathbf{t}) = P_{G-x}(\mathbf{t}) + \frac{\mathbf{t}}{1-\mathbf{t}}P_{G-N(x)-x}(\mathbf{t})$$

with

$$P_0(\mathbf{t}) = 1.$$

Here $G - x$ denotes the graph obtained from G by deleting the vertex x, and $G - N(x) - x$ the graph from which x and all its neighbors $N(x)$ have been deleted.

Proof. Let $I = \{m_1,\ldots,m_s, x_{i_1}x_n,\ldots,x_{i_k}x_n\}$ where the $x_{i_1},\ldots,x_{i_k}$ are the neighbors of the vertex x_n and the $m_1,\ldots,m_s$ are the remaining edges. We write $I = (x_n L, J)$, where $L = \{x_{i_1},\ldots,x_{i_k}\}$ and $J = \{m_1,\ldots,m_s\}$.

From the exact sequence

$$0 \to (x_n,J)/I \longrightarrow R/I \longrightarrow R/(x_n,J) \to 0,$$

since $I:x_n = (L,J)$, we obtain the exact sequence

$$0 \to R/(L,J)(-1) \longrightarrow k[G] \longrightarrow k[G - x_n] \to 0$$

from which we have the equality of series

$$P_G(\mathbf{t}) = \mathbf{t} \cdot P_{R/(L,J)}(\mathbf{t}) + P_{G-x_n}(\mathbf{t}).$$

Finally, note that x_n is not a vertex of the graph represented by (L,J), so that

$$R/(L,J) = k[G - N(x_n) - x_n][x_n],$$

and therefore

$$P_{R/(L,J)}(\mathbf{t}) = \frac{1}{1-\mathbf{t}} \cdot P_{k[G-N(x_n)-x_n]}(\mathbf{t}),$$

to complete the proof. □

This device may also be used for more general ideals, like those generated by squarefree monomials of arbitrary degrees. For example, if Δ is a simplicial complex and x is one of its vertices, and

$$\Delta = x * \Delta_1 \cup \Delta_2$$

is a disjoint decomposition, then one has an exact sequence

$$0 \to k[\Delta_1 \cup \Delta_2][-1] \longrightarrow k[\Delta] \longrightarrow k[\Delta_2] \to 0,$$

of face rings, with a corresponding relation of their Hilbert functions.

General Monomials

The broad algorithms of [BSb92] and [BCR93] remove all of these restrictions, even by allowing arbitrary grading of the ring of polynomials. Their settings are however the same: Given a monomial ideal I and a monomial f there is the exact sequence

$$0 \to R/(I:f)(-d) \longrightarrow R/I \longrightarrow R/(I,f) \to 0,$$

where $d = \deg f$. It follows that

$$P_{R/I}(\mathbf{t}) = P_{R/(I,f)}(\mathbf{t}) + \mathbf{t}^d P_{R/(I:f)}(\mathbf{t}).$$

The researchers have an abbreviated notation for these series:

$$\langle I \rangle = \langle I, f \rangle + \mathbf{t}^d \langle I:f \rangle.$$

Here are two approaches that have been used. They further differ in the way that corner cases are handled.

(a) (Bayer-Stillman) The equality above can be used backwards: If J is an ideal and $J = (I, f)$, then

$$\langle J \rangle = \langle I \rangle - \mathbf{t}^d \langle I : f \rangle,$$

where both I and $I : f$ have fewer generators that J. This is the approach of [BSb92] that was originally implemented in *Macaulay*.

(b) (Bigatti-Caboara-Robbiano) In this approach, used in *CoCoA*, f is chosen to be a variable that occurs in the monomials: $R/(I, f)$ is defined over fewer variables, while the ideal $I : f$ is given as follows. If $I = (fL, J)$, and f does not occur in the monomials of J, then $I : f = (L, J)$.

Taking for f the highest power of a variable x_n that occur in the monomials strips that variable from all monomials of $I : f$, but may complicate the handling of (I, f), except when the degree is very low.

Stanley Decompositions

Finally we consider in greater detail an approach that is not greatly useful computationally but it is often nice theoretically.

In order to make the graded structure of $A = R/I$ more visible, we introduce Stanley decompositions [SWh91]. Not only can the Hilbert series be read directly from the decomposition, but the proof that such decompositions exist yields an algorithm for their computation in the case where $R = k[x_1, \ldots, x_n]/I$, and I is a monomial ideal (e.g. [Wa92], which we follow here).

Definition 1.21. *A* Stanley decomposition *of the k-algebra A is a representation as a finite direct sum of k-vector spaces*

$$A = \bigoplus_{\alpha \in F} \mathbf{x}^\alpha k[\mathbf{X}_\alpha]$$

where F is a finite subset of $\mathbf{N}^n$, α is the vector $(\alpha_1, \ldots, \alpha_n)$, $\mathbf{X}_\alpha$ is a subset of $\{x_1, \ldots, x_n\}$ and $\mathbf{x}^\alpha = x_1^{\alpha_1} x_2^{\alpha_2} \cdots x_n^{\alpha_n}$. Note that we are identifying each x_i with its image $\bar{x}_i$ in the residue ring.

Example 1.22. Let $A = k[x, y, z]/(xy, yz, xz)$. Then A has Stanley decomposition

$$A = k[y] \oplus xk[x] \oplus zk[z].$$

That this is a Stanley decomposition is clear because

$$\{1, y, y^2, \ldots\} \subseteq k[y], \ \{x, x^2, \ldots\} \subseteq x \cdot k[x], \ \{z, z^2, \ldots\} \subseteq z \cdot k[z],$$

and these monomials form a basis for A since there are no cross terms.

We now prove our claim that the Stanley decomposition enables us to find the Hilbert series.

Proposition 1.23. *Let R be a standard k-algebra with a Stanley decomposition*

$$R = \bigoplus_{\alpha \in F} \mathbf{x}^\alpha k[\mathbf{X}_\alpha]$$

and let $|\alpha| = \deg \mathbf{x}^\alpha = \sum_{i=1}^n \alpha_n$ *and* $d = \max_{\alpha \in F} |\mathbf{X}_\alpha|$ *where* $|\mathbf{X}_\alpha|$ *is the cardinality of* $\mathbf{X}_\alpha$. *Then*

(a) *The Krull dimension of R is d, and*
(b) *The Hilbert series of R is*

$$P_R(\mathbf{t}) = \sum_{\alpha \in F} \frac{\mathbf{t}^{|\alpha|}}{(1-\mathbf{t})^{|\mathbf{X}_\alpha|}}.$$

Proof. Let $V_\alpha = \mathbf{x}^\alpha \cdot k[\mathbf{X}_\alpha]$ so that $R = \bigoplus_{\alpha \in F} V_\alpha$. We need only calculate the Hilbert series for each V_α since $P_R = \sum_{\alpha \in F} P_{V_\alpha}$. The monomial $\mathbf{x}^\alpha$ in V_α corresponds to a shift in grading so $H_{V_\alpha}(m) = H_{k[\mathbf{X}_\alpha]}(m - |\alpha|)$ and $P_{V_\alpha}(\mathbf{t}) = \mathbf{t}^{|\alpha|} P_{k[\mathbf{X}_\alpha]}(\mathbf{t}) = \mathbf{t}^{|\alpha|}/(1-\mathbf{t})^{|\mathbf{X}_\alpha|}$. The dimension is the order of the pole of P_R at 1 and is therefore d. □

Existence of Stanley Decompositions

The following results show that every k-algebra has a Stanley decomposition. When the ideal I is generated by monomials an inductive proof of existence can be given as follows[2].

Proposition 1.24. *If* $R = k[x_1, \ldots, x_n]/I$, *where I is a monomial ideal, then R has a Stanley decomposition.*

Proof. Write the generators of I as monomials in $k[x_1, \ldots, x_{n-1}][x_n]$. Thus $I = (m_1 \cdot x_n^{d_1}, \ldots, m_l \cdot x_n^{d_l})$, with $m_1, \ldots, m_l \in k[x_1, \ldots, x_{n-1}]$ and $0 \le d_1 \le d_2 \le \ldots \le d_l$. The Stanley decomposition, denoted $SD\{R\}$, is defined inductively on the number of indeterminates n.
If $n = 1$, then set

$$SD\{R\} = \begin{cases} k \oplus x_1 \cdot k \oplus x_1^2 \cdot k \oplus \ldots \oplus x_1^{d_1 - 1} \cdot k & I = (x_1^{d_1}) \\ k[x_1] & I = (0). \end{cases}$$

If $n \ge 2$, then define

$$SD\{R\} = \bigoplus_{j=0}^{d_l - 1} x_n^j \cdot SD\{k[x_1, \ldots, x_{n-1}]/(m_i \mid d_i \le j)\}$$

$$\bigoplus x_n^{d_l} \cdot k[x_n] \cdot SD\{k[x_1, \ldots, x_{n-1}]/(m_1, \ldots, m_l)\}.$$

---

[2] Francesco Piras has pointed out another algorithm, due to M. Janet, to find these decompositions: J. de Math. **8** (1920), 65–151; *Leçons sur les Systèmes d'Équations aux Dérivées Partielles.* Gauthier-Villars, Paris, 1929.

In the above formula we distribute monomials and polynomial rings over direct sums and collect terms by taking $\mathbf{x}^\alpha \cdot k[\mathbf{X}_\alpha] \cdot \mathbf{x}^\beta \cdot k[\mathbf{X}_\beta] = \mathbf{x}^\alpha \cdot \mathbf{x}^\beta \cdot k[\mathbf{X}_\alpha \cup \mathbf{X}_\beta]$. For example $x \cdot k[x,y] \cdot (y \cdot k[z] \oplus x^3 \cdot k) = xy \cdot k[x,y,z] \oplus x^4 \cdot k[x,y]$.

We now show that this is indeed a Stanley decomposition of R. It is sufficient to show that all monomials, which form a basis for R, can be expressed uniquely. We proceed by induction on the number of indeterminates.

For $n = 1$ this is obvious; for $n \geq 2$ assume $\mathbf{x}^\alpha = x_1^{\alpha_1} \cdots x_{n-1}^{\alpha_{n-1}} \cdot x_n^{\alpha_n} \in R$ is a nonzero monomial. The term $x_n^{\alpha_n}$ occurs in exactly one of $1, x_n, x_n^2, \ldots, x_n^{d_l - 1}, x_n^{d_l} \cdot k[x_n]$. Thus we need only show $x_1^{\alpha_1} \cdots x_{n-1}^{\alpha_{n-1}}$ is non-zero in the corresponding quotient ring in equation 1.14 and thus by induction occurs uniquely in its Stanley decomposition. But if $x_1^{\alpha_1} \cdots x_{n-1}^{\alpha_{n-1}} = 0$ in $k[x_1, \ldots, x_{n-1}]/(m_i \mid d_i \leq \alpha_n)$ then $x^\alpha = 0$ in R, contradicting our assumption.   □

Exercise 1.25. Let $G = K_{m,n}$ be the complete bipartite graph on $\{m,n\}$ vertices. Show that

$$P_G(\mathbf{t}) = \frac{1}{(1-\mathbf{t})^m} + \frac{1}{(1-\mathbf{t})^n} - 1.$$

Problem 1.26. A more delicate problem regarding Hilbert functions of rings associated to graphs is the following. For a graph G on the vertex set $\{x_1, \ldots, x_n\}$, denote by A the subring of $k[x_1, \ldots, x_n]$ generated by all products $x_i x_j$ such that (x_i, x_j) is an edge of G. How to find the Hilbert function of this ring?

1.5 Computer Algebra Systems

The division algorithms above, addressing so quickly membership and syzygy problems, give an early indication of the usefulness of this theory. These algorithms have been programmed into general purpose routines in various computer algebra systems, or in specialized programs with embedded knowledge of the algebraic ideas, often taking into account features present in special cases. Among the implementations, we single out the following. (It is of interest to keep in mind that some systems seek to simultaneously fill several roles and may therefore fail to take advantage of the specificity of the computation.)

General CAS

The following are large, commercial computer algebra systems, that seek to address symbolic computational needs of many communities. They differ on availability on various kinds of platforms.

- **Macsyma**
- **Reduce**
- **Maple**
- **Mathematica**

- **Axiom**

Among the commercial CAS, **Maple** ([CGW91]) and **Reduce** ([Hea87]) are full–fledged computer algebra systems whose Gröbner basis algorithms are periodically updated; the latter has an extremely useful forum/bulletin board, with a worldwide community of contributors. **Macsyma** has also a Gröbner basis algorithm but it does not incorporate recent optimizations. **Mathematica** [Wol96] has a limited implementation thus far, but its user's community is beginning to generate more efficient packages. **Axiom** ([JS92]) promises to pick up where **Scratchpad** left off, and its advanced factorization techniques should be very useful.

The ideal environment might consist of a system such as **Mathematica** or **Maple**, from which one could call on fast Gröbner *engines*, lean (but knowledgeable!) programs tailored for the computation of Gröbner bases and the direct information that is embedded in them. It may then be necessary to return to the fuller environment, to carry out a possibly numerical phase of the computation.

Specialized CAS

These are systems specifically designed for computation in commutative algebra and algebraic geometry that can be mediated by Gröbner basis techniques. They are freely available and instructions on how to obtain them are given in ([BS92], [CNR95], [GS96], [GPS95]). Those with worldwide followings are:

- *CoCoA*
- *Macaulay*
- *Singular*
- *Macaulay 2*

Macaulay ([BS92]) was in 1996 the most widely used computer algebra system in commutative algebra and algebraic geometry. It lacks a programming language and but has a rich library of scripts. *CoCoA* ([CNR95]) is a developing system, with a friendly user interface. *Singular* ([GPS95]) is another recent player, particularly focused on singularity theory. Finally, late in 1996 a new player, *Macaulay 2* ([GS96]), a successor to *Macaulay*, entered the playing field. It has a full-fledged programming language and incorporates many recently discovered efficiencies.

A significant aspect of these programs is the role played by working mathematicians on the choice of algorithms and algebraic knowledge in their implementations.

The methods discussed in this text require a range of operations that cannot yet be found in any single computer algebra system. In our own practice, we have had to mix and choose from several capabilities of the systems mentioned above, a task that is made harder by the difficulty these systems have in communicating with one another.

2

Toolkit

In this chapter we assemble some fundamental devices to manipulate ideals of rings of polynomials to form new rings and to set up the conditions to help ascertain the presence of certain properties in rings, modules and their morphisms.

- Nuts and bolts
- Rings of endomorphisms
- Noether normalization
- Fitting ideals
- Integral extensions
- Flatness and Cohen–Macaulayness testing

Elimination techniques have been in the forefront of applications of Gröbner bases to ideal theory from very early. The required adjunction of new variables is often very natural and appealing despite the potential threat of combinatorial explosion.

There are related two operations with ideals which are pervasive in the constructions. They are:

- The formation of ideal quotients,

$$I:J = \{\, x \in R \mid x \cdot J \subset I \,\}$$

in a ring of polynomials R

- The construction of the ring of endomorphisms $\operatorname{Hom}_R(I, I)$ of an ideal I in an affine domain R.

Their key role occurs since (i) through $I:J$ one perturbs the primary decomposition of I in a reasonably controlled form, and (ii) $\operatorname{Hom}_R(I, I)$ leads to a new algebra which is an integral extension of R. They represent basic manipulations with the functor Ext of homological algebra.

Another major process, necessary to study morphisms of rings, is Noether normalization. It provides a baseline, adequate for Gröbner basis computation, from which to convert problems into others that may be amenable to linear algebra techniques.

2.1 Elimination Techniques

Elimination Theory is concerned with the determination of the image of a morphism between algebraic varieties

$$\varphi : Y \mapsto X.$$

Its computational aspect consists in the development of techniques to solve the following problem. Given a homomorphism of rings

$$\psi : A \mapsto B,$$

and an ideal $L \subset B$, determine the ideal

$$I = \psi^{-1}(L) \subset A.$$

Strictly speaking these two formulations are not equivalent except under conditions controlled by the fundamental:

Theorem 2.1 (Main Theorem of Elimination Theory). *Let R be a Noetherian ring and let $\mathbb{P}_R^n$ be a projective space over $\mathrm{Spec}(R)$. The projection*

$$p : \mathbb{P}_R^n \mapsto \mathrm{Spec}(R)$$

is a closed mapping.

In actual practice, A and B are affine rings and the issue is to find a description of the image of the corresponding morphism of affine varieties. The most important case is that of a ring B which is a polynomial ring over A, $B = A[\mathbf{T}] = A[T_1, \dots, T_m]$.

In this section we consider common uses of Gröbner bases to effect elimination of variables. They will permit the construction of the intersection of ideals and of some subrings. As a rule, these tricks require many additional variables and therefore keep the overhead very high.

Elimination of Variables

Let R be the polynomial ring $k[x_1, \dots, x_n]$, let t be an indeterminate over R, and put $B = R[t]$. Let I be an ideal of B. The technique of elimination of variables is based on the following:

Proposition 2.2. *Let T be an ordering of the variables such that $t >_T \mathbf{x}^a$ for any monomial in the x_i's. Let F be a Gröbner basis of I. Then $F \cap R$ is a Gröbner basis of $I \cap R$.*

Proof. Follows immediately from the division algorithm. $\square$

Replacing $t \mapsto \mathbf{T}$ in the Gröbner basis calculation, the full set $\mathbf{T}$ can be eliminated. Alternatively the T_i can be successively eliminated.

Corollary 2.3. *If I and J are two ideals of R, then $I \cap J$ can be computed.*

Proof. To apply Proposition 2.2 we must show how $I \cap J$ can be obtained as the contraction of some ideal $L \subset R[t]$.

We claim that if t is a new variable, then $I \cap J = (I \cdot t, J \cdot (1 - t)) \cap R$. Indeed, if $a \in I \cap J$, then

$$a = at + (1 - t)a.$$

On the other hand, any element of $(I \cdot t, J \cdot (1 - t)) \cap R$,

$$b = \sum a_i h_i(t)t + \sum b_j g_j(t)(1 - t), \ a_i \in I, \ b_j \in J, \ h_i(t), g_j(t) \in R[t],$$

evaluates to itself if $t \mapsto 0$ or $t \mapsto 1$, but it is mapped into J in one case and into I in the other. $\qquad\qquad\square$

The ability to compute syzygies gives a distinct advantage in carrying out the ideal theoretic operations mentioned earlier. For instance, the computation of the intersection of two ideals $I = (a_1, \ldots, a_m)$ and $J = (b_1, \ldots, b_n)$ is now handled as that of finding the syzygies of the matrix

$$\begin{bmatrix} 1 & a_1 & \cdots & a_m & 0 & \cdots & 0 \\ 1 & 0 & \cdots & 0 & b_1 & \cdots & b_n \end{bmatrix}.$$

The desired intersection is the ideal generated by the entries at $(1, 1)$.

Corollary 2.4 (Radical Membership). *If $f \in R$ and I is an ideal, then $f \in \sqrt{I}$ can be decided.*

Proof. If t is a variable, and R is any commutative ring, then $f \in \sqrt{I}$ if and only if $(I, 1 - tf) = R[t]$. (Details left as an exercise.) $\qquad\qquad\square$

Homomorphism of Affine Rings

A k–homomorphism

$$\psi : A = k[x_1, \ldots, x_s]/(f_1, \ldots, f_m) \mapsto B = k[y_1, \ldots, y_r]/J,$$

of two affine k–algebras is the assignment

$$x_i \mapsto g_i(y_1, \ldots, y_r) \in k[y_1, \ldots, y_r], \ i = 1, \ldots, s$$

such that $f_j(g_1, \ldots, g_s) \in J$, for all i.

Image of a Morphism

Another application of elimination is to determine the image of a mapping between affine spaces. Let $\phi: \mathbb{A}^r \mapsto \mathbb{A}^s$ be a polynomial mapping between two affine spaces defined over a field k. Denote by I the ideal of all polynomials $h(Y_1, \ldots, Y_s) \in k[Y_1, \ldots, Y_s]$ that vanish on the image of ϕ:

$$\{x \in k^r \mid h(\phi_1(x), \ldots, \phi_s(x)) = 0\}.$$

Proposition 2.5. *Let k be an infinite field and let ϕ be as above. Then*

$$I = (Y_i - \phi_i(X), \ i = 1, \ldots, s) \cap k[Y].$$

Proof. For $h(Y) \in I$ we have

$$
\begin{aligned}
h(Y_1, \ldots, Y_s) &= h(\phi_1(X) + (Y_1 - \phi_1(X)), \ldots, \phi_s(X) + (Y_s - \phi_s(X))) \\
&= h(\phi_1(X), \ldots, \phi_s(X)) + \sum_i h_i(X, Y)(Y_i - \phi_i(X)).
\end{aligned}
$$

In one direction the assertion follows since $h(\phi_1(X), \ldots, \phi_s(X))$ vanishes identically. The converse is clear.   □

More generally, suppose $\phi: V \mapsto W$ is a morphism of affine subvarieties of $\mathbb{A}^r$ and $\mathbb{A}^s$ respectively.

Proposition 2.6. *Let I_V and I_W denote the ideals defining V and W. Then*

(a) *The image of V lies in W if and only if $h(\phi(x)) = 0$ for any $h \in I_W$.*
(b) *The closure of the image of V is defined by the ideal*

$$I = (I_V + (Y_i - \phi_i(X), \ i = 1, \ldots, s)) \cap k[Y].$$

Proof. We leave the verifications to the reader.   □

Example 2.7. A motivation for deciding the membership question in rings of polynomials is provided by the following formulation by D. Bayer of the 4-color question. Let M be a map made up of the regions R_i, $i \in A$, that we want to color with, say, 4 colors. The 'colors' we use will be the four roots of 1. In the ring of polynomials

$$R = k[x_i, \ i \in A],$$

let I be the ideal generated by all $f_i = x_i^4 - 1$ and the polynomials

$$h_{ij} = x_i^3 + x_i^2 x_j + x_i x_j^2 + x_j^3,$$

associated to each neighboring pair of regions R_i, R_j. Noting that h_{ij} only vanishes along with f_i, f_j when the roots of these polynomials are chosen to be distinct, it follows from the Nullstellensatz that $I \neq R$ if and only if M is 4-colorable.

Regular Elements and Ideal Quotients

Two of the most common manipulations with ideals concern the underlying primary decompositions.

Definition 2.8. *Given two ideals I and J of a ring R, the* ideal quotient *of I by J is the ideal*

$$I:J = \{r \in R \mid r \cdot J \subset I\}.$$

It will figure prominently in our constructions, so that we must have several ways to find it.

Proposition 2.9. *An element $f \in R$ is regular modulo the ideal I if and only if one of the following conditions hold in $R[t]$:*

(a) $((I \cdot t, (1-t) \cdot f) \cap R) \cdot f^{-1} = I.$
(b) $(I, 1 - f \cdot t) \cap R = I.$

The first formula computes $I :_R f$, whereas the second determines

$$I : \langle f \rangle = \bigcup_{n \geq 1} (I :_R f^n).$$

Perhaps the most direct approach is: If $J = (a_1, \ldots, a_n)$, it can be computed as

$$I:J = \bigcap_{i=1}^{n} (I : a_i).$$

An alternative is the following construction.

Proposition 2.10. *Let t be a variable over the ring R and let*

$$f = a_1 + a_2 t + \cdots + a_n t^{n-1}.$$

Then

$$I:J = (I \cdot R[t] : f) \cap R.$$

Definition 2.11. *Let I and J be two ideal of the ring R. The* saturation *of I with respect to J is the limit ideal quotient*

$$I : \langle J \rangle = I : J^{\infty} = \bigcup_{k} (I : J^k).$$

This operation strips away from any primary decomposition of I all components whose radicals contain J. In [BSa87] the following device is given to compute it.

Proposition 2.12. *Let f be the polynomial defined above and let y be a fresh variable. Then*

$$I:J^{\infty} = ((I, y - f) : y^{\infty}) \cap R.$$

Comparison of Ideals

One of the most common tasks faced is that of comparing two ideals I and J for containment. It is usually set up by assuming $I \subset J$, after replacing $J \mapsto I + J$. Since several systems have implemented the quotient of ideals operation,

$$I = J \Longleftrightarrow I{:}J = (1). \tag{2.1}$$

When I and J are homogeneous ideals, one can also just compare their Hilbert functions:

$$I = J \Leftrightarrow H_{R/I}(\mathbf{t}) = H_{R/J}(\mathbf{t}) \Leftrightarrow H_I(\mathbf{t}) = H_J(\mathbf{t}).$$

Homogenization

Homogeneous ideals and modules have attached numerical data that is useful in controlling a computation and benefit of the efficiencies of certain monomial orderings. It is therefore quite common to homogenize a module and later capture information of the original module. This is often carried out in the following manner.

Let $f \in R = k[x_1, \ldots, x_n]$, and denote by T a fresh variable. The homogenization of f is:

$$f^H = T^{\deg f} f\left(\frac{x_1}{T}, \ldots, \frac{x_n}{T}\right).$$

For an ideal $I = (f_1, \ldots, f_m)$, we have two ways to define its homogenization:

$$J = (f_1^H, \ldots, f_m^H) \quad \text{or as}$$
$$L = (f_1^H, \ldots, f_m^H){:}T^\infty.$$

The first depends on the generating set but the other doesn't. The key property of both is that $T - 1$ is a regular element and they deform into the original ideal: $J_{T=1} = L_{T=1} = I$. Whenever we speak of homogenizing we refer to the second process. The same steps can be carried out on a module by acting on its presentation matrix.

Exercise 2.13. Let ψ be an endomorphism of an affine algebraic variety V. Describe how to set the computation that yields the fixed points of ψ.

Exercise 2.14. (Bayer-Stillman) Let $I \subset k[x_1, \ldots, x_n]$ be a homogeneous ideal, and consider a graded, reverse lexicographic order that makes x_n the lowest variable. Prove that

$$I{:}x_n = I \Longleftrightarrow in(I){:}x_n = in(I).$$

Exercise 2.15. Let I be an ideal and let E be an R–module. Prove the following statements:

- I is a prime ideal if and only if I^H is a prime ideal.
- $(\sqrt{I^H})_{T=1} = \sqrt{I}$.
- $(\text{annihilator}(E^H))_{T=1} = \text{annihilator}(E)$.
- Find an example of a regular sequence f, g, h such that the homogenization f^H, g^H, h^H is not of codimension 3.

2.2 Rings of Endomorphisms

Let D be a Noetherian ring and let Q denote its total ring of fractions. For two ideals J and I, we use the notation

$$J:_Q I = \{q \in Q \mid q \cdot I \subset J\}.$$

This operation should be distinguished from the ordinary quotient $I : J$. When they are used in the same setting, the notations $I :_D J$ and $I :_Q J$ are used.

The significance lies in the following two identifications: If I contains regular elements then

$$\operatorname{Hom}_D(I,D) \simeq D:_Q I$$
$$\operatorname{Hom}_D(I,I) \simeq I:_Q I.$$

The first of these modules is the *inverse* or *dual ideal* of I; it is denoted by I^{-1}. The ring of endomorphisms is also called the *multiplier ring* of I.

The actual determination can be carried out as follows:

Proposition 2.16. *Let* $L = (a_1, \ldots, a_n)$ *be an ideal with each* a_i *a regular element of* D. *Then:*

$$D:_Q L = ((a_1):_D (a_2, \ldots, a_n)) a_1^{-1} \tag{2.2}$$

$$L:_Q L = (\bigcap_{i=1}^{n} L \cdot (a_1 \cdots \widehat{a_i} \cdots a_n))(a_1 \cdots a_n)^{-1}. \tag{2.3}$$

Proof. We only verify the second assertion. Denote $a = \prod_{i=1}^{n} a_i$. We have

$$L:_Q (a_1, \ldots, a_n) = \bigcap_{i=1}^{n} L:_Q (a_i)$$

$$= \bigcap_{i=1}^{n} L a_i^{-1},$$

which upon multiplying and dividing by a yields the desired expression. $\square$

Let us sketch out how the ring

$$I:_Q I = \operatorname{Hom}_A(I,I)$$

can be represented. Suppose $I \subset A = R/\mathfrak{p}$, where $R = k[x_1, \ldots, x_n]$. Write $I = L/\mathfrak{p}$, and pick a representation $L = (\mathfrak{p}, a_1, \ldots, a_n)$, with $a_i \notin \mathfrak{p}$.

Proposition 2.17. *Let* $a = \prod_{i=1}^{n} a_i$ *and set*

$$(\bigcap_{i=1}^{n} L(a/a_i)) a^{-1} = (b_1, \ldots, b_m) a^{-1}.$$

For each b_j pick a new variable y_j and consider the ideal

$$J = (\mathfrak{p}, ay_1 - b_1, \ldots, ay_m - b_m).$$

Then $P = J : a^\infty$ is the presentation ideal of $I : I$, that is

$$I : I = k[x_1, \ldots, x_n, y_1, \ldots, y_m]/P.$$

Proof. The mapping

$$\varphi : k[x_1, \ldots, x_n, y_1, \ldots, y_m] \longmapsto I : I$$

where the x_i are mapped to their homomorphic images in $A \subset I : I$ and the y_j are mapped to $b_i a^{-1}$ is a surjection. The ideal J is obviously contained in the kernel K of φ, and therefore P is also contained in K. On the other hand, if we localize the mapping at the powers of a, $A_a = (I : I)_a$, since $I_a = A_a$. This implies that J_a is the kernel of φ_a, that is $J_a = K_a$, which establishes the assertion. $\square$

Remark 2.18. In Section 6.2, we will discuss a particular kind of endomorphism ring that has a more amenable approach to its calculation: when the ideal I is a dual, $I = \operatorname{Hom}_D(J, D)$.

To an ideal I of a Noetherian domain D one can attach another integral extension of D,

$$B = \bigcup_{n \geq 0} \operatorname{Hom}_D(I^n, I^n).$$

This ring has a well-known interpretation: If $A = D[It]$ is the Rees algebra of I and $X = \operatorname{Proj}(A)$, then $B = H^0(X, \mathcal{O}_X)$ and therefore is a finitely generated D–module by Serre's theorem (see [Har77, Theorem 5.2]). There does not seem to exist any particularly simple approach to the calculation of B.

Remark 2.19. A very delicate question, with no known natural approach, is that of finding the minimal number of generators of an ideal I of an affine ring R. Even deciding whether I is a principal ideal does not seem to be easy. Of course, if R is a domain, checking that I is invertible has a direct formulation: $I \cdot I^{-1} = R$.

Exercise 2.20. Let I and J be ideals of the affine integral domain R. Prove that the ideals are isomorphic in every localization $R_\mathfrak{p}$ if and only if $\operatorname{Hom}_R(J, I)$ is an invertible ideal of the ring $\operatorname{Hom}_R(I, I)$. Explain how this property can be verified.

Problem 2.21 (The Isomorphism Problem). Let A and B be two algebraic structures defined over a field k. The isomorphism problem is to decide whether they are isomorphic. If A and B are affine algebras given by generators and relations, how can one proceed? An instance that is already interesting is that of zero–dimensional rings. If A and B are graded modules over a ring of polynomials R, David Eisenbud has pointed out the following approach to determine whether A and B are isomorphic (as graded modules, that is the isomorphism must preserve the grading). Compute the graded module $C = \operatorname{Hom}_R(A, B)$ and test a generic linear combination of a basis of C_0, the vector space of forms of degree 0 of C.

2.3 Noether Normalization

One of the most ubiquitous procedures in computational commutative algebra is Noether normalization. Its role cannot be understated. Given an algebra A there are two natural frameworks to study it, through a presentation as an affine algebra, $A = R/I$, $R = k[x_1, \ldots, x_n]$, or viewing A as a finite module over a smaller ring of polynomials:

$$
\begin{array}{c}
R = k[x_1, \ldots, x_n] \\
\downarrow \\
k[z_1, \ldots, z_d] \longrightarrow A = R/I
\end{array}
$$

In this section we are concerned with the second representation. It is of great importance that the process be carried out very efficiently. It might be fair to say that almost anything that may be obtained from generic projections can be obtained through Noether normalizations (see Theorem A.60).

Definition 2.22. *Let $R = k[x_1, \ldots, x_n]$ be a polynomial ring and let I be an ideal of codimension $n - d$. A Noether normalization of $A = R/I$ is a polynomial ring*

$$
k[\mathbf{z}] \hookrightarrow A
$$

over which A is finitely generated as a module. At its best, $\mathbf{z}$ should be a subset of the $\mathbf{x}$'s:

$$
\{z_1, \ldots, z_d\} = \{x_{i_1}, \ldots, x_{i_d}\} \subset \{x_1, \ldots, x_n\}.
$$

In this case the ideal I is said to be in Noether position. *If R is a graded algebra generated by elements of degree 1, a Noether normalization $k[\mathbf{z}]$ generated by forms of degree 1 will be said to be* standard.

To be more useful it is important to have some sense of the structure of A as a $k[\mathbf{z}]$–module, particularly when it comes in the form of a presentation

$$
k[\mathbf{z}]^p \xrightarrow{\varphi} k[\mathbf{z}]^q \longrightarrow A \longrightarrow 0.
$$

Example 2.23. Let $R = k[\mathbf{x}^{\alpha_1}, \ldots, \mathbf{x}^{\alpha_m}]$ be a two–dimensional monomial subring of the ring of polynomials $k[x_1, \ldots, x_n]$. A Noether normalization generated by two of the $\mathbf{x}^{\alpha_i}$'s can be found as follows.

Pick two exponent vectors, α_1 and α_2 for simplicity, which are linearly independent. This implies that for any other vector α_3 (see Section 7.3) there are rational numbers p, q such that

$$
\alpha_3 = p\alpha_1 + q\alpha_2,
$$

and therefore integers a, b, c, with $a > 0$, such that

$$a\alpha_3 = b\alpha_1 + c\alpha_2.$$

If both $b, c \geq 0$ this means that $\mathbf{x}^{\alpha_3}$ is integral over $k[\mathbf{x}^{\alpha_1}, \mathbf{x}^{\alpha_2}]$, and we go on to consider α_4. If instead $b < 0$, then $c > 0$ and we have

$$c\alpha_2 = a\alpha_3 + (-b)\alpha_1,$$

and we keep $\mathbf{x}^{\alpha_1}, \mathbf{x}^{\alpha_3}$, and go on to the next monomial. It is clear that in the end we will realize a Noether normalization.

Picking Variables for Noether Normalization

A method to find $k[\mathbf{z}]$ ([BSV88]) is based on the following observation (see [Log89] for a detailed analysis).

Let $F = \{f_1, \ldots, f_m\}$ be a set of generators of the ideal I of the polynomial ring $k[x_1, \ldots, x_n]$. Suppose $f_1 \in F$ is monic in one of the variables, say x_n (see Theorem A.60 on how to achieve this condition). Let $J = I \cap k[x_1, \ldots, x_{n-1}]$. Then

$$k[x_1, \ldots, x_{n-1}]/J \hookrightarrow k[x_1, \ldots, x_n]/I \qquad (2.4)$$

is an integral extension. In particular, height $I = 1 + $ height J.

It will provide for systems of parameters and projective dimensions as well. We blur the distinction between an ideal and a set of generators.

Algorithm 2.24 *Let $F = \{f_1, \ldots, f_m\}$ be a set of generators of the ideal I.*

$V := \{x_1, \ldots, x_n\}$
$G := F$
while $G \neq \emptyset$ *do*
  if there is a monic polynomial in *the variable* x_i in G *then*
    $V := \{x_1, \ldots, \widehat{x_i}, \ldots, x_n\}$
    $G := G \cap k[x_1, \ldots, \widehat{x_i}, \ldots, x_n]$
  else effect a change of variables φ so that a monic polynomial occurs in $\varphi(G)$
  and $G := \varphi(G)$.

Note that when all the changes of variables are taken into account, we have a sequence of polynomials

$$f_i(y_1, \ldots, y_i) \in k[\mathbf{z}][y_1, \ldots, y_{n-d}] = k[x_1, \ldots, x_n],$$

monic in y_i, lying in the ideal I. This leads immediately to a presentation of A as a $k[\mathbf{z}]$-module.

Noether Complexity

Finding the monic polynomial requested by the algorithm requires a change of variables (which is always possible according to classical arguments; see Appendix A). If k is a sufficiently large field, this means effecting a coordinate change

$$y_i = c_{i1}x_1 + \cdots + c_{in}x_n, \ 1 \le i \le n,$$

such that the first new d variables $y_1, \ldots, y_d$ are in Noether position for A.

A measure of costs that has been used is that of *Noether complexity*: the minimum number of nonzero coefficients c_{ij}, for $i \le d$, that is needed to achieve Noether normalization. If I is a homogeneous ideal, it has been studied in conjunction with the Chow form of A, but mainly for theoretical purposes (see the comments in [EiS94, p. 155]).

When the codimension of I is low, it is often practical to find the variables in Noether position one at a time. Later in this section, we discuss a general guide on selecting the variables.

Parameters and Superficial Elements

The case of a homogeneous integral domain has an added interest for the role it plays in invariant theory. Here $A = R/I$, where I is a homogeneous ideal of codimension d. Noether normalization of A can proceed more speedily than outlined in the general case. Suppose A is a graded k–algebra generated by elements of degree ≥ 1. One then has the following criterion:

Proposition 2.25. *If the Krull dimension of A is d and $h_1, \ldots, h_d$ are homogeneous elements of degree > 0, then*

$$k[h_1, \ldots, h_d] \hookrightarrow A$$

is a Noether normalization if and only if $\dim_k(A/(h_1, \ldots, h_d)) < \infty$.

Such sets of elements are called *systems of parameters* of the algebra A. If $\mathfrak{p}_1, \ldots, \mathfrak{p}_r$ are the minimal primes of A with $\dim A/\mathfrak{p}_i = d$, it suffices to pick h_1 homogeneous lying outside of each $\mathfrak{p}_i$ to obtain $\dim A/(h_1) = d - 1$. The other elements are obtained by iteration.

Additional control over the choice of the h_i's can be exercised as follows. Suppose the A is generated by elements of degree 1 and $d > 0$. Let

$$H_A(n) = e_0 \binom{n+d-1}{d-1} - e_1 \binom{n+d-2}{d-2} + \cdots + (-1)^{d-1} e_{d-1}$$

be the Hilbert polynomial of A (see Appendix B). Assume now $\mathfrak{q}_1, \ldots, \mathfrak{q}_s$ is the set of all associated prime ideals of A. Pick h in A_1 but not lying in any $\mathfrak{q}_i$ which is distinct from the maximal ideal (A_1) (this may require k to be large enough).

Proposition 2.26. *The following hold:*

(a) $0:h$ *is annihilated by a power of* (A_1).
(b) *The Hilbert polynomial of* $A/(h)$ *is*

$$H_{A/(h)}(n) = e_0 \binom{n+d-2}{d-2} - e_1 \binom{n+d-3}{d-3} + \cdots + (-1)^{d-2} e_{d-2}.$$

Proof. The first assertion (see Proposition A.41) follows since for any prime ideal $\mathfrak{p} \neq (A_1)$, in the localization $A_{\mathfrak{p}}$ the image of h is a regular element and therefore $(0:h)_{\mathfrak{p}} = 0$. This means that the support of the module $0:h$ consists at most of the maximal ideal (A_1).

For (b), we consider the exact sequence of graded modules

$$0 \to 0:h \longrightarrow A(-1) \xrightarrow{h} A \longrightarrow A/hA \to 0.$$

Since from (a) $0:h$ is an Artinian module it vanishes for sufficiently large degrees. This implies that

$$H_{A/(h)}(n) = H_A(n) - H_A(n-1),$$

from which the claim follows. $\qquad\square$

The element h is said to be *superficial*.[1] If h is a regular element the entire Hilbert functions of A and $A/(h)$ are derivable from one another.

Rings Defined by Monomial Ideals

Noether normalization of an affine algebra $A = k[x_1, \ldots, x_n]/I$, where I is generated by monomials, has a very direct solution, according to the following.

Proposition 2.27 (Stanley). *Suppose* $\dim A = d$ *and denote by*

$$z_r = \sum x_{i_1} x_{i_2} \cdots x_{i_r}$$

the image in A of the r-th elementary symmetric function of the x_i's. Then the subring $k[z_1, \ldots, z_d]$ is a Noether normalization of A.

Proof. It suffices to prove that the homogeneous ideal $(z_1, \ldots, z_d)$ is primary for the maximal ideal of A. Let P be a homogeneous prime ideal containing I and these d symmetric polynomials. Since the minimal primes of I are generated by subsets of $\{x_1, \ldots, x_n\}$ and $\dim A = d$, P contains a subset of s indeterminates, $s \geq n-d$. Along with these variables, by hypothesis P contains the symmetric functions of all the remaining variables, in particular P contains all symmetric polynomials in the x_i's. It follows that $P = (x_1, \ldots, x_n)$. $\qquad\square$

When Noether normalizations by forms of degree 1 are required, the fact that the defining ideal I is monomial does not offer any special simplification.

---

[1] Someone's idea of a joke.

Ideal of Leading Coefficients

A strategy to find monic polynomials in Noether normalization is sometimes enhanced by the following observations.

Let S be a Noetherian ring and I be an ideal of the polynomial ring $S[x]$. Denote by $c(I)$ the ideal of S generated by the leading coefficients of all the elements of I. This ideal is interesting because it is at least as 'large' as I:

$$\text{height } c(I) \geq \text{height } I,$$

according to Theorem A.64; for the purpose of the statement we say that (1) has infinite height.

Suppose $S = k[x_1,\ldots,x_n]$ and I is an ideal of $S[x]$. Choose an ordering for $S[x]$ that is the lexicographic product of the degree order on $k[x]$ by some term ordering on $k[x_1,\ldots,x_n]$. We leave the proof of the following to the reader:

Proposition 2.28. *If $\{g_1(x),\ldots,g_s(x)\}$ is a Gröbner basis of I then the leading coefficients of the $g_i(x)$ generate $c(I)$.*

Exercise 2.29. Let I be an ideal of a polynomial ring $k[x_1,\ldots,x_n]$. Show that I is in Noether position if and only if $\sqrt{I}$ is also in Noether position.

2.4 Fitting Ideals

Let A be a ring and let E be a finitely generated A–module. We are going to attach to E a sequence of ideals which generalize the elementary divisors of modules over PID's. We discuss some circumstances under which these ideals extend the power of decision of the classical invariants.

Let

$$A^m \xrightarrow{\ \varphi\ } A^n \longrightarrow E \to 0$$

be a presentation of E. We assume that bases have been chosen in A^m and A^n so that φ is represented by a $n \times m$ matrix.

Definition 2.30. *For an integer $0 \leq j < n$, the jth Fitting ideal of E is the ideal $I_{n-j}(\varphi)$ generated by the minors of order $n - j$ of the matrix φ. They are denoted $F_j(E)$, and the notation extended by setting $F_j(E) = A$ if $j \geq n$.*

The notation $F_j(E)$ highlights the fact that these ideals do not depend on the matrix φ itself, a fact that follows easily from the Schanuel's lemma in homological algebra (or by determinantal calculations as originally done by Fitting).

The processing of these ideals is obviously very expensive. On some occasions, we are going to get away with the computation of only a few of their generators.

Remark 2.31. We point out some of the basic properties of the Fitting ideals of E, all readily derived from the definition.

- (annihilator E)$^n \subset F_0(E) \subset$ annihilator E.
- If $h : A \mapsto B$ is a ring homomorphism and E is a finitely generated module, then $F_j(E \otimes_A B) = h(F_j(E)) \cdot B$.
- For a prime ideal $\mathfrak{p}$, the localization $E_\mathfrak{p}$ is minimally generated by r elements if and only if $F_{r-1}(E) \subset \mathfrak{p}$ but $F_r(E) \not\subset \mathfrak{p}$. In particular, the Fitting ideals of E detect whether $E_\mathfrak{p}$ is a free $A_\mathfrak{p}$-module.
- If $E \to E'$ is a surjection then $F_j(E) \subset F_j(E') \ \forall j$.

Some features of the module are captured by their Fitting ideals and in the case of the classical:

Proposition 2.32. *Let R be a Noetherian ring of Krull dimension 1 and E a finitely generated R-module. If $F_0(E)$ is an invertible ideal of R, then E is an Artinian module of length $\ell(E) = \ell(R/F_0(E))$.*

Of more recent vintage is (see [Bru86] for this extension and its history):

Proposition 2.33. *Let φ be an $m \times n$ matrix with entries in the Cohen–Macaulay local ring R of dimension d. Suppose $m \leq n$ and denote $E = \mathrm{coker}\ \varphi$. If $F_0(E)$ is an ideal of height d then $\ell(E) = \ell(R/F_0(E))$.*

Re-assembling the Module

Let A be a PID and let E be a finitely generated A-module, with

$$\underbrace{0,\ldots,0}_{r},D_1,\ldots,D_s,\ D_1 \neq 0,$$

as its sequence of Fitting ideals. The module can be reconstructed as

$$E \simeq A^r \oplus \bigoplus_{j=1}^{s} A/(D_j : D_{j+1}).$$

In other words, one can reconstruct the module from its Fitting ideals. This property is not always valid however according to the following example. Let $A = \mathbb{Z}[2i]$ and let E be the module defined by the single relation $(2, 2i)$ and E' the quotient module defined by the relations $(2, 2i)$ and $(2i, -2)$. Note that E and E' are not isomorphic (prove it) but have the same Fitting ideals.

A delicate problem that occurs often is whether a homomorphism $\varphi : E \to E'$ of R-modules is an isomorphism. If φ is a homomorphism of graded modules that is either injective or surjective, one can use the Hilbert functions of E and E' to test for the property. For 'abstract' modules it is not very clear how to proceed and the circumstances that must be paid attention to, as indicated in the following test.

Proposition 2.34. *Let R be an integrally closed Noetherian domain and let E be a finitely generated R–module whose associated primes have height at most 1. If $\varphi : E \to E'$ is a surjection and E and E' have the same Fitting ideals then φ is an isomorphism.*

Proof. Let

$$0 \to L \longrightarrow R^n \longrightarrow E \to 0$$

be a presentation of E. Any quotient module E' is obtained by adding to L extra elements from R^n; thus a quotient submodule E' is defined by module of relations $L \subset L' \subset R^n$.

If $\mathfrak{p}$ is a prime ideal of height 1, $R_\mathfrak{p}$ is a discrete valuation domain and $F_j(E_\mathfrak{p}) = F_j(E'_\mathfrak{p})$ implies that $E_\mathfrak{p}$ and $E'_\mathfrak{p}$ are isomorphic, which in turn implies that $\varphi \otimes R_\mathfrak{p}$ is an isomorphism. This means that $L'_\mathfrak{p} = L_\mathfrak{p}$.

With the hypothesis on the associated primes of E, note that if x, y is a regular sequence on R, then (x, y)–depth of E is at least 1 and therefore the (x, y)–depth of L is 2, in other words, (x, y) is (locally) a regular sequence on L.

We claim that $L = \bigcap_\mathfrak{p} L_\mathfrak{p}$, for height $\mathfrak{p} = 1$. This will follow from the following basic facts about reflexive modules. $\qquad\qquad\square$

Let R be a ring and E an R–module. The abelian group $E^* = \mathrm{Hom}_R(E, R)$ has a natural structure of R–module; it is called the *dual module* of E. Its own dual is denoted E^{**}, and there is a natural mapping $\varphi_E : E \to E^{**}$. The module E is *reflexive* if φ_E is an isomorphism.

Proposition 2.35. *Let R be an integrally closed Noetherian domain and let E be a finitely generated torsionfree R–module. The following conditions are equivalent:*

(a) *E is reflexive;*
(b) *every (locally or globally) regular sequence x, y on R is a regular sequence on E;*
(c) *$E = \bigcap_\mathfrak{p} E_\mathfrak{p}$, height $\mathfrak{p} = 1$.*

Proof. See [BH93, Proposition 1.4.1]. $\qquad\qquad\square$

Splitting–off the Torsion Submodule

We discuss a version of the classical decomposition of modules over PID's [Lip69] that plays a role in our approach to regularity criteria.

Let E be a finitely generated module and let $F_0(E), F_1(E), \ldots$ be the sequence of its Fitting ideals. Suppose that $F_r(E)$ is the first non–vanishing of these ideals. If $F_r(E)$ contains regular elements, passing to the total ring of quotients Q of R, the module $E \otimes Q$ is projective (actually free) of rank r. In this case we shall say that E has *generic rank r*.

Proposition 2.36. *Let A be a Noetherian ring and E a finitely generated A–module. If $F_r(E)$ is an invertible ideal then proj dim $E = 1$ and its torsion submodule splits.*

Proof. We may assume that A is a local ring and if

$$\varphi = \begin{bmatrix} a_{11} & \cdots & a_{1m} \\ \vdots & \ddots & \vdots \\ a_{n1} & \cdots & a_{nm} \end{bmatrix}$$

is the presentation matrix of E, then $F_r(E)$ is generated by the determinant D of the submatrix ψ of size $n - r$ located in the upper left corner. Let $e_1, \ldots, e_n$ be the corresponding generators of E; from the relations

$$\sum_{j=1}^{n} a_{ij} e_j = 0 \ i = 1, \ldots, n - r,$$

and for each $1 \leq h \leq n - r$, multiplying by the cofactors of the hth column and adding we get

$$D e_h + \sum_{j=n-r+1}^{n} D_j e_j = 0.$$

Since D_j is divisible by D,

$$e_h + \sum_{j=n-r+1}^{n} (D_j/D) e_j$$

is annihilated by D and hence lies in the torsion submodule $T(E)$ of E. It follows that the images $e'_{n-r+1}, \ldots, e'_n$ of the last r elements generate $E/T(E)$. Since this module is generated by r elements and has the same rank as E, it is free.

Finally, it follows from Cramer's rule that the matrix ψ gives a presentation of the module $T(E)$, to establish the claim. □

Syzygies

Similar considerations lead to the explicit syzygies of special matrices. Let us assume that R is an integral domain and that the image of φ is a module of projective dimension at most 1. A general question is whether there is an identifiable set of elements in R^n that generate $\ker \varphi$. Let us at least identify some of these syzygies.

Suppose that rank $\varphi = r$ and denote by $v_1, \ldots, v_m$ the columns of φ. This means that given any submatrix ψ of $r+1$ columns has the property that $I_{r+1}(\psi) = 0$. To fix notation, suppose the chosen columns are $v_1, v_2, \ldots, v_{r+1}$. For a subset of $r+1$ rows, denote by $D_1, \ldots, D_{r+1}$ the (signed) cofactors of one of its rows. It follows that

$$\sum_{j=1}^{r+1} D_j v_j = 0, \tag{2.5}$$

and therefore the n-tuple $u = (D_1, D_2, \ldots, D_r, D_{r+1}, 0, \ldots, 0)$ lies in the kernel of φ. For each such u, consider the ideal $\mathfrak{D} = (D_1, \ldots, D_{r+1})$. Let $\mathfrak{D}^{-1} = \mathrm{Hom}_R(\mathfrak{D}, R)$ be the divisorial dual of $\mathfrak{D}$. Then $\mathfrak{D}^{-1}u \subset R^n$ and lies in $\ker\varphi$. Note that if R is a factorial domain then $\mathfrak{D}^{-1}u$ is simply the submodule generated by the element obtained by stripping from the vector u the gcd of all of its entries. We denote by $\mathfrak{K}$ the submodule generated by all these elements.

Proposition 2.37. *If* image φ *is a projective module then* $\ker\varphi = \mathfrak{K}$.

Proof. It is enough to prove the assertion at each localization $R_{\mathfrak{p}}$. Note that the similarly defined $\mathfrak{K}(\mathfrak{p})$ is just the localization of the globally defined $\mathfrak{K}$, since the same matrix elements were used and the construction of the $\mathfrak{D}^{-1}$ commutes with localization.

To prove the assertion in the case of a local ring, we may assume by Nakayama lemma that a subset of r column vectors, say the first r columns, generate image φ. For say the $(r+1)^{\text{th}}$ column v_{r+1} we must have a representation

$$v_{r+1} = a_1 v_1 + \cdots + a_r v_r$$

which when compared to equation (2.5) yields $a_i = -\frac{D_i}{D_{r+1}}$. In turn this shows that

$$w = (a_1, \ldots, a_r, -1, 0, \ldots, 0) \in \mathfrak{D}_{\mathfrak{p}}^{-1}u.$$

But the kernel of $\varphi_{\mathfrak{p}}$ is generated by elements such as w. $\square$

The following consequence follows a similar result of [AS95].

Corollary 2.38. *In addition, if R is a ring of polynomials and the entries of φ have degrees bounded by s then $\ker\varphi$ is generated by elements whose entries are of degrees at most $r \cdot s$.*

The torsion submodule of a module

Let $R = k[x_1, \ldots, x_n]$ be a ring of polynomials over the field k and E a finitely generated module defined over the integral domain $A = R/\mathfrak{p}$. We assume that E is given with a presentation over the ring R,

$$R^m \xrightarrow{\varphi} R^n \longrightarrow E \longrightarrow 0,$$

and set ourselves the task of determining the torsion submodule of E as a module over A.

We use the Fitting ideals of E as an R–module. Let $F_i(E) = I_{n-i}(\varphi)$, $i \geq 0$, be these ideals. First, the rank of E as an A–module is r, if r is the smallest integer such that $F_r(E)$ is not contained in $\mathfrak{p}$. Pick $f \in F_r(E) \setminus \mathfrak{p}$, and let Z be the image of φ in R^n. Set

$$L = Z :_{R^n} f^\infty = \bigcup_{s \geq 1} (Z :_{R^n} f^s).$$

Then $T = L/Z$ is the torsion submodule of E. This is clear since T/Z is just the kernel of the localization mapping

$$R^n/Z = E \longrightarrow E_f,$$

and E_f is a projective module over A_f, and therefore it is a torsionfree module over A.

We note that the operations are standard computation commands in several of the computer algebra systems; their outputs are usually in the form of a presentation of T/Z.

2.5 Finite and Quasi–Finite Morphisms

Given a homomorphism $\psi : A \mapsto B$ of affine domains, there are two very direct questions about its nature:

- When is ψ an integral, birational, or quasi-finite morphism?
- If B is contained in the integral closure of A, what is the conductor $\mathfrak{c}(B/A)$?

After replacing A by its image in B, let us recall that ψ is *finite* if B is finitely generated as a module over A, *birational* if B lies in the total ring of fractions of A, and *quasi–finite* if for each prime ideal $\mathfrak{p} \subset A$ the ring $B_\mathfrak{p}/\mathfrak{p}B_\mathfrak{p}$ is a finite dimensional vector space over $R_\mathfrak{p}/\mathfrak{p}R_\mathfrak{p}$.

Finite and Birational Morphisms

Some of these questions are easy to address. Let us consider the general one first. Suppose

$$\psi : A = k[x_1,\ldots,x_s]/(f_1,\ldots,f_m) \mapsto B = k[x_1,\ldots,x_s,y_1,\ldots,y_r]/J = k[\mathbf{x},\mathbf{y}]/J,$$

and we further assume that $J \cap k[x_1,\ldots,x_s] = (f_1,\ldots,f_m)$.

We bring in the technique of *tag variables* used in the study of morphisms (see [ShS86], [CT91]). If $g \in k[\mathbf{x},\mathbf{y}]$, we denote still by g its image in B, if no confusion arises.

Proposition 2.39. *Let $g \in k[\mathbf{y}]$, let t be a new indeterminate. Pick $x_1 < \cdots < x_s < t < y_1 < \cdots < y_r$ an elimination order of $k[\mathbf{x},\mathbf{y},t]$, and let G be a Gröbner basis of $(J,t-g)$. Then g is an element in the field of fractions of A if and only if G contains a polynomial of $k[\mathbf{x},t]$, linear in t, and is an element integral over A if and only if G contains a polynomial in $k[\mathbf{x},t]$, monic in t.*

Proof. Let us show the second assertion. If g is integral over A, there exists a polynomial

$$F(T) = T^n + a_{n-1}T^{n-1} + \cdots + a_0 \in A[T],$$

such that $F(g) \in J$. This implies that $F(t) - F(g) \in (J,t-g)$, and thus $F(t)$ will indeed occur in the Gröbner basis of $(J,t-g)$ for the given ordering. $\square$

Corollary 2.40. *If g lies in the field of fractions of A and $\{a_i t + b_i,\ i = 1,\ldots,m\}$ are the linear polynomials in $G \cap k[x_1,\ldots,x_s,t]$, then $A:_A g = (a_1,\ldots,a_m)$.*

Repeating the computation for each of the other y_i's, possibly with a new term order, will deal with the issue. It would be better to have it done with a single ordering.

Quasi–finite Morphisms

The structure of a quasi–finite morphism of finite type, $\psi : A \mapsto B$, is given by the Zariski Main Theorem:

Theorem 2.41 (ZMT). *Let C be the integral closure of A in B. Then B is quasi–finite over A if and only if B is a flat epimorphism over C.*

To avoid misunderstanding, we recall that the assertion means that B is flat over C and the canonical mapping

$$B \otimes_C B \longrightarrow B$$

is an isomorphism. The condition is straightforward to set up in simple cases.

Proposition 2.42. *Let $A = k[x_1,\ldots,x_r]/I \subset k[x_1,\ldots,x_r,y]/J$.*

(a) *B is quasi-finite over A if and only if the A–content ideal of J is A. More concretely, for each generator g of J, $g = a_n y^n + \cdots + a_0$, let $c(g) = (a_n,\ldots,a_0)$. Then B is quasi-finite if and only if*

$$(c(g),\ g \in J) = (1).$$

(b) *If $B_i = A[y_i] \subset C,\ i = 1,\ldots,s$ is a sequence of quasi–finite extensions of A, then $B = A[y_1,\ldots,y_s]$ is quasi–finite over A.*

Conductor

The conductor of a ring extension $A \hookrightarrow B$ is the annihilator of B/A. It many ways it mediates the relationship between A and B, and it would be interesting to have ways to compute it.

If B lies in the integral closure of A, picking a product order as above, but with a degree ordering for the y_i's, gives a set of module generators of B relative to A. It will consist of all

$$\mathbf{y}^\alpha = y_1^{\alpha_1} \cdots y_r^{\alpha_r},$$

which are not leading monomials of the elements in G.

The conductor is the ideal

$$\mathfrak{c}(B/A) = \bigcap_\alpha A :_A \mathbf{y}^\alpha$$

for all monomials $\mathbf{y}^\alpha = y_1^{\alpha_1} \cdots y_r^{\alpha_r}$. The equations of integral dependence give bounds for the exponents α_i, which is enormously inefficient!

Lacking a more direct determination of the conductor, in terms of the ideal J, it will be to interest to: (i) find elements that sit naturally in the conductor (discriminants, differents, etc.); (ii) give special classes of extensions whose conductors are more amenable to computation.

Let us introduce the following outrageous definition. If $\psi : A \mapsto B$ is a homomorphism, we say that an ideal $D \subset A$ is a *semiconductor* if it defines the same locus as $\mathfrak{c}(B/A)$, in other words, if $\sqrt{D} = \sqrt{\mathfrak{c}(B/A)}$.

Here are some examples (or near examples!) of such ideals.

- **Fitting ideal:** If the module structure of B as an A–module is known, there will be a presentation

$$A^m \xrightarrow{\varphi} A^n \longrightarrow B/A \to 0.$$

The Fitting ideal $F_0(B/A)$ is generated by the $n \times n$ minors of φ. One has

$$\mathfrak{c}(B/A)^n \subset F_0(B/A) \subset \mathfrak{c}(B/A).$$

- **Denominators:** Suppose it is already known that B lies in the integral closure of A. Then any of the two ideals

$$\prod_{i=1}^r (A : y_i) \subset \bigcap_{i=1}^r (A : y_i)$$

is a semiconductor.
- **Noether different:** It is defined for a general morphism $\psi : A \mapsto B$ as follows (see [Kun86, p. 383]). Let

$$0 \to \Delta \longrightarrow B \otimes_A B \xrightarrow{\mu} B \to 0$$

be the natural mapping given by multiplication, $\mu(u \otimes v) = uv$, and let $\mathfrak{J} = ann(\Delta)$. The Noether *different* is the ideal

$$\mathfrak{N}(B/A) = \mu(\mathfrak{J}).$$

Note that the ring $B \otimes_A B$ has a presentation

$$B \otimes_A B = k[x_1, \ldots, x_s, y_1, \ldots, y_r, Y_1, \ldots, Y_r]/(J(\mathbf{y}), J(\mathbf{Y})),$$

where $J(\mathbf{y}) = J$ and $J(\mathbf{Y})$ is another copy of J with the variables y_i's replaced by corresponding Y_i's. The ideal Δ is generated by the images of the $y_i - Y_i$, so that

$$\mathfrak{J} \equiv (J(\mathbf{y}), J(\mathbf{Y})) : (y_i - Y_i, i = 1, \ldots, r).$$

The quotient Δ/Δ^2 is the module of relative differentials $\Omega_{B/A}$. It easy to see that, it vanishes if and only if $\mathfrak{N}(B/A) = B$, in which case B is unramified over A. This makes the ideal $\mathfrak{n}(B/A) = A \cap \mathfrak{N}(B/A)$ a good predictor for the comparison $A = B$.

More precisely, it is clear that $\mathfrak{c}(B/A) \subset \mathfrak{n}(B/A)$. It is not true that $\mathfrak{n}(B/A)$ is always a semiconductor, according to the following example of Kunz. If $B = \mathbb{C}[x]$ and $A = \mathbb{R} + x\mathbb{C}[x]$, then $\mathfrak{n}(B/A) = A$ but $\mathfrak{c}(B/A) = x\mathbb{C}[x]$.

Nevertheless, the two ideals $\mathfrak{c}(B/A) \subset \mathfrak{n}(B/A)$ are often close. (Is there a cohomological description of the quotient?) The point is that by introducing a small set of tag variables, the Y_i's, we can carry out the computation with the algebra presentation instead of the much larger module presentation in order to obtain a sharp upper bound for $\mathfrak{c}(B/A)$.

Problem 2.43. Let A be an affine domain over the field k. How to decide whether A is a rational extension of a ring of polynomials $R = k[z_1, \ldots, z_d]$?

2.6 Flat Morphisms

Let A be a commutative ring and B an A–module. We recall that B is a *flat module* if whenever

$$0 \to E \xrightarrow{\varphi} F$$

is an injective homomorphism then the induced homomorphism

$$E \otimes_A B \xrightarrow{\varphi \otimes 1_B} F \otimes_A B$$

is also injective. Given the well-known properties of tensor products, this means that the functor $E \mapsto E \otimes_A B$ is exact. More concretely, to test whether an A-module B is flat, it suffices to test it on embeddings $0 \to I \longrightarrow A$, where I is an ideal. When A is Noetherian, one may just take prime ideals, an observation we use often.

Example 2.44. The simplest examples of such modules are the free A–modules (e.g. the ring of polynomials $A[x]$), followed by the localizations of A at a multiplicative set S. Often they arise out of constructions that combine those two processes. Let us consider an example of this.

Let R be a commutative ring and let $\varphi : J \to J$ be a mapping of the set J into itself. In the free R–module

$$E = \bigoplus_{\alpha \in J} Re_\alpha,$$

let us define a structure of $R[x]$–module by $x \cdot e_\alpha = e_{\varphi(\alpha)}$.

It is not difficult to describe the conditions under which E is a flat $R[x]$–module. This is the case if and only if: (i) φ is injective and (ii) no power φ^n, $n > 0$, has fixed points. That (i) and (ii) are necessary follow from the fact that x and $x^n - 1$ are regular elements of $R[x]$ and must remain nonzero divisors on E.

For the converse, define a relation "$\sim$" on J as follows: $\alpha \sim \beta$ if and only if there exists an integer $n \geq 0$ such that $\varphi^n(\alpha) = \beta$ or $\varphi^n(\beta) = \alpha$. It is clear that this is an equivalence relation. We write

$$E = \bigoplus E_{(\alpha)}, \quad \text{where } E_{(\alpha)} = \bigoplus_{\beta \in (\alpha)} R_\beta.$$

It is now easy to see that

$$E_{(\alpha)} \simeq \begin{cases} R[x] & \text{if } (\alpha) = \{\varphi^n(\beta),\, n \geq 0\} \text{ for some } \beta \in J \\ R[x, x^{-1}] & \text{if } \varphi \text{ acts transitively on the subset defined by } \alpha \end{cases}$$

Let $\psi : A \mapsto B$ be a homomorphism of finitely generated algebras over a field k or over $\mathbb{Z}$. In this section we are concerned with methods and tools to ascertain whether ψ makes B a flat module over A. Morphisms with such properties are very common and desirable in the study of mappings between algebraic varieties.

The aim is more specific than that of discussing general flatness criteria. The literature is very rich in details to look at this question but we require that the underlying algebra structure of B play a enhanced role beyond that of its module structure over A.

The standard Noetherian setting to discuss flatness is the following: A is a Noetherian ring, B is a finitely generated A–algebra and M is a finitely generated B–module, and one asks whether M is A–flat. Through the use of the idealization principle it is almost always possible to reduce to the case $M = B$.

Three legs in any computable test of flatness are efficient versions of

- Generic flatness
- Local criterion
- Reduction of Krull dimension

and strategies on combining them. After treating the first two, we show several instances of using them together following methods in the literature and [Vas97].

Torsionfreeness and Generic Flatness

Suppose R is an integral domain—and we have $\mathbb{Z}$ or a polynomial ring over a field in mind—and $A = R[T_1, \ldots, T_n]/I$. How does one decide whether A is a torsionfree R–module? The answer passes through a constructive and considerably shorter proof of the theorem of generic flatness ([Mat80, p. 156]).

Theorem 2.45 (Theorem of Generic Flatness). *Let R be a Noetherian domain and A an R–algebra of finite type. Then there exists $0 \neq f \in R$ such that A_f is R_f–free.*

Proof. Let $A = R[\mathbf{T}]/I = R[T_1, \ldots, T_n]/I$ and pick a term ordering $<$ for the monomials in the T_i's and denote by $in(I)$ the ideal of $R[\mathbf{T}]$ generated by the leading terms of all polynomials in I.

Let $a_i \cdot \mathbf{T}^{\alpha_i}$, $i = 1, \ldots, m$ be a generating set of monomials of $in(I)$. Setting $f = \prod_{i=1}^{m} a_i$, we claim that A_f is R_f-free. For the same monomial order, since f is a regular element of R, the initial ideal of the localization I_f is just $in(I)_f$ (i.e. $in(I_f) = in(I)_f$), and therefore its generators can be taken to be the monomials $\mathbf{T}^{\alpha_i}$. This means that $R_f[\mathbf{T}]/in(I_f)$ is now a free R_f-module on the standard monomials for I (i.e. those which are not multiples of the $\mathbf{T}^{\alpha_i}$'s). In this setting the normal form algorithm (Theorem 1.5) provides an isomorphism of R_f-modules

$$\text{NormalForm} : R_f[\mathbf{T}]/I_f \simeq R_f[\mathbf{T}]/in(I_f), \qquad (2.6)$$

so that these same monomials clearly form a R_f-basis of $R_f[\mathbf{T}]/I_f$. □

Corollary 2.46. *Let R be a Noetherian domain, $A = R[T_1, \ldots, T_n]/I$, $<$ a monomial term order, $in(I) = (a_1 \mathbf{T}^{\alpha_1}, \ldots, a_m \mathbf{T}^{\alpha_m})$ the associated initial ideal and set $f = \prod_{i=1}^{m} a_i$. Then*

$$A \text{ is } R\text{-torsionfree if and only if } I : f = I. \qquad (2.7)$$

Proof. The previous proof shows that A_f is R_f-free, while $I : f = I$ implies that the powers of f are regular on A, leading to the embedding $A \hookrightarrow A_f$. The converse is clear. □

This corollary can be used to determine when a $\mathbb{Z}$-algebra is flat. This is in view of the fact that Buchberger's algorithm to compute Gröbner bases works in nearly the same way over $\mathbb{Z}$ as over a field (see [AL90, Section 4.5]).

Local Criterion of Flatness

Let $\psi : A \mapsto B$ be a homomorphism of graded algebras. Denote by $\mathfrak{m}$ the maximal homogeneous ideal of A. To test that B is A-flat one appeals to the basic criterion of flatness ([Mat80, Theorem 49]):

Theorem 2.47. *Let A be a Noetherian ring, I an ideal of A and M an A-module. Suppose that $\bigcap_{n \geq 0} I^n M = 0$. Then M is A-flat if and only if M/IM is A/I-flat and $I \otimes_A M \overset{\varphi}{\simeq} IM$ where φ is the natural map.*

We prove the main special case of this theorem.

Theorem 2.48. *Let $\psi : (A, \mathfrak{m}) \mapsto (B, \mathfrak{n})$ be a local homomorphism of the local Noetherian ring $(A, \mathfrak{m})$ into the Noetherian local ring $(B, \mathfrak{n})$. A finitely generated B-module E is flat over A if and only if $\mathrm{Tor}_1^A(A/\mathfrak{m}, E) = 0$.*

Proof. To begin note that if M is a finitely generated A-module, then $\mathrm{Tor}_1^A(M, E)$ is a finitely generated B-module. Indeed, if

$$\mathbb{F}_\bullet : \qquad \cdots \longrightarrow F_n \longrightarrow \cdots \longrightarrow F_1 \longrightarrow F_0$$

is a projective resolution of M with the F_i finitely generated free A-modules, then $\mathbb{F}_\bullet \otimes_A E$ is a complex of finitely generated B-modules and thus its homology groups are finitely generated.

It will be enough to show that $\mathrm{Tor}_1(A/\mathfrak{p}, E) = 0$ for each prime ideal. If $\dim A/\mathfrak{p} = 0$ then $\mathfrak{p} = \mathfrak{m}$ and the hypothesis applies. Suppose that $\dim A/\mathfrak{p} > 0$ and that the statement holds for all primes of lower dimension. Let $a \in \mathfrak{m} \setminus \mathfrak{p}$, and consider the sequence

$$0 \to A/\mathfrak{p} \xrightarrow{a} A/\mathfrak{p} \longrightarrow A/(\mathfrak{p}, a) \to 0.$$

Tensoring with E yields the exact sequence

$$\mathrm{Tor}_1^A(A/\mathfrak{p}, E) \xrightarrow{a} \mathrm{Tor}_1^A(A/\mathfrak{p}, E) \longrightarrow \mathrm{Tor}_1^A(A/(\mathfrak{p}, a), E).$$

Since $\dim A/(\mathfrak{p}, a) < \dim A/\mathfrak{p}$, the module $A/(\mathfrak{p}, a)$ admits a filtration with factors $A/\mathfrak{q}$, with $\mathfrak{q}$ a prime properly containing $\mathfrak{p}$ and $\mathrm{Tor}_1^A(A/(\mathfrak{p}, a), E) = 0$ by induction. But then $\mathrm{Tor}_1^A(A/\mathfrak{p}, E) = a \cdot \mathrm{Tor}_1^A(A/\mathfrak{p}, E)$, which by Nakayama lemma implies that $\mathrm{Tor}_1^A(A/\mathfrak{p}, E) = 0$. $\qquad\square$

The point of this proof is that it does not require that B be Noetherian, B coherent and E finitely presented would suffice.

This result can be rephrased as a numerical criterion in the following manner:

Proposition 2.49. *Let $(A, \mathfrak{m})$ be a graded algebra with maximal homogeneous ideal $\mathfrak{m}$ and let $\psi : A \mapsto B$ be a homomorphism of graded algebras. Then B is a flat A-algebra if and only if the Hilbert series of $\mathfrak{m}B$ and $\mathfrak{m} \otimes_A B$ are identical, $H_{\mathfrak{m}B}(\mathbf{t}) = H_{\mathfrak{m} \otimes_A B}(\mathbf{t})$.*

Proof. We use Theorem 2.47 with $I = \mathfrak{m}$, $M = B$, so the preconditions of the criterion are realized. B is flat over A if and only if $\mathrm{Tor}_1^A(A/\mathfrak{m}, B) = 0$, in other words precisely when the canonical surjection

$$\varphi : \mathfrak{m} \otimes_A B \longrightarrow \mathfrak{m}B \to 0$$

is an isomorphism, a fact that will be detected by the Hilbert series.

The ideal $\mathfrak{m}B$ is handled in the obvious manner. For the module $\mathfrak{m} \otimes_A B$, we make use of a presentation of $\mathfrak{m}$,

$$A^m \xrightarrow{\varphi} A^n \longrightarrow \mathfrak{m} \to 0,$$

as A-module and get $\mathfrak{m} \otimes_A B$ as coker $(\varphi \otimes 1_B)$. These are the objects that are handed over to *Macaulay* or *CoCoA* for the computation of Hilbert-Poincaré series. $\qquad\square$

Algebras over Regular Rings

Let $R = k[x_1, \ldots, x_d]$ be a ring of polynomials and let A be a finitely generated R–algebra. We give now a method to decide whether A is R–flat.

We first require the following formulation of flatness:

Proposition 2.50. *Let R be a ring of global dimension d and let M be an R–module. Then M is R–flat if and only if for each $\mathfrak{p} \in \mathrm{Spec}(R)$ there exists a regular sequence $x_1, \ldots, x_g \in \mathfrak{p}$, $g = $ height $\mathfrak{p}$, which is also a regular sequence on M.*

Proof. For each prime ideal $\mathfrak{p}$, we have exact sequences

$$0 \to R/\mathfrak{p} \longrightarrow R/(\mathbf{x}) \longrightarrow C \to 0.$$

Note that the support of C consist of primes of codimension at least g. Tensoring with M yields the long exact sequence

$$\longrightarrow \mathrm{Tor}_{d+1}(C,M)$$
$$\longrightarrow \mathrm{Tor}_d(R/\mathfrak{p},M) \longrightarrow \mathrm{Tor}_d(R/(\mathbf{x}),M) \longrightarrow \mathrm{Tor}_d(C,M)$$
$$\longrightarrow \mathrm{Tor}_{d-1}(R/\mathfrak{p},M) \longrightarrow \mathrm{Tor}_{d-1}(R/(\mathbf{x}),M) \longrightarrow \mathrm{Tor}_{d-1}(C,M)$$
$$\cdots \qquad\qquad \cdots \qquad\qquad \cdots$$
$$\longrightarrow \mathrm{Tor}_1(R/\mathfrak{p},M) \longrightarrow \mathrm{Tor}_1(R/(\mathbf{x}),M) \longrightarrow \mathrm{Tor}_1(C,M).$$

Since $\mathrm{Tor}_{d+1}(\cdot,\cdot) = 0$, $\mathrm{Tor}_d(R/\mathfrak{p},M)$ vanishes along with $\mathrm{Tor}_d(R/(\mathbf{x}),M)$. Since C has a filtration with factors of the form $R/\mathfrak{q}$, for $\mathfrak{q}$ prime, $\mathrm{Tor}_d(C,M) = 0$ as well. We move down the next row to show that $\mathrm{Tor}_{d-1}(R/\mathfrak{p},M) = 0$ and complete the proof using induction. $\qquad\square$

Theorem 2.51. *Let R be a ring of polynomials in d variables over a field k and let A be a finitely generated R–algebra. Let $0 \neq f \in R$ be such that A_f is R–flat and let S be a Noether normalization of $R/(f)$. Then A is R–flat if and only if f is regular on A and A/fA is flat over S.*

Proof. Suppose for a given f the module A satisfies the stated conditions. For a given $\mathfrak{p} \in \mathrm{Spec}(R)$ we seek to build the regular sequence required by Proposition 2.50. There will be two cases to consider, according to whether f lies in $\mathfrak{p}$ or not.

If $f \in \mathfrak{p}$, the prime ideal $\mathfrak{p}/(f) \subset R/(f)$ lies over a prime ideal $\mathfrak{q} \subset S$ of the same height since $R/(f)$ is finite over S (actually $R/(f)$ is a free S–module). We build a regular sequence on A that begins with f and the lifts of the regular sequence on A/fA obtained from $\mathfrak{q}$ that the hypothesis guarantees.

Note that if we make sure, in the argument given in the proof of Proposition 2.50, that for all primes such that $f \in \mathfrak{p}$ we have f in the regular sequence, we can ensure here that $\mathrm{Tor}_i^R(R/\mathfrak{p},A) = 0$ for $i \geq 1$ and all such primes.

Suppose now that $f \notin \mathfrak{p}$ and consider the exact sequence

$$0 \to R/\mathfrak{p} \xrightarrow{\ f\ } R/\mathfrak{p} \longrightarrow R/(\mathfrak{p},f) \to 0.$$

Tensoring with A we have the exact sequence

$$\operatorname{Tor}_2^R(R/(\mathfrak{p},f),A) \to \operatorname{Tor}_1^R(R/\mathfrak{p},A) \xrightarrow{f} \operatorname{Tor}_1^R(R/\mathfrak{p},A) \to \operatorname{Tor}_1^R(R/(\mathfrak{p},f),A),$$

where the terms with $R/(\mathfrak{p},f)$ vanish by the previous observation. In turn this says that multiplication by f on the module $\operatorname{Tor}_1^R(R/\mathfrak{p},A)$ is an isomorphism and therefore

$$\operatorname{Tor}_1^R(R/\mathfrak{p},A) = \operatorname{Tor}_1^R(R/\mathfrak{p},A)_f = \operatorname{Tor}_1^{R_f}(R_f/\mathfrak{p}_f,A_f) = 0.$$

This argument says that we do not need to check the condition of Proposition 2.50 on the primes containing f.

For the converse, since A/fA is flat over $R/(f)$, and the latter is free over S, A/fA must be S–flat.   □

Remark 2.52. Here is a variation on a standard test of flatness that could be stated as follows: Let M be a module over a commutative ring R and let $f \in R$ be regular on M and on R. Then M is R-flat if and only if M_f is R_f–flat and M/fM is $R/(f)$–flat.

Proof. Setting $E = M/fM$ in the change of rings spectral sequence ([Wei94, Theorem 5.6.6])

$$E_2^{p,q} = \operatorname{Tor}_p^{R/(f)}(E, \operatorname{Tor}_q^R(R/(f),N)) \Longrightarrow \operatorname{Tor}_n^R(E,N),$$

we get that $\operatorname{Tor}_2^R(M/fM,N) = 0$ for any R–module N. This means that tensoring the exact sequence

$$0 \to M \xrightarrow{f} M \longrightarrow M/fM \to 0$$

by N gives a monomorphism induced by multiplication by f,

$$0 \to \operatorname{Tor}_1^R(M,N) \xrightarrow{f} \operatorname{Tor}_1^R(M,N).$$

On the other hand, tensoring $\operatorname{Tor}_1^R(M,N)$ by R_f we get the embedding

$$\operatorname{Tor}_1^R(M,N) \hookrightarrow \operatorname{Tor}_1^R(M,N)_f = \operatorname{Tor}_1^{R_f}(M_f,N_f) = 0,$$

and therefore $\operatorname{Tor}_1^R(M,N) = 0$, to establish the claim.   □

Corollary 2.53. *Let A be a finitely generated algebra over $R = k[x_1,\dots,x_n]$. If k is an infinite field there exists a linear change of the x_i's, $\mathbf{x} \mapsto \mathbf{z}$, and an R–regular triangular sequence*

$$\mathbf{f} = \{f_1(z_1,\dots,z_n), f_2(z_2,\dots,z_n),\dots,f_n(z_n)\},$$

such that A is R–flat if and only if $\mathbf{f}$ is a regular sequence on A.

General Algebras

Let $A = k[x_1,\ldots,x_d]/J$ and $B = k[x_1,\ldots,x_d,T_1,\ldots,T_n]/I$ with $J \subset I$. If the algebra A is not regular or A and B are both not graded, it is not clear how to test for flatness from operations effected on J and I as in the previous cases. We consider a few very special cases here using a mix of known results.

Proposition 2.54. *Let* $\varphi : R \mapsto A$ *be an injective homomorphism of affine algebras over a field k and let R_0 be a Noether normalization of R. Suppose R is an integral domain and there exists a regular algebra $R \hookrightarrow S$ which is finite over R. Then A is R–flat if and only if $A \otimes_R S$ is R_0–flat.*

Proof. By the result of [Fer69], it suffices to prove that $A \otimes_R S$ is S-flat. Since R_0 is also a Noether normalization of S, S is a flat R_0–module. By Proposition 2.50 the required regular sequences in S might as well be found in R_0.   □

Example 2.55. If R is an affine domain of dimension 1, we could take its integral closure for S. As an illustration, if $R = k[t^2,t^3]$, then an R–module A is (i) torsionfree if and only if it is torsionfree over $k[t^2]$, and (ii) it is flat if and only if $A \otimes_R k[t]$ is torsionfree over $k[t^2]$.

Proposition 2.56. *Let* $\varphi : A \mapsto B$ *be a homomorphism of affine algebras over a field k of characteristic zero, let A_0 be a Noether normalization of A. Suppose A is an integral domain which has isolated singularities, denote by J its Jacobian ideal and let $(f_1,\ldots,f_s) = J \cap A_0$. Then B is A–flat if and only if*

(i) B_{f_i} *is* $(A_0)_{f_i}$*–flat for* $1 \leq i \leq s$*, and*
(ii) $\sqrt{J} \otimes_A B$ *is a torsionfree A_0–module.*

Proof. At this point we may assume that A is a local ring. The first condition assures that $\mathrm{Tor}_A^1(A/\mathfrak{p},B) = 0$ for the regular primes of A, while (ii) takes care of the primes in the singular locus.   □

Remark 2.57. Combining this result and descent of flatness by a finite morphism, one has the means to test flatness over a two–dimensional domain A: If A is an affine algebra defined over a field of characteristic zero, pass to the integral closure of A, which has isolated singularities.

In characteristic zero it is possible to test flatness through repeated bouts of radicals computations and Noether normalizations as follows.

Proposition 2.58. *Let* $\varphi : A \mapsto B$ *be a homomorphism of affine algebras over a field k of characteristic zero. Suppose A is reduced and denote by J its Jacobian ideal. Let $f \in J$ be a regular element of A. Then B is A-flat if and only if*

(i) f *is a regular element on B,*
(ii) B_f *is* A_f*–flat,*
(iii) $\sqrt{fA} \otimes_A B \hookrightarrow B$*, and*

(iv) $B/\sqrt{fAB}$ is $A/\sqrt{fA}$–flat.

Proof. Conditions (iii) and (iv) enable the use of Theorem 2.47. To test (iii), it suffices to check whether f is regular on $\sqrt{fA} \otimes_A B$. $\square$

Remark 2.59. The reduction to the general case requires the embedding

$$\sqrt{(0)} \otimes_A B \hookrightarrow B.$$

Torsionfreeness in Tensor Powers

The methods described above are too much dependent on Noether normalizations. Inspired by the results and techniques of [Au61], we look for flatness in the torsion-freeness of powers of morphisms $\psi : A \mapsto B$.

Let A be an integral domain and let M be an A–module. If M is A–flat, any tensor power

$$M^{\otimes n} = \underbrace{M \otimes_A \cdots \otimes_A M}_{n}$$

is also A–flat and therefore torsionfree. Auslander proved that if A is an unramified regular local ring and M is a finitely generated A–module and $M^{\otimes n}$ is torsionfree, $n \geq \dim A$, then M must be A-free (later extended in [Lic66] to all regular rings).

In our case $M = B$ will be a Noetherian ring which is an A–algebra, so it will not be in general finite over A. As a cautionary note, suppose $A = k[x]$ and $M = k[x,x^{-1}] \oplus (k[x,x^{-1}]/k[x])$. It is easy to see that $M \otimes_A M = k[x,x^{-1}]$, but M is not torsionfree.

Proposition 2.60. *Let $(A, \mathfrak{m})$ be a regular local ring of dimension 2 and let B be a Noetherian A–algebra. Then B is A-flat if and only if $B \otimes_A B$ is a torsionfree A–module.*

Proof. It is not hard to see that we may assume that B is a local ring with maximal ideal $\mathfrak{M}$ with $\psi^{-1}(\mathfrak{M}) = \mathfrak{m}$. We are first going to show that B is torsionfree. Denote by B_0 its torsionfree part; B_0 is an ideal of B and defines the algebra

$$0 \to B_0 \longrightarrow B \longrightarrow \overline{B} \to 0. \tag{2.8}$$

Tensoring this sequence with B, we get

$$B_0 \otimes_A B \longrightarrow B \otimes_A B \longrightarrow \overline{B} \otimes_A B \to 0,$$

and therefore $B \otimes_A B = \overline{B} \otimes_A B$ since $B_0 \otimes_A B$ is a torsion module so that its image in $B \otimes_A B$ must vanish. Iterating we get $B \otimes_A B = \overline{B} \otimes_A \overline{B}$.

Now embed $\overline{B}$ into a flat module,

$$0 \to \overline{B} \longrightarrow F \longrightarrow C \to 0,$$

and tensor with $\overline{B}$,

$$0 \to \operatorname{Tor}_1^A(C,\overline{B}) \longrightarrow \overline{B} \otimes_A \overline{B} \longrightarrow F \otimes_A \overline{B} \longrightarrow C \otimes_A \overline{B} \to 0,$$

and therefore $\operatorname{Tor}_1^A(C,\overline{B}) = 0$, since it is a torsion module.

$\overline{B}$ is a torsionfree module and A is a regular local ring of dimension 2, therefore $\overline{B}$ has flat dimension at most 1 and therefore $\operatorname{Tor}_2^A(C,\overline{B}) = 0$. But this last module is also $\operatorname{Tor}_1^A(\overline{B},\overline{B})$. Tensoring (2.8) by $\overline{B}$, we have the exact sequence

$$\operatorname{Tor}_1^A(\overline{B},\overline{B}) \longrightarrow B_0 \otimes_A \overline{B} \longrightarrow B \otimes_A \overline{B} \longrightarrow \overline{B} \otimes_A \overline{B} \to 0,$$

from which we have that $B_0 \otimes_A \overline{B} = 0$. In view of the natural surjection of abelian groups

$$B_0 \otimes_A \overline{B} \longrightarrow B_0 \otimes_B \overline{B},$$

one has $B_0 \otimes_B B/B_0 = 0$. But this module is just B_0/B_0^2 and therefore $B_0 = B_0^2$ and $B_0 = 0$ by Nakayama lemma.

Let

$$0 \to F \xrightarrow{\alpha} G \longrightarrow B \to 0 \tag{2.9}$$

be a flat presentation of B. It suffices to prove that $\operatorname{Tor}_1^A(A/\mathfrak{m}, B) = 0$. Tensoring this sequence by $A/\mathfrak{m}$, we get an exact sequence of vector spaces over $A/\mathfrak{m}$,

$$0 \to \operatorname{Tor}_1^A(A/\mathfrak{m}, B) \longrightarrow \overline{F} \xrightarrow{\overline{\alpha}} \overline{G} \longrightarrow \overline{B} \to 0.$$

On the other hand, if we tensor the complex (2.9) by itself, and use that $\operatorname{Tor}_1^A(B,B) = 0$, we get an exact flat presentation

$$0 \to F \otimes_A F \xrightarrow{\varphi} F \otimes_A G \oplus G \otimes_A F \longrightarrow G \otimes_A G \longrightarrow B \otimes_A B \to 0, \tag{2.10}$$

where the cokernel of φ is flat since $B \otimes_A B$ is torsionfree and A has dimension 2. In particular, tensoring (2.10) with $A/\mathfrak{m}$ gives a complex of vector spaces which is exact at the left end. However, if $0 \neq u \in \operatorname{Tor}_1^A(A/\mathfrak{m}, B)$, $u \otimes u$ is a nonzero element of $\overline{F} \otimes_A \overline{F}$ that under the action of $\overline{\varphi}$

$$\overline{\varphi}(u \otimes u) = (u \otimes \overline{\alpha}(u), \overline{\alpha}(u) \otimes u) = 0.$$

This contradiction establishes the claim.   □

Conjecture 2.61. Let A be a regular ring of dimension d and let B be an algebra essentially of finite type over A. Then B is A–flat if and only if for some $n \geq d$ the algebra $B^{\otimes n}$ is torsionfree over A.[2]

---

[2] M. Kwieciński has since proved ([Kwi98]) that if A is an affine normal domain over $\mathbb{C}$ and B is a finitely generated A–algebra, then B is A–flat if and only if *all* tensor powers $B^{\otimes n}$ are torsionfree over A. A stronger assertion, that only requires that $B^{\otimes \dim A}$ be torsionfree over A has been proved by Galligo and Kwieciński ([GKi0]); it still requires that B is pure equidimensional.

Remark 2.62. The rigidity of Tor can be used to construct examples when $n < d$ does not suffice. Suppose R is a regular local ring of dimension d and $x_1, \ldots, x_d$ is a system of parameters. The module E defined by the exact sequence

$$0 \longrightarrow R \xrightarrow{(x_1, \ldots, x_d)} R^d \longrightarrow E \longrightarrow 0,$$

has the property that $E^{\otimes n}$ is torsionfree for all $n < d$ but not for $n = d$ (see [Au61]). It follows that the algebra $B = R \oplus E$ (the idealizer algebra of E) has the property that all $B^{\otimes n}$ are torsionfree for $n < d$ but not for $n \geq d$.

Exercise 2.63. Let A be a regular algebra finitely generated over a field. Establish Conjecture 2.61 when B is a birational extension of A.

Exercise 2.64. Let R be an Artinian local ring. Prove that R is Gorenstein if and only if the following property always holds: the tensor product of two faithful modules is faithful. Give an extension to arbitrary Noetherian rings.

Exercise 2.65. Let $(R, \mathfrak{m})$ be a Noetherian local ring, let φ be a $m \times n$, $m \geq n$, matrix with entries in the ring of polynomials $A = R[x_1, \ldots, x_d]$ and denote by I the ideal generated by the maximal sized minors of φ. If

$$\text{grade}(I) = \text{height}(L) = m - n + 1,$$

where L is ideal generated by the image of I in $R/\mathfrak{m}[x_1, \ldots, x_d]$, show that A/I is R–flat. For the case $m = n + 1$, the Hilbert–Burch theorem is useful.

2.7 Cohen–Macaulay Algebras

Let $A = k[x_1, \ldots, x_n]/I$ be an equidimensional algebra over the field k (that is, the minimal primes of A have the same dimension). Here we examine some tests to check whether A is a Cohen–Macaulay algebra using the tools developed in the previous two sections and Hilbert function theory discussed in the Appendix B. Some of these tests only check for the signature of the Cohen–Macaulay property in a manner of speaking.

Flatness

Suppose that $\dim A = d$, and let $R = k[z_1, \ldots, x_d] \hookrightarrow A$ be a Noether normalization.

Proposition 2.66. *A is a Cohen–Macaulay ring if and only if A is a flat R–module.*

Proof. For finitely generated modules over a Noetherian ring, "flat" and "projective" are equivalent. We then use Theorem A.98. $\square$

To test we may apply Proposition 2.51.

Fitting Ideals

Suppose the Noether normalization $R \hookrightarrow A$ comes detailed enough to provided a module presentation

$$R^m \longrightarrow R^n \longrightarrow A \rightarrow 0.$$

Proposition 2.67. *A is a Cohen–Macaulay ring if and only if all the Fitting ideals $F_i(A)$ are either (0) or (1).*

Hilbert Functions

Let us assume that A is a graded algebra generated over k by elements of degree 1 (such algebras are said to be standard; more generally if A is integral over $k[A_1]$ we say that A is semistandard; in the sequel the assertions apply to these extensions as well)

$$A = k + A_1 + \cdots + A_k + \cdots,$$

and let

$$H_A(t) = \frac{h_0 + h_1 t + \cdots + h_r t^r}{(1-t)^d}, \ h_r \neq 0,$$

be its Hilbert series. The integral vector $\mathbf{h}(A) = (h_0, h_1, \ldots, h_r)$ is the h-vector of A.

The comparison between the Hilbert function of an algebra A and some of its hypersurface sections $B = A/(f)$ is a time–honored approach to test the Cohen–Macaulay property (of modules as well).

Proposition 2.68. *Let A be a graded ring with Hilbert function $H_A(t)$ and let f be a homogeneous element of degree $s > 0$. Then f is regular on A if and only if*

$$H_{A/(f)}(t) = (1 - t^s)H_A(t).$$

In particular, if $s = 1$ then f is regular on A if and only if $\mathbf{h}(A) = \mathbf{h}(A/(f))$.

Proof. It suffices to consider the exact sequence

$$0 \rightarrow (0:_A f) \longrightarrow A(-s) \xrightarrow{f} A \longrightarrow A/(f) \rightarrow 0,$$

and observe that

$$H_{A/(f)}(t) = (1 - t^s)H_A(t) + H_{(0:_A f)}(t),$$

as we have done earlier. $\square$

Proposition 2.69. *Suppose that $z_1, \ldots, z_d$ is a homogeneous system of parameters. Then A is a Cohen–Macaulay ring if and only if $\mathbf{z}$ is a regular sequence on A. If the z_i's are of degree 1, the polynomial $h_0 + h_1 t + \cdots + h_r t^r$ is the Hilbert series of the Artin ring $A/(\mathbf{z})A$. In particular $h_i \geq 0$.*

Proof. The first assertion follows from Theorem 2.47. The other assertions follow from the exact sequence induced by an element of degree 1 which is regular on A,

$$0 \to A(-1) \xrightarrow{z} A \longrightarrow A/(z) \to 0$$

which yields

$$H_{A/(z)}(t) = (1-t)H_A(t)$$

by a standard calculation.   □

Remark 2.70. It is the very last assertion, that the $h_i \geq 0$, that can be used as a pretest for Cohen-Macaulayness. Of course, the same proof works for modules.

Here is another control for the degrees of certain extensions of Cohen–Macaulay modules ([BVV97]). The perceptive reader will recognize it as yet another version of the fundamental theorem for finitely generated modules over PIDs.

Proposition 2.71. *Let R be a standard graded algebra and let $A \subset B$ be finitely generated graded R–modules of the same dimension d and multiplicity e. If A and B are Cohen–Macaulay and*

$$H_A(t) = \frac{f(t)}{(1-t)^d}$$

$$H_B(t) = \frac{g(t)}{(1-t)^d}$$

are their Hilbert series, then $\deg f(t) \geq \deg g(t)$.

Proof. Consider the exact sequence

$$0 \to A \longrightarrow B \longrightarrow C \to 0$$

of graded modules. If $A \neq B$, C is a module of dimension $< d$ since A and B have the same multiplicity. Since A and B are assumed Cohen–Macaulay, by depth chasing C is also Cohen–Macaulay of dimension $d-1$.

We have the equality of Hilbert series,

$$\frac{g(t)}{(1-t)^d} = \frac{f(t)}{(1-t)^d} + \frac{h(t)}{(1-t)^{d-1}},$$

and therefore

$$g(t) - f(t) = (1-t)h(t).$$

The assertion follows since the h–vector of these modules are positive.   □

Additional constraints on the h_is occur if A is special, as in the following ([Sta91]) whose proof we just sketch:

Theorem 2.72. *Suppose A is a semistandard graded Cohen–Macaulay domain with* $\mathbf{h}(A) = (h_0, \ldots, h_s)$, $h_s \neq 0$. *Then*

$$h_0 + h_1 + \cdots + h_i \leq h_s + h_{s-1} + \cdots + h_{s-i}$$

for all $0 \leq i \leq s$.

Proof. Let ω_A be a canonical module of A: ω_A is a graded A–module and for convenience we shift degrees so that

$$\omega_A = \omega_0 + \omega_1 + \cdots, \quad \omega_0 \neq 0.$$

The Hilbert series of ω_A is

$$H_{\omega_A}(\mathbf{t}) = \frac{h_s + h_{s-1}\mathbf{t} + \cdots + h_0\mathbf{t}^s}{(1-\mathbf{t})^d}.$$

Since A is an integral domain, ω_A can be identified to an ideal of A, so that for any $0 \neq u \in \omega_0$, we have an exact sequence

$$0 \rightarrow A \simeq Au \longrightarrow \omega_A \longrightarrow C \rightarrow 0,$$

in which C has Krull dimension at most $d-1$. Actually, since both A and ω_A are Cohen–Macaulay of dimension d, C must have depth at least $d-1$, so that C is a Cohen–Macaulay module of dimension $d-1$.

From the equality

$$H_{\omega_A}(\mathbf{t}) = H_A(\mathbf{t}) + H_C(\mathbf{t}), \tag{2.11}$$

and the fact that C is Cohen–Macaulay,

$$H_C(\mathbf{t}) = \frac{c_0 + c_1\mathbf{t} + \cdots + c_s\mathbf{t}^s}{(1-\mathbf{t})^{d-1}}$$

has the property (according to Remark 2.70) that $c_i \geq 0$. A simple calculation in (2.11) will show that

$$c_i = (h_s + h_{s-1} + \cdots + h_{s-i}) - (h_0 + h_1 + \cdots + h_i),$$

to prove the assertion. $\qquad\square$

Multiplicities

A very useful Cohen–Macaulayness test is Theorem A.103. Let us phrase it for the case of a graded algebra. Let $A = k[x_1, \ldots, x_n]/I$ be a graded ring of dimension d. A system of parameters of degree 1 is a set of linear forms $z_1, \ldots, z_d$ such that $\dim A/(I, z_1, \ldots, z_d) = 0$. This means that the subring $R = k[z_1, \ldots, z_d]$ of A generated by the images of the z_i's is a Noether normalization of A.

Theorem 2.73. *Let $z_1, \ldots, z_d$ be a system of parameters for A. Then*

$$\dim_k(A/(I, z_1, \ldots, z_d)) \geq \deg A, \tag{2.12}$$

and equality holds if and only if A is a Cohen–Macaulay algebra. (If the z_i's have degrees different from 1, $\deg A$ is changed to $\deg A \prod_{i=1}^{d} \deg z_i$.)

Let us give a proof of a slightly changed version.

Proposition 2.74. *Let $B = k[\mathbf{z}] \hookrightarrow A = k[\mathbf{z}, \mathbf{x}]/I$ be a Noether normalization of A. Then*

$$\deg(B/A) \leq \dim_k(k[\mathbf{z}, \mathbf{x}]/(I, \mathbf{z})),$$

with equality holding if and only if $A_{(\mathbf{z})}$ is a free $B_{(\mathbf{z})}$–module.

Proof. This is special enough that we can argue directly. The assertion is that we have a local domain $(R, \mathfrak{m})$ $(= B_{(\mathbf{z})})$ of field of fractions K and a finitely generated R–module M $(= A_{(\mathbf{z})})$ such that

$$\dim_K M = \dim_{R/\mathfrak{m}} M/\mathfrak{m}M.$$

Let

$$0 \to L \longrightarrow R^n \longrightarrow M \to 0$$

be a minimal presentation of M, that is, $L \subset \mathfrak{m}R^n$. Tensoring with K, the assumption implies that $L \otimes_R K = 0$, and therefore $L = 0$ as desired. $\qquad\square$

Gorenstein Ideals

We formulate some of these criteria for the special case of Gorenstein rings.

Let $A = k[x_1, \ldots, x_n]/I$ be a homogeneous graded algebra. If height $I = m$, let $J = (f_1, \ldots, f_m)$ be a regular sequence of homogeneous elements of I. On the other hand, suppose $k[x_1, \ldots, x_d] \hookrightarrow A$, $d = n - m$, is a Noether normalization of A.

Proposition 2.75. *The algebra A is Gorenstein if and only if one of the following holds:*

(a) *I is a Cohen–Macaulay ideal and $J : I = (J, h)$, where h is a homogeneous polynomial.*

(b) *$\{x_1, \ldots, x_d\}$ is a regular sequence on A and if I_0 is the image of I in*

$$k[x_1, \ldots, x_n]/(x_1, \ldots, x_d),$$

then $I_0 : (x_{d+1}, \ldots, x_n) = (I_0, h)$.

Question 2.76. Let A be an affine graded algebra over the field k and denote by B its Segre square:

$$B = \bigoplus B_n, \quad B_n = A_n \otimes_k A_n.$$

If $H_A(\mathbf{t}) = \dfrac{f(\mathbf{t})}{(1 - \mathbf{t})^d}$ is the Hilbert series of A, how does one find the Hilbert series of B? In particular, if the h–vector $h(A)$ of A is positive, will $h(B)$ also be positive?

Problem 2.77 (Commuting Varieties). Let $\mathbf{X}$, $\mathbf{Y}$ be two generic $n \times n$ matrices over a field k, and denote by I the ideal of the ring of polynomials in $2n^2$ indeterminates generated by the entries of

$$\mathbf{X} \cdot \mathbf{Y} - \mathbf{Y} \cdot \mathbf{X}.$$

It is known that the variety $V(I)$ is irreducible (see [Vas94, Chapter 9] for a discussion). The expectation is that I itself is a prime ideal and that the variety $V(I)$ is normal.

The primality assertion would follow by showing that I is a Cohen–Macaulay ideal. This is a consequence of the fact that the equations define the symmetric algebra of a module over the ring of polynomials in one set, say $k[\mathbf{X}]$, generated by the $\mathbf{Y}$. Any symmetric algebra over an integral domain is an integral domain itself if there is a unique associated prime for its trivial ideal. To prove the Cohen–Macaulayness requires, according to Theorem 2.73, that for some system of parameters $\mathbf{z}$ of degree 1 forms,

$$\deg(k[\mathbf{X}, \mathbf{Y}]/I) = \ell(k[\mathbf{X}, \mathbf{Y}]/(I, \mathbf{z})).$$

For a clever solution when $n = 4$, see [Hre94].

3

Principles of Primary Decomposition

A turning point in the history of Modern Algebra was the abstract proof by Emmy Noether of primary decomposition of ideals ([Noe21]). It has the freshness and directness of Hilbert's proofs in invariant theory.

Let A be a Noetherian ring and let I be an ideal. The argument is composed of two steps.

- **Irreducible decomposition:** I is a finite intersection of irreducible ideals

$$I = L_1 \cap \cdots \cap L_r, \qquad (3.1)$$

and each irreducible ideal L_i is primary.
- **Primary (de)composition:** Collecting together those L_i with the same radical yields a primary decomposition of I:

$$I = Q_1 \cap \cdots \cap Q_s. \qquad (3.2)$$

It bears recalling that this representation arises simply from playing with the following: given $f \in R$ let n be such that $I : f^n = I : f^{n+1}$; then

$$I = (I, f^n) \cap (I : f). \qquad (3.3)$$

There is a measure of uniqueness in some of the Q_i but almost none amongst the L_j, beyond their numbers. The irreducible components may, on occasion, have additional structure that make (3.1) more appealing. For instance, if $R = k[x,y]$, k a field, then every irreducible ideal can be generated by two elements.

Each of these representations presents formidable difficulties for explicitly computing them. In this chapter we focus on another representation, which can be enabled along by the radical formulas of the next chapter. We consider approaches to primary decomposition that at worst require only Gröbner bases calculations over the field of rational functions over a ground field k. However this is not good enough as one would like to perform *all* the arithmetic over k and to be fully deterministic.

It can be argued, why do we care to find the primary decomposition of an ideal I and not leave it alone? There is great weight on this argument particularly because what makes I interesting often depends on how its components hang together. On the other hand, many properties of I are indeed connected to its primary components and we must develop the means to access them directly, or derive numbers indicative of their sizes, such as their multiplicities.

A facetious justification for studying primary decomposition is the following. Let $Q_1, \dots, Q_s$ be primary ideals whose generators are used to code some data and let I be their intersection. Suppose that some of these primary ideals have comparable radicals, in other words, the ideal I has embedded components. Given that the outcome of effecting a primary decomposition process on it may not return all the Q_i—as it depends on the how the process is put together—it makes likely that it can be used to hide the data from those without knowledge of the actual process.

Effective primary decomposition of ideals has received systematic study in classical [Her26], neo–classical [Sei74], [Sei84] and modern treatments [GH91], [GTZ88]. Our aim here will be to show how certain tools from Homological Algebra can be engaged in carrying out part of the task. It is an approach that cannot be up to the full problem. It will only lead the solution up to the point where refined factorization techniques come into play.

The major aim of this chapter is to bring to the fore many different aspects of primary decomposition of ideals and modules. From the outset the importance of prime ideals is highlighted, as the main goal of any theory of primary decomposition is to identify the prime ideals attached to the given ideal.

There are several models of primary decomposition. A scheme pursued here towards the primary decomposition of I,

$$I = Q_1 \cap \cdots \cap Q_s,$$

runs as follows. Let I be an ideal of the polynomial ring $R = k[x_1, \dots, x_n]$.

(i)   Express I as an intersection

$$I = I_1 \cap \cdots \cap I_r$$

   of equidimensional ideals I_i, with codim $I_i \neq$ codim I_{i+1}, whose associated primes are also associated primes of I.

(ii) For each I_i compute its radical $L_i = \sqrt{I_i}$.

(iii) For each radical ideal L_i effect its irredundant primary decomposition

$$L_i = P_{i1} \cap \cdots \cap P_{ir_i}.$$

(iv) For each prime P_{ij}, set I'_{ij} for the stable value of the quotient

$$I'_{ij} = I_i : P^{\infty}_{ij} = \bigcup I_i : P^p_{ij}, \qquad p = 1, 2, \ldots.$$

Setting $Q_{ij} = I_i : I'_{ij}$, it will follow that

$$\boxed{I = \bigcap_{ij} Q_{ij}.} \tag{3.4}$$

is an irredundant primary decomposition.

In this chapter we deal with step (i), and in Chapter 5 we address (ii). The last step is theoretically straightforward. It is (iii) that is most bothersome as it usually requires factorization over function fields; some aspects are discussed later in this chapter and in Chapter 4.

There are other models of primary decomposition, in which step (i) above is entirely bypassed (see [ShY96]). We shall discuss them in sections 3.3 and 3.4.

In Chapter 9, we use semi–numerical measures to look at the difficulties to be found in the computation of the radical of I, and for the purposes of predicting outcomes in the process of Noether normalization.

3.1 Associated Primes and Irreducible Decomposition

Throughout R is a Noetherian ring and modules are finitely generated. Attached to a module a small set of primes, the so-called *associated primes*, play a role similar to the decomposition of finitely generated abelian groups into torsion-free and torsion components. One of the goals is to identify the elements of R that act as injective endomorphisms of M and thereby mesh wonderfully with the theory of derived functors in homological algebra.

Zerodivisors and Associated Primes

Let M be a module over the Noetherian ring R. For $0 \neq m \in M$, the set (the annihilator of m)

$$0 : m = 0 :_R m = \{ r \in R \mid rm = 0 \}$$

is an ideal of R. Its elements are said to be *zerodivisors* of M. If R is a graded ring, M is a graded module and m is a homogeneous element, $0 : m$ is a homogeneous ideal.

More generally, we use the following notation: Given a module M, a submodule N and an ideal I,

$$N :_M I = \{ m \in M \mid Im \subset N \}.$$

Definition 3.1. *A prime ideal* $\mathfrak{p}$ *is associated to a module M if there exists an element* $e \in M$ *whose annihilator is* $\mathfrak{p}$. *The set of all these primes is denoted* $\mathrm{Ass}(M)$.

In a flagrant abuse of terminology we set:

Definition 3.2. *Let I be an ideal. The associated primes of the module* R/I *are called the* associated primes *of I.*

A related set of prime ideals is the *support* of M

$$\mathrm{Supp}(M) = \{\, \mathfrak{p} \in \mathrm{Spec}(R) \mid M_{\mathfrak{p}} \neq 0 \,\}.$$

It is naturally identified to $\mathrm{Spec}(R/I), I = \mathrm{annihilator}(M)$. One has

$$\text{minimal primes of } I \subset \mathrm{Ass}(M) \subset \mathrm{Supp}(M).$$

The following lists some of the properties of these sets of prime ideals.

Proposition 3.3. *Let M be a Noetherian R–module. Then*

(a) $\mathrm{Ass}(M) = \emptyset$ *if and only if* $M = 0$.
(b) $\bigcup_{\mathfrak{p} \in \mathrm{Ass}(M)} \mathfrak{p}$ *is the set of zero divisors on M.*
(c) $\mathrm{Ass}(M)$ *is finite.*
(d) *If S is a multiplicative set in R, then* $\mathrm{Ass}(M_S)$ *consists of prime ideals* $\mathfrak{p}_S$ *such that* $\mathfrak{p} \in \mathrm{Ass}(M)$ *and* $\mathfrak{p} \cap S = \emptyset$.

Proof. The first two and the last assertion are clear, since for $0 \neq e \in M$, if $I = 0 :_R e$, the annihilator of e, and $x \cdot y \in I$, then the annihilator of xe contains (I, y). From this it follows that if I is not prime, then it can be enlarged into a bigger annihilator.

To prove (c), consider first an exact sequence of R–modules,

$$0 \to M_1 \longrightarrow M \overset{\psi}{\longrightarrow} M_2 \to 0. \tag{3.5}$$

We claim that $\mathrm{Ass}(M) \subset \mathrm{Ass}(M_1) \cup \mathrm{Ass}(M_2)$. Indeed, suppose $\mathfrak{p}$ is the annihilator of $e \in M$. If $Re \cap M_1 = 0$, $\psi(e) \in M_2$ has for annihilator precisely $\mathfrak{p}$. On the other hand, if $Le = Re \cap M_1 \neq 0$, for $x \in L \setminus \mathfrak{p}$, the annihilator of the element xe is $\mathfrak{p}$.

If $\mathrm{Ass}(M)$ is not finite, pick a maximal submodule M_1 such that $\mathrm{Ass}(M_1)$ is finite. In $M_2 = M/M_1$, pick by (a) an element e' whose annihilator is the prime ideal $\mathfrak{p}$ and lift it to $e \in M$. We have the exact sequence

$$0 \to M_1 \longrightarrow (M_1, e) \longrightarrow R/\mathfrak{p} \to 0,$$

which by the preceding proves that $\mathrm{Ass}(M_1, e)$ is also finite. □

Let us highlight the following test:

Corollary 3.4. *Let R be a Noetherian ring and let I be an ideal. A prime ideal* $\mathfrak{p}$ *is associated to I if and only if*

$$I : (I : \mathfrak{p}) \subset \mathfrak{p}. \tag{3.6}$$

Proof. We may assume that $\mathfrak{p}$ is the unique maximal ideal of R. We have that $\mathfrak{p}$ is an associated prime of I if and only if $I \neq I{:}\mathfrak{p}$, a condition that is equivalent to the assertion.   □

The relationship between the minimal primes of the support of a module and associated primes can be expressed in terms of certain Fitting ideals as follows.

Proposition 3.5. *Suppose R is a regular ring and let M be a finitely generated module. For each integer $i \geq 0$, let I_i be the Fitting ideal defined in* Theorem A.85 *from a projective resolution of M. Then*

$$\mathrm{Ass}(M) = \bigcup_{i \geq 0} \{\text{minimal primes of } I_i \text{ of codimension } i\}.$$

Proof. We leave the proof as an exercise to the reader, who may want to peek ahead at Corollary 3.42.   □

The following abuse of terminology has become standard.

Definition 3.6. *An R–submodule S of M is said to be* primary *if $\mathrm{Ass}(M/S) = \{\mathfrak{p}\}$. It will then be referred as a $\mathfrak{p}$–primary module. An ideal I is* primary *if the module R/I is primary.*

It will follow that the intersection of primary submodules for the same prime $\mathfrak{p}$ is $\mathfrak{p}$–primary. It is easy to see that the annihilator of M/S is a primary ideal $\mathfrak{q}$ in the standard dressing: If $x \cdot y \in \mathfrak{q}$ and $x \notin \mathfrak{p}$ then $y \in \mathfrak{q}$.

Corollary 3.7. *Let $I \subset J$ be ideals of a Noetherian ring R. Then $I = J$ if and only if for every associated prime P of I, $I_P = J_P$.*

Proof. Apply Proposition 3.3 to the submodule J/I of R/I. Note that $\mathrm{Ass}(J/I) \subset \mathrm{Ass}(R/I)$.   □

Associated Primes and Projective Dimension

There are properties of Cohen–Macaulay rings which are useful in controlling primary decomposition.

Proposition 3.8. *Let I be an ideal of a Cohen–Macaulay ring R and let $\mathfrak{p}$ be an associated prime of I. Then* height $\mathfrak{p} \leq \mathrm{proj}\, \dim_R(R/I)$.

Proof. We may assume that the projective dimension of R/I is finite, since otherwise height $\mathfrak{p} < \infty$ proves the assertion. Localizing at $\mathfrak{p}$ can only lower the projective dimension of $(R/I)_{\mathfrak{p}}$ as an $R_{\mathfrak{p}}$–module while the height of its associated prime $\mathfrak{p}R_{\mathfrak{p}}$ is maintained. Now we use Theorem A.83.   □

Corollary 3.9. *Let $\mathbf{f} = \{f_1, \dots, f_s\}$ be a set of monomials in $R = k[x_1, \dots, x_n]$. Then any associated prime of $(\mathbf{f})$ has height at most s.*

Proof. By Theorem A.81, proj dim $_R R/(\mathbf{f}) \leq s$, and the assertion follows from the formula above.   □

Another way to control the codimension of associated primes is through their initial ideals.

Proposition 3.10. *Let R be a ring of polynomials with a term ordering $<$, let I be an ideal and let $I' = \mathrm{in}_<(I)$. Then any associated prime of I has codimension bounded by* proj $\dim_R(R/I')$.

Proof. It follows from Proposition 3.8 and Corollary B.12.   □

Irreducible Decomposition

Definition 3.11. *A submodule S of M is* irreducible *if it is not the intersection of two properly larger submodules.*

Theorem 3.12. *Every submodule of a Noetherian module is a finite intersection of irreducible submodules. Every irreducible submodule is primary.*

Proof. We just remark on the second assertion in the setting of modules. We can assume that we are dealing with the null submodule of a module M. If $\mathfrak{p}, \mathfrak{q}$ are distinct elements of $\mathrm{Ass}(M)$ and $\mathfrak{p} \not\subset \mathfrak{q}$, let e and f be elements of M with annihilators $\mathfrak{p}$ and $\mathfrak{q}$, respectively. If $Re \cap Rf \neq 0$, say $xe = yf \neq 0$, we have that $\mathfrak{p}xe = \mathfrak{p}yf = 0$ and therefore $y\mathfrak{p} \subset \mathfrak{q}$, so that $y \in \mathfrak{q}$ and thus $yf = 0$, which is a contradiction.   □

Proposition 3.13. *Let M be a Noetherian module. Then every irredundant representation of the null submodule as a finite intersection of irreducible submodules contains the same number of components.*

Proof. Let $0 = X_1 \cap \cdots \cap X_n = Y_1 \cap \cdots \cap Y_m$ be two such decompositions. Denote $L = X_2 \cap \cdots \cap X_n$ and set $N_j = L \cap Y_j$. Then

$$N_1 \cap \cdots \cap N_m \subset Y_1 \cap \cdots \cap Y_m = 0.$$

On the other hand, $(L + X_1)/X_1 \simeq L/(X_1 \cap L) \simeq L$, and so the null submodule of L is irreducible. But each $N_j \subset L$, which means that some $N_{j_0} = 0$. We have exchanged X_1 by Y_{j_0} in the decomposition:

$$0 = Y_{j_0} \cap X_2 \cap \cdots \cap X_n = Y_1 \cap \cdots \cap Y_m.$$

We can apply the same argument with respect to X_2, and so on, to eventually arrive at $m \leq n$.   □

Let us examine this invariant in some special cases and its behavior under hyperplane section.

Proposition 3.14. *Let $(R, \mathfrak{m})$ be a Noetherian local ring and let M be an R–module of finite length. Denote by M_0 the set of elements of M annihilated by $\mathfrak{m}$. M_0 is a vector space over $R/\mathfrak{m}$ whose dimension is the number of irreducible components of the null module of M.*

Proof. Let $J_1 \cap \cdots \cap J_n$ be an irredundant decomposition of the null submodule where each J_i is irreducible. Consider the natural injection

$$0 \to M \longrightarrow \bigoplus_{i=1}^{n} (M/J_i). \tag{3.7}$$

Since Ass $(M/J_i) = \{\mathfrak{m}\}$, there is an element $e_i \in M/J_i$ whose annihilator is $\mathfrak{m}$. Moreover, since the null submodule of M/J_i is irreducible, for any nonzero submodule M_i of M/J_i, $M_i \cap Re_i \neq 0$, and therefore $Re_i \subset M_i$, as the former is a simple module.

In the diagram (3.7), it is clear that M_0 maps into $\bigoplus_{i=1}^{n} (M/J_i)_0$. As $(M/J_i)_0 \simeq R/\mathfrak{m}$, our claim amounts to showing that M_0 maps onto $\bigoplus_{i=1}^{n} Re_i$. To show this to be the case, for each i, consider the submodule of M/J_i given by $N_i = (J_i + J_1 \cap \cdots \cap \widehat{J_i} \cap \cdots \cap J_n)/J_i$. As $e_i \in N_i$, we can choose an element $x_i \in J_1 \cap \cdots \cap \widehat{J_i} \cap \cdots \cap J_n$ as its coset representative. Since each x_i will vanish in M/J_k, for $k \neq i$, it will follow that $\sum_{i=1}^{n} Rx_i$, which is contained in M_0, maps onto $\sum_{i=1}^{n} Re_i$. $\qquad\square$

For a given Noetherian R–module, we now link its set of associated primes,

$$\mathrm{Ass}(M) = \{\mathfrak{p}_1, \ldots, \mathfrak{p}_m\},$$

to the number of $\mathfrak{p}$–primary components in an irredundant irreducible decomposition of the null submodule of M,

$$0 = J_1 \cap \cdots \cap J_n. \tag{3.8}$$

There are two sets of primes, $\mathrm{Ass}(M)$ and the associated primes to the modules M/J_i. In view of the natural embedding (3.7), we have that $\mathrm{Ass}(M) \subset \bigcup \mathrm{Ass}(M/J_i)$, where each of the latter consists of a single element. To show that these two sets are identical we can argue as follows. It is enough to consider a single J_i. Since M/J_1 is irreducible, of associated prime, say $\mathfrak{p}$, its submodule $(J_1 + J_2 \cap \cdots \cap J_n)/J_1 \subset M/J_1$ must also be $\mathfrak{p}$–primary. Since $0 = \cap_{i=1}^{n} J_i$ is irredundant, we can choose $f \neq 0$ in $(J_1 + J_2 \cap \cdots \cap J_n)/J_1$ with preimage $e \in J_2 \cap \cdots \cap J_n$ and $\mathrm{ann}(f) = \mathfrak{p}$. Thus $\mathfrak{p}e \subseteq J_1 \cap \cdots \cap J_n = 0$. As the annihilator of e also kills f, we conclude that $\mathfrak{p} \in \mathrm{Ass}(M)$.

We can now use Proposition 3.14 to describe the number of irreducible components in (3.8) that have the same associated prime.

Proposition 3.15. *Let $\mathfrak{p} \in \mathrm{Ass}(M)$ and denote by $\Delta_\mathfrak{p}(M)$ the submodule of M whose elements are annihilated by $\mathfrak{p}$. The number of irreducible $\mathfrak{p}$–primary components in (3.8) is $\dim_{k(\mathfrak{p})}(\Delta_\mathfrak{p}(M))_\mathfrak{p}$.*

Proof. Note that the module $(\Delta_{\mathfrak{p}}(M))_{\mathfrak{p}}$ is indeed a vector space over the residue field $k(\mathfrak{p})$ of the ring $R/\mathfrak{p}$. $\Delta_{\mathfrak{p}}(\cdot)_{\mathfrak{p}}$ is actually a functor that can be applied to any module. Applied to a component M/J_i, of associated prime $\mathfrak{q}$, we find that

$$\Delta_{\mathfrak{p}}(M)_{\mathfrak{p}} = \begin{cases} 0, & \text{if } \mathfrak{p} \neq \mathfrak{q}, \\ k(\mathfrak{p}), & \text{if } \mathfrak{p} = \mathfrak{q}. \end{cases}$$

This shows that

$$(\Delta_{\mathfrak{p}}(\bigoplus_{i=1}^{n} M/J_i))_{\mathfrak{p}} = k(\mathfrak{p})^s,$$

one copy of $k(\mathfrak{p})$ for each J_i such that $\mathrm{Ass}(M/J_i) = \{\mathfrak{p}\}$. It is now clear that the argument of Proposition 3.14 gives an isomorphism

$$(\Delta_{\mathfrak{p}}(M))_{\mathfrak{p}} \simeq k(\mathfrak{p})^s,$$

which proves the assertion. □

Definition 3.16. *Let M be a module and $\mathfrak{p}$ a prime ideal. The integer*

$$r_M(\mathfrak{p}) = \dim_{k(\mathfrak{p})} \Delta_{\mathfrak{p}}(M)_{\mathfrak{p}}$$

is the type *of M at $\mathfrak{p}$.*

Hyperplane Section

A point of interest in irreducible decompositions consists in comparing the decomposition of the null submodules of M and M/xM, where x is a regular element on M.

Proposition 3.17. *Let M be a finitely generated R–module and let $\mathfrak{p}$ be one of its associated prime ideals. Suppose x is a regular element on M and $\mathfrak{m}$ is a prime minimal over $(\mathfrak{p},x)$. Then $\mathfrak{m} \in \mathrm{Ass}(M/xM)$ and $r_{M/xM}(\mathfrak{m}) \geq r_M(\mathfrak{p})$.*

Proof. The assertion follows directly from the theory of the canonical module; the knowledgeable reader is invited to provide his own proof. We offer a proof involving a counting argument.

We may assume that $(R,\mathfrak{m})$ is a local ring. Let L be the submodule of M annihilated by $\mathfrak{p}$. Since $\mathfrak{m}$ is not associated to L, this module can be viewed as a torsion free module over the one–dimensional local domain $R/\mathfrak{p}$. Note that $r_M(\mathfrak{p}) = s$ is the dimension of the $k(\mathfrak{p})$–vector space $L_{\mathfrak{p}}$, in other words, s is the rank of L as an $R/\mathfrak{p}$–module. Denote by $\mathfrak{p}^*$ the image of $\mathfrak{p}$ in $R^* = R/(x)$. Finally let $0 : \mathfrak{p}^*$ be the submodule of $M^* = M/xM$ annihilated by $\mathfrak{p}$. In the canonical surjection

$$M \longrightarrow M^* \to 0,$$

the elements of L map into $0 : \mathfrak{p}^*$. On the other hand, if $f \in L$ is mapped to 0, then $f = x \cdot g$, and since $\mathfrak{p}f = 0$, and x is a regular element on M, $g \in L$. This means that we have an injection

$$0 \to L/xL \longrightarrow 0 : \mathfrak{p}^* \subset M/xM,$$

so that all we have to do is to show that $r_{L/xL}(\mathfrak{m}) \geq s$.

Lemma 3.18. *Let $(R, \mathfrak{m})$ be a local ring, let I be an $\mathfrak{m}$–primary ideal and let E be a finitely generated R–module. Then the modules $(0 :_E I)$ and $(0 :_E \mathfrak{m})$ have finite length and*

$$\ell(0 :_E I) \leq \ell(R/I) \cdot \ell(0 :_E \mathfrak{m}).$$

Proof. The modules defined are Artinian. To establish the length estimate, we argue by induction on $\ell(R/I)$, the formula being clear if $I = \mathfrak{m}$. In any event, the module R/I contains a submodule isomorphic to $R/\mathfrak{m}$,

$$0 \to R/\mathfrak{m} \longrightarrow R/I \longrightarrow R/J \to 0,$$

with $\ell(R/J) = \ell(R/I) - 1$. Applying the functor $\operatorname{Hom}_R(\cdot, E)$ to the exact sequence, we get the exact sequence,

$$0 \to \operatorname{Hom}_R(R/J, E) \longrightarrow \operatorname{Hom}_R(R/I, E) \longrightarrow \operatorname{Hom}_R(R/\mathfrak{m}, E),$$

or

$$0 \to (0 :_E J) \longrightarrow (0 :_E I) \longrightarrow (0 :_E \mathfrak{m}),$$

and thus $\ell(0 :_E I) \leq \ell(0 :_E J) + \ell(0 :_E \mathfrak{m})$, to prove the assertion. $\qquad\square$

We apply this to the proof of the Proposition, with $E = L/xL$ and $I = (\mathfrak{p}, x)$, so that we have $\ell(L/xL) \leq \ell(R/(\mathfrak{p}, x)) \cdot r_{L/xL}(\mathfrak{m})$. To complete the proof, one determines $\ell(L/xL)$. L is a torsion free $R/\mathfrak{p}$–module of rank s, so there exists an exact sequence

$$0 \to (R/\mathfrak{p})^s \longrightarrow L \longrightarrow C \to 0,$$

where C is a torsion module of the one–dimensional ring $R/\mathfrak{p}$, and therefore has finite length. Multiplication by x on this sequence, by the snake lemma, induces an exact sequence of modules of finite length

$$0 \to {}_xC \longrightarrow (R/(\mathfrak{p}, x))^s \longrightarrow L/xL \longrightarrow C/xC \to 0,$$

where ${}_xC$ is the kernel of the multiplication induced by x on C:

$$0 \to {}_xC \longrightarrow C \xrightarrow{x} C \longrightarrow C/xC \to 0.$$

Counting lengths in the first sequence, we obtain $\ell(L/xL) = s \cdot \ell(R/(\mathfrak{p}, x))$, since the second sequence implies that $\ell({}_xC) = \ell(C/xC)$. $\qquad\square$

Irredundant Primary Decomposition

There is, obviously, not much in the way of uniqueness in irreducible representations of submodules. A measure of invariance is obtained by collecting together the irreducible components with the same associated primes and dropping irrelevant factors from the intersection.

Definition 3.19. *A primary decomposition of the null submodule of M is an irredundant representation*

$$0 = J_1 \cap \cdots \cap J_m, \tag{3.9}$$

where the J_i are primary modules for distinct prime ideals.

Remark 3.20. We observe that if R is a graded ring and I is a homogeneous ideal then the constructions considered here ensure that I has a primary decomposition by homogeneous ideals.

Uniqueness and Non–uniqueness

The analysis of uniqueness in these decompositions proceeds nicely when we use Theorem A.12.

Proposition 3.21. *Let $\mathfrak{p}$ be a prime ideal. Then*

$$L = \bigcap_{n \geq 1} \mathfrak{p}^{(n)}$$

is the intersection of the primary components of (0) that are contained in $\mathfrak{p}$.

Proof. The intersection J of the components contained in $\mathfrak{p}$ is also the kernel of the canonical homomorphism $R \to R_{\mathfrak{p}}$. On the other hand, J is contained in each $\mathfrak{p}^{(n)}$. But by Theorem A.12, the image of L in $R_{\mathfrak{p}}$ is zero. □

Corollary 3.22. *Let J_1 and J_2 be primary components relative to the prime ideals $\mathfrak{q}$ and $\mathfrak{p}$. If $\mathfrak{q} \subset \mathfrak{p}$, there exists a $\mathfrak{p}$–primary ideal $J_2' \subsetneq J_2$ such that $J_1 \cap J_2' = J_1 \cap J_2$.*

Proof. In the localization $R_{\mathfrak{p}}$, for $n \gg 0$, one has that $\mathfrak{p}^n R_{\mathfrak{p}}$ is properly contained in $(J_2)_{\mathfrak{p}}$. Since the $\mathfrak{p}$–primary ideals in R and $R_{\mathfrak{p}}$ correspond bijectively, we have $J_1 \cap J_2 \subset \mathfrak{p}^{(n)} \subsetneq J_2$, and thus $J_2' = \mathfrak{p}^{(n)}$ will serve the purpose. □

Definition 3.23. *Let M be a module and $\mathfrak{p} \in \mathrm{Ass}(M)$. The ideal $\mathfrak{p}$ is an* embedded prime *of M if it contains properly another element of $\mathrm{Ass}(M)$, otherwise it is called a* minimal prime *of M.*

Proposition 3.24. *Let M be a Noetherian module and let $0 = J_1 \cap \cdots \cap J_m$ be an irredundant primary decomposition. If $\mathfrak{p}_i$ is a minimal prime of M and I is the product of the other associated primes of M, then $J_i = \Gamma_I(M) = H_I^0(M)$.*

Proof. Left as an exercise to the reader.

Remark 3.25. We could restrict the search of primary decomposition of ideals with embedded components by focusing on 'maximal' components instead of the choices above. Thus in a decomposition of primary ideals it may look more natural to increase L into a 'maximal' primary ideal (for the same radical)

$$I = J \cap L = J \cap L', \quad L \subset L'.$$

But the properties of such decompositions have not been exploited particularly with respect to the degrees of generating sets (in the case of rings of polynomials).

Monomial Ideals

It is straightforward—but may still be time consuming—to carry out primary decomposition on monomial ideals. It requires no Gröbner basis computation. It is based on the following elementary description of primary monomial ideals.

Proposition 3.26. *A monomial ideal $I \subset k[x_1,\ldots,x_n]$ is primary if and only if there exists a subset $\{y_1,\ldots,y_r\} \subset \{x_1,\ldots,x_n\}$ such that I is generated by monomials in the y_i and contains a power of each of these variables.*

We leave its proof to the reader. We point out two consequences:

Corollary 3.27. *Let I be a primary monomial ideal. Then:*

(a) *I is a Cohen–Macaulay ideal, and*
(b) *I is Gorenstein if and only if $I = (y_1^{a_1},\ldots,y_r^{a_r})$ for some subset of the variables as above.*

In particular, every irreducible monomial ideal is generated by powers of a subset of the indeterminates.

Proof. Part (a) is clear since I will be the extension to $k[x_1,\ldots,x_n]$ of a Cohen–Macaulay ideal of the ring $k[y_1,\ldots,y_r]$. As for (b), it will again be left to the reader as an exercise. $\square$

An outline of how to find the primary decomposition of a monomial ideal goes as follows (see [DEP82, p. 25]).

Algorithm 3.28 *Let I be an ideal generated by the set of monomials*

$$L = \{f_1,\ldots,f_m\} \subset k[x_1,\ldots,x_n].$$

- *If there is a power of each x_i in L, then I is primary.*
- *If a variable, say x_n, occurs in some f_i but does not occur as a simple power in L, let x_n^s be the largest power of x_n that appears as a factor. Then*

$$I = (I,x_n^s) \bigcap (I:x_n^s).$$

- *Repeat the previous two steps on (I,x_n^s) and on $(I:x_n^s)$.*
- *Returns a primary decomposition of I.*

The decomposition in the third step follows since s is the first integer such that $I:x_n^s = I:x_n^{s+1}$. Note that $(I:x_n^s)$ has been stripped of the variable x_n.

Finitely Generated Algebras

There are relative versions of the notion of sets of associated primes that apply to non–finitely generated modules as well. We briefly examine an instance.

Proposition 3.29. *Let R be a Noetherian ring and let A be a finitely generated R–algebra. For any finitely generated A–module M the set $\text{Ass}_R(M)$ is finite.*

Proof. The assertion is clear when $M \simeq A/Q$ for some prime ideal Q; in this case $\text{Ass}(M) = \{\mathfrak{q}\}$ where $\mathfrak{q} = R \cap Q$. (More precisely the inverse image of Q.)

Now observe that for any exact sequence of A–modules

$$0 \to L \longrightarrow N \longrightarrow A/Q \to 0,$$

$$\text{Ass}_R(N) \subset \text{Ass}_R(L) \cup \text{Ass}_R(A/Q).$$

Indeed, for any embedding $R/\mathfrak{p} \hookrightarrow N$, its image in A/Q either vanishes–in which case $R/\mathfrak{p}$ embeds into L–or embeds into A/Q–in which case $\mathfrak{p} = R \cap Q$–or there is a nontrivial submodule E of $R/\mathfrak{p}$ that embeds into L.

In the general case, we make use of the fact that M admits a filtration of A–submodules

$$M = M_1 \supset M_2 \supset \cdots \supset M_r = 0,$$

whose factors M_i/M_{i+1} are isomorphic to A/Q_i, for Q_i a prime ideal. We now induct on the length of the filtration. □

A graded algebra A over a ring R is *homogeneous* if A is generated over R by elements of degree 1, $A = R[A_1]$. The ideal generated by all elements of degree > 0 is denoted A_+.

Proposition 3.30. *Let R be a Noetherian ring and let A be a homogeneous finitely generated graded R–algebra. Let $M = \oplus M_n$ be a finitely generated graded A–module. Then for all large n*

$$\text{Ass}_R(M_n) = \text{Ass}_R(M_{n+1}).$$

Proof. Let $L = H^0_{A_+}(M)$ be the submodule of elements of M annihilated by some power of A_+. L is finitely generated as an R–module so that the quotient M/L has the property that $(M/L)_k = M_k$ for sufficiently large k. Replacing M by M/L we may thus assume that $H^0_{A_+}(M) = 0$.

Let $f_1, \ldots, f_s$ be a set of generators of A_1. Consider the mapping

$$M_n \xrightarrow{\psi} M^s_{n+1}, \quad \psi(x) = (f_1 x, \ldots, f_s x).$$

By assumption ψ is an embedding and we have

$$\text{Ass}_R(M_n) \subset \text{Ass}_R(M^s_{n+1}) = \text{Ass}(M_{n+1}).$$

Now apply the previous proposition. □

Problem 3.31. The use Proposition 3.21 led to the lack of uniqueness in primary decompositions for ideals with embedded components: Given such an ideal $I = J \cap L$, J and L primary, then one can replace L by a smaller primary ideal. It would be interesting to move in the opposite direction: Find a method that from $I = J \cap L$ produces a decomposition $I = J \cap L_0$, L_0 primary, but as large as possible.

Exercise 3.32. Let M be a finitely generated module over the commutative ring R. For any ideal I show that

$$\sqrt{\operatorname{ann}(M/IM)} = \sqrt{I + \operatorname{ann}(M)}.$$

Exercise 3.33. Describe how to test whether a prime ideal $\mathfrak{p}$ is a minimal prime of the ideal I.

3.2 Equidimensional Decomposition of an Ideal

Bits and pieces of Homological Algebra, particularly local duality theory (see [BH93, Chapter 3]) can be engaged to provide a decomposition of a polynomial ideal I as an intersection of known ideals of the same codimension.

Definition 3.34. *Let R be a Noetherian ring. An ideal I is said to be* height unmixed *if all the associated primes of I have the same codimension.*

When R is a polynomial ring, this means that the dimensions of the irreducible components of the algebraic variety $V(I)$ are the same. For such rings, I will be called *equidimensional* if the minimal associated primes have the same codimension.

Definition 3.35. *Let R be a Noetherian ring and let I be an ideal. The* grade *of I is the length of the longest regular sequence $z_1, \ldots, z_r \subset I$. If I is generated by a regular sequence we shall say that it is a* complete intersection. *A height unmixed ideal I is said to be* generically a complete intersection *if for each minimal prime P of I the ideal I_P is a complete intersection of R_P. A local algebra A is said to be a* complete intersection *if $A = R/I$, where R is a regular local ring and I is a complete intersection ideal.*

How does one test whether a given ideal I is a generic complete intersection? There are several devices of which perhaps the most direct is:

Proposition 3.36. *Let R be a Cohen–Macaulay ring and let I be an unmixed ideal of codimension g with a presentation*

$$R^m \xrightarrow{\varphi} R^n \longrightarrow I \to 0.$$

Then I is a generic complete intersection if and only if height $I_{n-g}(\varphi) \geq g + 1$.

Proof. This condition on the Fitting ideal expresses the fact that the I needs at most g generators at all primes of height g. Since by hypothesis I has codimension g, it will indeed be generated by g generators at each of its associated primes. $\qquad\square$

Definition 3.37. *Let I be an ideal of codimension m. Given a primary decomposition of I, let I_i be the intersection of those primary components of a given codimension i. If all the associated primes of I have the same codimension the ideal I is* height unmixed. *In general, in the representation*

$$I = I_m \cap I_{m+1} \cap \cdots \cap I_r \cap \cdots \cap I_q, \tag{3.10}$$

the ideal I_i is called the ith equidimensional component *of I.*

Remark 3.38. Although the ideals I_i are not uniquely defined for $i > m$, the initial intersections $I_m \cap I_{m+1} \cap \cdots \cap I_r$ have an intrinsic description:

$$I_m \cap I_{m+1} \cap \cdots \cap I_r = \{\, x \in R \mid \text{height } (I : x) \geq r + 1 \,\}.$$

More precisely, since some of the components may be missing, it is more convenient to interpret the notation as saying that

$$\text{height } I_{i+1} \geq 1 + \text{height } I_i,$$

and express the invariance accordingly.

Equidimensional Factors of an Ideal

We attach to an ideal I of a Gorenstein ring R a sequence of ideals J_r, $r \geq 0$, that are useful in obtaining an equidimensional decomposition for I. It is based on the notion of the hull of a module introduced in [EHV92].

Definition 3.39. *Let I be an ideal. For each non–negative integer r, the rth* equidimensional factor *of I is the ideal*

$$\text{hull}_r(I) = J_r = \text{annihilator}(\text{Ext}_R^r(\text{Ext}_R^r(R/I, R), R)). \tag{3.11}$$

This operation has several elementary properties that make verification of many assertions easy. For instance, for each integer r, the operation $\text{hull}_r(\cdot)$ commutes with localization. It may well occur that some of these ideals are improper, that is, equal to R. We discuss some of the properties of these ideals, for which we rely on [BH93, Chapter 3]. The point is to use the Ext functors as filters for the associated primes of modules.

Proposition 3.40. *Let R be a Gorenstein ring and E a finitely generated module. Let $\mathfrak{p}$ be a prime ideal of codimension r. Then $\mathfrak{p}$ is an associated prime of E if and only if it is a minimal prime of the support of $\text{Ext}_R^r(E, R)$.*

Proof. See Proposition 9.66 and the discussion preceding Proposition 9.7.   □

This changes the task of finding the associated primes of E into the task of finding the minimal primes among the associated primes of the modules $\text{Ext}_R^r(E,R)$. Iterating as given in the definition above has the effect of stripping from the modules $\text{Ext}_R^r(E,R)$ their other associated primes, some of which may not be among the associated primes of E.

Proposition 3.41. *Let R be a Gorenstein ring and let I be an ideal of codimension m. Let J_r denote the rth equidimensional factor of I. Then*

(a) $\text{hull}_r(I) = J_r = R$ *for* $0 \leq r < m$.
(b) $\text{hull}_r(I) = J_r$ *is either R or an equidimensional ideal of codimension r.*
(c) *If* $\mathbf{x} = \{x_1, \ldots, x_m\}$ *is a regular sequence contained in I, then*

$$\text{hull}_m(I) = J_m = (\mathbf{x}):((\mathbf{x}):I).$$

Proof. (a) The module $\text{Ext}_R^r(R/I,R)$ vanishes for $r < m$.

(b) First observe that the support of a module $\text{Ext}_R^r(M,R)$ has codimension at least r, since R is a Gorenstein ring and therefore for each prime ideal $\mathfrak{p}$ with height $\mathfrak{p} < r$, the injective dimension of $R_\mathfrak{p} < r$. We claim that J_r is the intersection of the primary components of codimension r of the annihilator of $\text{Ext}_R^r(R/I,R)$. Begin by observing that if the support of $\text{Ext}_R^r(R/I,R)$ has codimension $> r$ then $J_r = R$, because for the same reasons as above, the double Ext will vanish. More precisely, let E_0 be the subset of all elements of $E = R/I$ whose support has codimension $> r$; E_0 is a submodule of E and $\text{Ass}(F = E/E_0)$ consists of primes of codimension $\leq r$. Consider the long exact sequence of the functor Ext:

$$\text{Ext}_R^{r-1}(E_0,R) \longrightarrow \text{Ext}_R^r(F,R) \longrightarrow \text{Ext}_R^r(E,R) \longrightarrow \text{Ext}_R^r(E_0,R).$$

Since $\text{Ext}_R^{r-1}(E_0,R) = \text{Ext}_R^r(E_0,R) = 0$, $\text{Ext}_R^r(F,R) \simeq \text{Ext}_R^r(R/I,R)$. We already observed that $\text{Ext}_R^r(F,R)$ has no associated primes of codimension $< r$, let us show that it does not have associated primes of codimension $> r$. Let f be an element of R regular on F:

$$0 \to F \xrightarrow{f} F \longrightarrow F/fF \to 0.$$

In the cohomology sequence

$$\text{Ext}_R^r(F/fF,R) \longrightarrow \text{Ext}_R^r(F,R) \xrightarrow{f} \text{Ext}_R^r(F,R) \longrightarrow \text{Ext}_R^{r+1}(F/fF,R),$$

$\text{Ext}_R^r(F/fF,R) = 0$ since f is regular on F and therefore the support of F/fF has codimension $r+1$. This implies that f is regular on $\text{Ext}^r(F,R)$, which from the fact that F has no associated prime of codimension $> r$ proves that $\text{Ext}_R^r(F,R)$ is equidimensional (and similarly for the double Ext).

If I_0 is the intersection of the (unique) primary components of I of codimension m, as above we have

$$\text{Ext}_R^m(R/I_0,R) = \text{Ext}_R^m(R/I,R).$$

To show (c) we use the standard formula of Rees (see Proposition 9.71):

$$\text{Ext}_R^m(R/I, R) \simeq \text{Hom}(R/I, R/(\mathbf{x})) \simeq ((\mathbf{x}):I)/(\mathbf{x}) \simeq ((\mathbf{x}):I_0)/(\mathbf{x}).$$

This is the canonical module ω of R/I_0, so by local duality it follows that the annihilator of $\text{Ext}_R^m(\omega, R)$ is still I_0 since it is grade unmixed.   □

Corollary 3.42. *Let R be a Gorenstein ring and let I be an ideal of codimension m. Then*

$$\boxed{\text{Ass}(R/I) = \bigcup_{r \geq m} \text{Ass}(R/\text{hull}_r(I)).}$$ (3.12)

Unmixedness Test

We single out from Proposition 3.41.(c) a test to check whether an ideal has all of its associated primes of the same codimension.

Corollary 3.43. *Let R be a Gorenstein ring and I an ideal of codimension m. Let $\mathbf{x} = \{x_1, \ldots, x_m\} \subset I$ be a regular sequence. Then I is unmixed if and only if*

$$\boxed{I = (\mathbf{x}) : ((\mathbf{x}) : I).}$$

Simple Cryptography

The test above provides a simple system of cryptography in the following sense. We view it as saying that we have ideals J and I with the property

$$\underbrace{I}_{\text{wanted}} = \underbrace{J}_{\text{key}} : \underbrace{(J : I)}_{\text{given}}.$$

Actually $J : I$ must be modified before it is passed forward.

Let us consider a simple case. Suppose that J is an ideal generated by a set of powers of the variables of a ring of polynomials and I is an ideal generated by a set of monomials that contains J. Let L be the ideal generated by a generating set of monomials of $J : I$, with the *exclusion* of those generating J. We then have:

$$\boxed{I \rightsquigarrow J : I \rightsquigarrow L \mapsto L \rightsquigarrow J : L = J : (J : I) = I.}$$

It looks very hard to recover I from L without the knowledge of J. Note that in this case calculations with Gröbner bases are not required.

Equidimensional Decomposition

We show how the capability to compute the equidimensional factors leads to an explicit equidimensional decomposition.

We first illustrate the method with a simple example, an ideal I of a polynomial ring $k[x,y]$. Suppose $I = a \cdot J$, with J an ideal of codimension two. There are only two factors, $J_1 = (a)$ and $J_2 = J$. If a is a unit, I is already equidimensional. In general, set $I_1 = J_1$, and let I_2 be the stable value of $(I + J_2^n) : J_1$. The latter exists because each of these ideals contains J_2 so that the descending chain of ideals corresponds term wise to a descending chain in the Artinian ring R/J_2. I_2 is clearly equidimensional and we claim $I = I_1 \cap I_2$; for this we can localize and apply Krull intersection theorem.

In the general case the features that are automatic in this example, such as the stabilization step, must be made to happen. Let us indicate the steps. Let I be an ideal of codimension m and let J_r, for $r \geq m$, be its equidimensional factors.

- Set $I_m = J_m$; this is just the intersection of the primary components of I of codimension m.
- Suppose $I_m, \ldots, I_s$ are equidimensional ideals, height $I_j = j$, with the property that I and $L_s = I_m \cap \cdots \cap I_s$ coincide at any localization $R_{\mathfrak{p}}$ such that $\dim R_{\mathfrak{p}} \leq s$. By the previous step this already occurs for $s = m$.
- Consider the descending sequence of ideals

$$A_i = (I + J_{s+1}^i) : L_s.$$

We claim that the sequence of the $(s+1)$th factors of the A_i stabilizes. To prove this we are going to make several calculations.

Let J be any of these factors, and note $J_{s+1} \subset J$. Because they are both ideals with pure components of codimension $s+1$, we may assume that $(R, \mathfrak{p})$ is a local ring of dimension $s+1$ and both J and J_{s+1} are $\mathfrak{p}$–primary ideals.

From the previous step, we have that $I = L_s \cap L$, where L is an ideal of pure codimension $s+1$. Consider the cohomology sequence of the exact sequence

$$0 \to L_s/I \longrightarrow R/I \longrightarrow R/L_s \to 0.$$

We have the exact sequence

$$\mathrm{Ext}_R^{s+1}(R/L_s, R) \longrightarrow \mathrm{Ext}_R^{s+1}(R/I, R) \longrightarrow \mathrm{Ext}_R^{s+1}(L_s/I, R) \longrightarrow \mathrm{Ext}_R^{s+2}(R/L_s, R),$$

with the end terms of the sequence vanishing. This means that the annihilator of the module $\mathrm{Ext}_R^{s+1}(R/I, R)$, which is J_{s+1} by definition (we do not have to consider the double Ext at the top of the dimension), will also annihilate $\mathrm{Ext}_R^{s+1}(L_s/I, R)$ and thus its dual L_s/I. It shows that $J_{s+1} \cdot L_s \subset I$. This implies that the sequence

$$J_{s+1} \subset \cdots \subset J_{s+1}(A_n) \subset \cdots \subset J_{s+1}(A_1)$$

of equidimensional ideals of codimension $s+1$ will stabilize because it depends exclusively on what happens at the minimal primes of J_{s+1}.

The constructions prove the following:

Proposition 3.44. *Let q be the maximum of the two numbers, the exponent where the chain stabilizes and any integer that makes $J_{s+1}^q \cap L_s \subset J_{s+1} \cdot L_s$ to hold (such integers exist by the Artin–Rees lemma cf. Theorem A.11). Let I_{s+1} be the $(s+1)$th factor of $I + J_{s+1}^q$. Then I and $L_s \cap I_{s+1}$ agree in codimension $\leq s+1$.*

Example 3.45. To illustrate this whole process we look at the following ideal

$$I = (ax, ay + bx, az + by + cx, bz + cy, cz) \subset R = k[a, b, c, x, y, z].$$

The generators of I are the coefficients of the product of the polynomials $(a + bt + ct^2)(x + yt + zt^2)$. Let us go through the steps above, using *Macaulay*. We begin with the calculation of the equidimensional factors of I (height $I = 3$).

$$I_3 = J_3 = (a, b, c) \cap (x, y, z)$$
$$J_4 = J_5 = R$$
$$J_6 = (a^2, b^2 - ac, ab, c^2, ac, bc, z^2, cz, cx + by + az, -y^2 + xz,$$
$$ax, xy, bx + ay, x^2, cy + bz, yz, xz).$$

A calculation will show that $q = 3$ will satisfy the condition of the Proposition. The desired equidimensional factor is

$$
\begin{aligned}
I_6 = {}& (ax, bx + ay, cx + by + az, cy + bz, cz, a^6, a^4b^2 - a^5c, a^5b, \\
& a^4c^2, a^5c, a^4bc, a^2b^4 - 2a^3b^2c, a^3b^3, a^2b^2c^2 - a^3c^3, \\
& a^3b^2c, a^2b^3c - a^3bc^2, a^3bc^2, a^2c^4, a^3c^3, \\
& a^2bc^3b^6 - 3ab^4c, ab^5, b^4c^2 - 2ab^2c^3, ab^4c, b^5c - 2ab^3c^2, ab^3c^2, \\
& b^2c^4 - ac^5, ab^2c^3, b^3c^3 - abc^4, abc^4, c^6, \\
& ac^5, bc^5, z^6, y^2z^4 - xz^5, xyz^4, x^2z^4, yz^5, xz^5, y^4z^2 - 2xy^2z^3, \\
& xy^3z^2 - x^2yz^3, x^2y^2z^2 - x^3z^3, y^3z^3, xy^2z^3, x^3z^3, x^3yz^2, \\
& x^2yz^3, x^4z^2, y^6 - 3xy^4z, xy^5 - 2x^2y^3z, x^2y^4 - 2x^3y^2z, \\
& y^5z, xy^4z, x^3y^2z, x^3y^3 - x^4yz, x^2y^3z, \\
& x^4y^2 - x^5z, x^4yz, x^5z, x^5y, x^6).
\end{aligned}
$$

A final checking will certify that $I = I_3 \cap I_6$.

Remark 3.46. Motivated by this example and the regularity it displays this family of ideals was studied in [CVV98].

Let R be a commutative ring. For a polynomial $f = a_0 + a_1 t + \cdots + a_d t^d \in R[t]$, we define its *content* to be the ideal of R generated by its coefficients, $c(f) = (a_0, a_1, \ldots, a_d)$.

Let now R be a Noetherian integral domain and let

$$f = x_0 + x_1 t + \cdots + x_m t^m \quad \text{and} \quad g = y_0 + y_1 t + \cdots + y_n t^n,$$

be generic polynomials of degrees m and n over R. The Gaussian ideal $G(f,g) = c(fg)$ has a primary decomposition

$$c(fg) = c(f) \cap c(g) \cap [c(fg) + c(f)^{n+1} + c(g)^{m+1}].$$

Furthermore, if R is a Gorenstein ring then

$$L(f,g) = c(fg) + c(f)^{n+1} + c(g)^{m+1}$$

is a Gorenstein ideal.

We note that this was not the decomposition produced by the method. Instead primary ideals, smaller than $L(f,g)$, were obtained.

Exercise 3.47. Let f,g,h be 3 generic polynomials as above. Give a primary decomposition for the ideal $c(fgh)$.

3.3 Equidimensional Decomposition Without Exts

One drawback of the preceding approach to equidimensional decomposition lies in its over reliance on the functor Ext. Basically it was required to ensure that in the decomposition

$$I = I_m \cap I_{m+1} \cap \cdots \cap I_r \cap \cdots \cap I_q,$$

the associated primes of each I_r are also associated to I.

One can do away with this last requirement. Such representations are called *coarse* equidimensional decomposition. By giving us more room they are simpler to deal with, particularly when the dimension of R/I is small. In contrast, the other *strict* representation has the benefit that requires computation only at those codimensions where associated primes of I are present.

We sketch out the method.

Proposition 3.48. *Let R be a Gorenstein ring and let I be an ideal of codimension m and let I_m be its equidimensional component of codimension m. A coarse equidimensional decomposition of I can be iteratively found as follows:*

(i) *Let $\mathbf{x} = x_1,\ldots,x_m$ be a regular sequence in I. Set*

$$I_m = (\mathbf{x}) : ((\mathbf{x}):I).$$

(ii) *Set $K = I : I_m$ and let r be an integer such that $K^r \cap I_m \subset I$.*

Then

$$I = I_m \cap I', \quad I' = K^r + I,$$

where I_m is an equidimensional ideal of codimension m and I' is an ideal of codimension at least $m + 1$.

Proof. As observed earlier, I_m is the intersection of all primary components of I of codimension m. This implies that $I:I_m$ cannot be contained in any prime ideal of codimension m.

The existence of the integer r is guaranteed by the Artin–Rees lemma applied to the ideals $A = I_m/I$ and $B = K/I$ of the ring R/I: There exists c such that for all $n > c$,

$$B^n \cap A = B^{n-c}(B^c \cap A) = (0),$$

which suffices to establish the claim. $\square$

Irrelevant Components

The appeal to Artin–Rees stability here seems unavoidable. It creates several matters of concern:

(a) The representation $I = I_m \cap I'$ will not ensure that the associated primes of I' are also associated to I. This means that many other primes may be involved that are not related to I but just on the choice of I', a fact that may not be so significant if $\dim R/I$ is small but can become important when $\dim R/I$ is large.

(b) On the other hand, when applying the same process to I' we already have part of a regular sequence. This is a consequence of the fact that $K = I : I_m = I' : I_m$ and this ideal satisfies

$$\text{height}(I : I_m) = s > m, \quad \sqrt{(I : I_m)} = \sqrt{I'}.$$

Thus, if $s = m+1$ for instance, we pick $y \in K^r + I$ such that $\text{height}(x_1, \ldots, x_m, y) = m+1$. We use this sequence to determine the component of I' of codimension $m+1$. If s is larger we must find fresher elements to lengthen the sequence until we get one of codimension s.

(c) Suppose that for each I_i we have an irredundant primary decomposition

$$I_i = \bigcap_j Q_{ij}.$$

Stripping from the list of primary ideals those Q_{ij}'s whose radicals are not associated to I (for which end one can use earlier tests), gives the equidimensional ideal I'_i or R. It is now easy to see that

$$I = \bigcap_{i \geq m} I'_i = \bigcap_{i \geq m} \bigcap_j Q'_{i,j}, \tag{3.13}$$

is an irreducible decomposition of I.

(d) Suppose that R is a ring of polynomials and

$$\text{proj dim}_R R/I = t \geq m.$$

Then by Proposition 3.8, all the associated primes of I have codimension at most t. This implies that we can safely ignore all components I'_i, for $i > t$:

$$I = I_m \cap I'_{m+1} \cap \cdots \cap I'_t. \tag{3.14}$$

Exercise 3.49. Let x, y, u, v be independent indeterminates over a field k. Find an equidimensional decomposition of the ideal

$$I = (x^2 + y^2, xu + yv, u^2 + v^2).$$

Exercise 3.50. Let I be a perfect, Gorenstein ideal of codimension 3. Prove that I is generically a complete intersection if and only if $I^2 : I = I$.

Problem 3.51. Find a mechanism to capture the initial intersections $I_m \cap I_{m+1} \cap \cdots \cap I_r$ in Definition 3.2.1 without having to go through the whole process of finding equidimensional decompositions.

Problem 3.52. Suppose it is known that the ideal I has only 2 associated prime ideals. Develop an effective method to find its primary decomposition.

3.4 Mixed Primary Decomposition

There are approaches to primary decomposition do not emphasize equidimensional components. We discuss some of these now. First however we describe a tool that is common to all approaches–the technique of localization.

Other methods to look at the primary decomposition of an ideal I are an outgrowth of (3.3). Here is an instance:

Proposition 3.53. *Let I and J be ideals of a Noetherian ring R. Then for all large integers n,*

$$I = (I, J^n) \cap (I{:}J).$$

Proof. We apply the Artin–Rees lemma (Theorem A.11) to the filtration defined by the powers of J. There exists an integer r such that for all $n > r$,

$$J^n \cap (I{:}J) = J^{n-r}(J^r \cap (I{:}J)) \subset I,$$

from which the assertion follows. □

The issue is for a given I to make judicious choices for J that make both $I{:}J$ and (I, J^n) are more amenable.

Localization

The notion of *localization* gives the means to retrieve some information encoded in ideals, particularly with regard to their primary components. It is also a mechanism to generate non–trivial primary ideals associated to a given prime ideal.

Definition 3.54. *Let R be a ring, I an ideal and S a multiplicative set of R. The localization of I at S is the kernel $_SI$ of the natural homomorphism*

$$R \mapsto (R/I)_S.$$

In other words, $_SI = \{x \in R \mid \exists\, t \in S,\ tx \in I\}$. In particular if R is an integral domain then $_SI = R \cap IR_S$.

Another name for this notion is *saturation* of the ideal I with respect to the multiplicative set S. We follow the most current usage in dealing with primary decomposition.

For the purpose of studying primary decomposition, the following examples are the most significant:

$$S = \begin{cases} R \setminus \mathfrak{P}, & \text{for some prime ideal } \mathfrak{P}; \\ \{f^n,\ n \geq 0\}; \\ 1 + J, & \text{for some ideal } J. \end{cases}$$

The difficulty with the first example lies in the fact that the multiplicative set is only given indirectly, as the complement of $\mathfrak{P}$. We consider a few special cases of interest.

In [GTZ88], the following method is given to determine some localizations. Let R be a principal ideal domain of quotient field K and suppose an algorithm is given to find Gröbner bases of ideals over of polynomial ring $R[\mathbf{x}]$ (e.g. $R = \mathbb{Z}$ or $R = k[t]$).

Proposition 3.55. *For an ideal $I \subset k[\mathbf{x}]$ it is possible to find an element $f \in R$ such that*

$$I \cdot K[\mathbf{x}] \cap R[\mathbf{x}] = I \cdot R_f[\mathbf{x}] \cap R[\mathbf{x}].$$

Proof. The proposed f is obtained as follows. Let $\{g_1, \ldots, g_r\}$ be the Gröbner basis of I *computed with respect to R*. The initial term of g_i has the form $\text{in}(g_i) = s_i \mathbf{x}^{\mathbf{a}_i}$. Take f to be the product of the s_i's. $\qquad\square$

Detaching Components

The following properties of localization provide us with tools to isolate special primary components of ideals.

Let R be a Noetherian ring and let I and J be ideals. There is one elementary property of the quotient operation $I{:}J$ *vis-à-vis* the primary decomposition of I that is useful to keep in mind in the sequel. Let

$$I = Q_1 \cap \cdots \cap Q_s$$

be a primary decomposition of I. For any ideal J we have

$$I:J = (Q_1:J)\cap\cdots\cap(Q_s:J).$$

A quotient $Q_i:J$ evaluates as follows. Denote by P_i the radical of Q_i. Then

$$Q_i:J = \begin{cases} Q_i, & \text{if } J \not\subset P_i \\ R, & \text{if } J \subset Q_i \\ P_i\text{-primary ideal properly containing } Q_i, & \text{if } J \subset P_i,\ J \not\subset Q_i \end{cases}$$

Proposition 3.56. *Let R be a commutative ring, let S be a multiplicative set and let I be an ideal.*

(a) *If I is a primary ideal, then $_SI = I$ or R, according to whether $I\cap S = \emptyset$ or not.*

(b) *If S is generated by a single element f (or more generally if S is finitely generated by elements $a_1,\ldots,a_n$ and $f = \prod a_i$), then*

$$_SI = I:f^\infty.$$

(c) *If I and J are ideals of R then $_S(I\cap J) = {}_SI\cap {}_SJ$.*

(d) *Let I be an ideal and let $\mathfrak{P}$ be a prime ideal. If I has a primary decomposition*

$$I = Q_1\cap\cdots\cap Q_s\cap Q_{s+1}\cap\cdots\cap Q_r,$$

in which the first s components are contained in $\mathfrak{P}$ but the others lie outside, then

$$_\mathfrak{P}I = Q_1\cap\cdots\cap Q_s.$$

Furthermore, if

$$f \in \sqrt{Q_{s+1}\cap\cdots\cap Q_r}\setminus P,$$

then

$$_\mathfrak{P}I = I:f^\infty.$$

(e) *If R is a Gorenstein ring, I is an ideal of codimension m and $\mathfrak{P}$ is an associated prime of the same codimension, then the $\mathfrak{P}$–primary component of I is given by $\mathrm{hull}_m(I,\mathfrak{P}^n)$, where n is the first integer for which*

$$\mathrm{hull}_m(I,\mathfrak{P}^n) = \mathrm{hull}_m(I,\mathfrak{P}^{n+1}).$$

(f) *Let $I = Q_1\cap\cdots\cap Q_r$ be the primary decomposition of an ideal without embedded components, and let $\mathfrak{P}$ be one of its minimal primes, say $\mathfrak{P} = \sqrt{Q_1}$. Suppose n is large enough so that*

$$L = I:\mathfrak{P}^n = I:\mathfrak{P}^{n+1}.$$

Then

$$Q_1 = {}_\mathfrak{P}I = I:L.$$

Proof. (a) and (b) follow directly from the definition, while (c) is a consequence of flatness of the morphism $R \to R_S$: It suffices to note that the natural embedding of R–modules

$$0 \to R/(I\cap J) \longrightarrow R/I\oplus R/J$$

remains exact upon tensoring by R_S

The proof of (d) follows from (a), (b) and (c).

(e) Given that the ideal $\operatorname{hull}_m(I,\mathfrak{P}^n)$ is unmixed, of codimension m, by localization we see that it must be a $\mathfrak{P}$–primary ideal. By Theorem A.12 we must have $\operatorname{hull}_m(I,\mathfrak{P}^n)_\mathfrak{P} = I_\mathfrak{P}$.

(f) Note that under the conditions on I, L is simply the intersection of the components Q_i, $i \geq 2$.   □

Geometrically Irreducible Decomposition

We assemble the elements of the primary decomposition described in [ShY96], and which is the basis for algorithms developed by its authors. It is modeled on the Chinese Remainder Theorem.

Proposition 3.57. *Let I be an ideal of a Noetherian ring R and suppose J is another ideal generated by $f_1,\ldots,f_n$ which have the following properties:*

(i) *For each f_i, $I\!:\!f_i = I\!:\!f_i^2$;*
(ii) *$f_i f_j \in I$ for $i \neq j$.*

Then

$$I = (I,J) \cap (I\!:\!J). \tag{3.15}$$

Proof. We must show that if

$$f = a_1 f_1 + \cdots + a_n f_n \in J \cap (I\!:\!J),$$

then $f \in I$. Note that for each f_i, $f f_i \in I$, and by condition (ii), we have that $a_i f_i^2 \in I$ also. Now use (i) to conclude that $a_i \in I\!:\!f_i$, and establish the assertion.   □

Corollary 3.58. *Let R be a Noetherian ring and let I be an ideal with an irredundant primary decomposition*

$$I = Q_1 \cap \cdots \cap Q_s \cap Q_{s+1} \cap \cdots \cap Q_r, \quad P_i = \sqrt{Q_i},$$

in which $\{P_1,\ldots,P_s\}$ are the minimal primes of I and the other primes are embedded. For each prime ideal P_i, let

$$g_i \in P_1 \cap \cdots \cap \widehat{P_i} \cap \cdots \cap P_s \setminus P_i.$$

For each g_i, let n_i be an integer such that

$$I\!:\!g_i^\infty = I\!:\!g_i^{n_i}.$$

Then the elements

$$f_i = g_i^{n_i}, \quad i = 1,\ldots,s,$$

satisfy the conditions of the previous proposition.

Proof. We only have to show that $h = f_i f_j \in I$ for $i \neq j$. Since by construction this element lies in the radical of I, there exists some n such that $h^n \in I$. We thus have

$$
\begin{aligned}
I = I{:}(f_i^n \cdot f_j^n) &= (I{:}f_i^n){:}f_j^n \\
&= (I{:}f_i){:}f_j^n = (I{:}f_j^n){:}f_i \\
&= (I{:}f_j){:}f_i = I{:}f_i f_j,
\end{aligned}
$$

using the ordinary rules of quotient formation and the fact that each f_i already satisfies (i) of Proposition 3.57. □

Suppose the primes P_i, for $1 \leq i \leq s$, are given by their Gröbner bases with respect to a term ordering. Then for each pair P_i, P_j, $i, j \leq s$, elements g_{ij} could be selected with the property $g_{ij} \in P_j \setminus P_i$. Setting

$$
g_i = \prod_{j \neq i} g_{ij},
$$

gives a collection of elements with the properties above.

Theorem 3.59. *Let R be a Noetherian ring and I is an ideal with a primary decomposition*

$$
I = Q_1 \cap \cdots \cap Q_s \cap Q_{s+1} \cap \cdots \cap Q_r,
$$

with associated data

$$
(P_1, f_1), \ldots, (P_s, f_s),
$$

as above. Then

$$
I = (I{:}f_1) \cap \cdots \cap (I{:}f_s) \cap (I, f_1, \ldots, f_s),
$$

and the ideals in this representation have the following properties:

(a) $\sqrt{(I{:}f_i)} = P_i$, *for* $i \leq s$.
(b) $\dim R/I > \dim R/(I, f_1, \ldots, f_s)$.

Proof. With J as defined,

$$
\begin{aligned}
I &= (I{:}(f_1, \ldots, f_s)) \cap (I, f_1, \ldots, f_s) \\
&= ((I{:}f_1) \cap \cdots \cap (I{:}f_s)) \cap (I, f_1, \ldots, f_s).
\end{aligned}
$$

The other assertions follow from the choices of the f_i's. □

Remark 3.60. In particular this provides a decomposition of I in which the ideals $(I{:}f_i)$ have a unique minimal prime and its associated primes are all associated primes of I. The associated primes of $(I, f_1, \ldots, f_n)$ may not all be among those associated to I. At some point the issues discussed in (3.13) and (3.14) come into play.

3.5 Elements of Factorizers

All the approaches to primary decomposition discussed thus far ultimately require decomposition of radical ideals. What is needed are *factorizers*. Informally, a *factorizer* for a ring of polynomials is a process that accepts a radical ideal,

$$I = \mathfrak{P}_1 \cap \cdots \cap \mathfrak{P}_s,$$

and outputs the collection

$$\{\mathfrak{P}_1, \ldots, \mathfrak{P}_s\},$$

of its associated primes. There has been a great deal of progress for the class of ideals generated by one single polynomial (see [Kal92]) but no breakthrough has been achieved for more general ideals. Already the question of deciding whether the ideal I is prime looms hard.

In this section we collect a number of ideas and facts about factorizers and primality criteria. The perspective followed is the following. Let $I \subset k[x_1, \ldots, x_n]$ be a radical ideal of dimension d and denote by

$$A = k[z_1, \ldots, z_d] \hookrightarrow B = k[x_1, \ldots, x_n]/I$$

some embedding obtained by picking the z_i as a set of variables independent modulo I, or perhaps even a a Noether normalization. The module structure of B over A is bound to have an impact on the process, particularly in case of the latter. One then expects the difficulty to be less when B is a free A–module (that is, when I is a Cohen–Macaulay ideal) or at least when B is a torsionfree A–module (that is, when I is height unmixed).

Reduction to Dimension Zero

Given the ideal I we sketch three reductions of the problem of primary decomposition to the zero–dimensional case (see also Remark 3.70).

Equidimensional Components

Let J_0 be an equidimensional component of I of dimension d. Since the associated primes of J_0 and of its radical J are the same, we may assume that J is a radical, equidimensional ideal. This means that for any equi–dimensional embedding, that is with A and B of the same dimension,

$$A = k[\mathbf{z}] = k[z_1, \ldots, z_d] \hookrightarrow B = R/J \tag{3.16}$$

R/J is a torsion free module over $k[\mathbf{z}]$. This means that the associated prime ideals of J correspond to the maximal ideals of the affine $k(\mathbf{z})$–algebra $A \otimes_{k[\mathbf{z}]} k(\mathbf{z})$. More precisely,

$$\{\mathfrak{p} \mid \mathfrak{p} \in \mathrm{Ass}\ (R/J)\} \Longleftrightarrow \{\mathfrak{P} \mid \mathfrak{P} \in \mathrm{Ass}\ (R/J) \otimes k(\mathbf{z})\},$$

where the second set consists of zero–dimensional ideals, with the correspondence being given by $\mathfrak{p} = \mathfrak{P} \cap R$.

This leads to the localization problem: Given a multiplicative set $S \subset R$ and an ideal $L \subset R$ find $R \cap S^{-1}L$. Some special cases are treated in [GTZ88]; see how Proposition 3.55 deals with the cases needed here.

One Step at a Time

There is no need to look for Noether normalization above by allowing for the computation of superfluous prime ideals that can be filtered out later.

Let $P_1, \dots, P_s$ be the minimal prime ideals of J. Pick one variable, say $x = x_1$, and consider $(f) = J \cap k[x]$.

- If $f = 0$, the associated prime ideals of J split into two subsets:

$$\{P_1, \dots, P_s\} = \{P_1, \dots, P_r\} \cup \{P_{r+1}, \dots, P_s\},$$

where $r \geq 1$. The primes in the first subset correspond to those of $J \otimes_{k[x]} k(x)$. After they are isolated, the remaining primes, if any, are obtained by taking the ideal quotient

$$J : (P_1 \cap \cdots \cap P_r)$$

into the next step.
- If $f \neq 0$, f is the product of the irreducible polynomials f_i:

$$f = f_1 \cdots f_r,$$

and the prime ideals of J correspond to the primes of

$$R/(f_i, J \cap k[x_2, \dots, x_n]).$$

If J is not zero–dimensional, then $J \cap k[x_j] = 0$ for one of the variables. It would be interesting to benefit from the second case more directly.

Integral Closure

The integral closure B of $A = R/J$ is a direct product of integral domains. If we succeed in computing B (see later) as an affine domain $B = D/L$, with $D = k[x_1, \dots, x_n, y_1, \dots, y_m]$,

$$B = S/L_1 \times \cdots \times S/L_r,$$

where L_i is generated by L and one idempotent e_i. The associated prime ideals of J are then the ideals $L_i \cap R$.

The case of a graded ring A is rather pleasing: The idempotents must be homogeneous of degree zero and therefore live in the finite dimensional algebra B_0.

Primitive Elements

With k infinite and a Noether normalization such as in (3.16), where the z_i is a subset of the set of indeterminates defining R, for any element $f \in R$ we have that in the ring of polynomials with the tag variable T,

$$(f - T, J)R[T] \cap k[z_1, \ldots, z_d, T)] = (F) \neq 0.$$

This polynomial $F(z_1, \ldots, z_d, T)$ may be used in finding the factorization of J. Indeed, the embedding

$$k[z_1, \ldots, z_d, T]/(F) \hookrightarrow R/J$$

makes F a squarefree polynomial. If F is not irreducible its factors can be used to detach some of the prime ideals of J. To ensure that this is effective the degree of F must be as close as possible to the torsionfree rank of R/J as an $k[z_1, \ldots, z_d]$–module.

A choice f that achieves maximal degree is gotten from a linear combination

$$f = \sum_{i=1}^{m} c_i x_i.$$

We pick linear combinations such that for any m–subset $u_1, \ldots, u_m$, the coefficient matrix (c_{ij}) is invertible. For instance, one could take $c_i = c^{i-1}$. Since the number of $k(\mathbf{z})$–subfields of $K = A \otimes k(\mathbf{z})$ is bounded by $\ell!$, by Steinitz's theorem ([Jac74, Theorem 4.25]), making sufficiently many choices of f, say

$$(m - 1) \times (\ell! - 1) + 1,$$

will assure that the question could be ascertained. Indeed, it is clear that any subfield of K that contains m linearly independent such combinations must equal K. The pigeon–hole principle guarantees that a simple extension such as S will occur if R/J is a domain. In fact, this also says that for most choices of u the expected extension $k(\mathbf{z}, f)$ is maximal, which opens up room for probabilistic approaches.

Primality Testing

Let $R = k[x_1, \ldots, x_n]$ be a polynomial ring over a field k. The question of deciding whether an ideal $I = (f_1, \ldots, f_m)$ of R is prime is simple to approach but often turns out difficult to verify. Because this problem occurs with some regularity, it is convenient to identify algebraic elements that might see their way into workable algorithms. Questions of certification being usually simpler that those of construction, in this section, we treat some known methods that do not involve primary decomposition.

Projection *cum* Reduction

To test primality of an ideal I in a polynomial ring $R = B[t]$, a natural approach is the following simple observation. We will leave most of the elementary proofs as exercises for the reader.

Proposition 3.61. *The ideal I is prime if an only if $J = I \cap B$ is a prime ideal, and the image I' of I in $(B/J)[t]$ is a prime ideal.*

Since B is Noetherian, the ideal I' must be either trivial (i.e., $I = J \cdot B[t]$), or it must be a prime ideal of height one. It is to deal with the second possibility that new techniques must be brought in. Note that if we have a generating set of I, $f_1(t), \ldots, f_m(t)$, we may decide, using pseudo-division, whether $I \cap B \neq 0$, or otherwise obtain a nonzero element of I of least degree in t.

Proposition 3.62. *Let B be a Noetherian integral domain with field of fractions K, and let I be an ideal of the polynomial ring $B[t]$ satisfying $I \cap B = 0$. Then I is prime if and only if the following conditions are satisfied:*

(i) *$IK[t]$ is generated by an irreducible polynomial $f(t) \in K[t]$—which may be taken to lie in I.*
(ii) *If α is the leading coefficient of $f(t)$ then $I : \alpha = I$.*

In this case, $I = \bigcup_s (f) : \alpha^s$.

The following has also a classical veneer:

Proposition 3.63. *Let R be a Noetherian integrally closed domain and let $f(t) \in R[t]$ be a polynomial irreducible over the field of fractions of R. If I is the ideal generated by the coefficients of $f(t)$ then $P = (I^{-1} f(t))$ is a prime ideal of $R[t]$.*

Proof. It suffices to verify that $R[t]/P$ is torsionfree as an R–module. We leave this checking to the reader. □

The main requirements here are factorization algorithms over various polynomial rings. Let us consider one case where factorization does not play a role.

Proposition 3.64. *Let R be a Noetherian domain satisfying the (S_2) condition of Serre's, and let u be an element in its field of fractions. Let D be the ideal that conducts u into R. Then the ideal $L = (D(U - u))$ is prime if and only if grade $(D + Du) \geq 2$.*

Proof. Note that L is a divisorial ideal of $R[U]$. Let P be one of its associated primes; P is an ideal of height 1. If $Q = P \cap R = 0$, $L_P = P_P$; otherwise, $L \subset QR[U]$, which is impossible as height $Q = 1$ contradicts the condition on the grade. This shows that L is prime. □

Computing Degrees

Another approach does not use projections, appealing to a composite of the Noether normalization theorem, with its consequent determination of the degree of the prime, and an eventual factorization over a ring of polynomials over a field.

This being a certification process, it is useful to have screening devices to weed out early ideals which are not prime.

Radical Test

One possible intermediate element would be a radical test. It may be used as a screening device.

Consider the ideal

$$I = (f_1, \ldots, f_m) \subset k[x_1, \ldots, x_n],$$

where k has characteristic zero. Assume height $I = g$ and let J be the Jacobian ideal of I. Suppose that I is unmixed (e.g. it has passed the test of Corollary 3.43).

Proposition 3.65. *Let f be an element of $J \setminus I$. If $I : f = I$, then I is a radical ideal. Conversely, if I is radical, such elements exist.*

Proof. This follows since the localization R_f is now a smooth k–algebra by Theorem 5.14, while the choice of f assures that the primary decomposition of I is not disturbed. The converse follows because radical ideals are generically smooth. □

The ideal I is then prime if and only if the localization I_f is prime as well. We may then replace I by $L = (I, 1 - Tf) \subset R[T]$. What this means is that we traded fI by the smooth ideal L, that is $R[T]/L$ is a regular ring. Since such rings decompose into direct products of integral domains, I is prime if and only if $R[T]/L$ has no non–trivial idempotent elements.

There are several instances of algebras where testing for reducedness already leads to integrality (see [HSV89]).

Cohen–Macaulay Ideals

We shall say that A is Cohen–Macaulay if it is *locally* Cohen–Macaulay. In the affine case, they decompose into direct products of equidimensional subrings.

We sketch a method that has given good results, particularly when R/I is Cohen–Macaulay, and its Krull dimension or the codimension of I is low.

Let

$$B = k[\mathbf{z}] \hookrightarrow A = k[\mathbf{z}, \mathbf{x}]/I = k[z_1, \ldots, z_d, x_1, \ldots, x_m]/I$$

be a Noether normalization of A. The *degree* of A over B is the dimension

$$\deg(A/B) = \dim_{k(\mathbf{z})}(A \otimes_B k(\mathbf{z})),$$

where $k(\mathbf{z})$ is the field of fractions of B. It equals the torsion–free rank of A as a B–module.

Proposition 3.66. *Suppose A is a Cohen–Macaulay, equidimensional ring. The degree of A over B is the dimension of the vector space*

$$\ell = \dim_k(k[\mathbf{z}, \mathbf{x}]/(I, \mathbf{z})).$$

A is an integral domain if and only if there exists a subring

$$B \hookrightarrow S = k[\mathbf{z}, U]/(f(\mathbf{z}, U)) \hookrightarrow A,$$

where $f(\mathbf{z}, U)$ is an irreducible polynomial of degree ℓ.

Proof. The assertions are clear from the fact that A is a free module of rank ℓ over B, and therefore lies in the field of fractions of S. □

Example 3.67. The ideal

$$I = (xw + y^3, yw + z^3 + xw^3, zw + x^2, w^3 + xy^2z^2 - y^2w^4)$$

is Cohen–Macaulay of codimension 3.

The subalgebra $k[w]$ is a Noether normalization of $k[x,y,z,w]/I$, and the irreducible polynomial

$$f(y,w) = y^{17} - w^{11}y^2 - w^{10}$$

lies in I. Since it is clear that

$$\dim_k(k[x,y,z,w]/(I,z)) = 17,$$

I is a prime ideal.

The computation of the degree in this manner uses Theorem 2.73 (see Corollary A.103 or [He78, Proposition 1.1]).

Simple Rational Extensions

The integer $\deg(A/B)$ is embedded in the details of a presentation of A as a B–module, that is, could be derived from a set of module generators and their relations.

View $A = k[\mathbf{z}, x_1, \dots, x_m]/I$ as the composite of the extensions $A_i = k[\mathbf{z}, x_i]/I_i$, where $I_i = I \cap k[\mathbf{z}, x_i]$. If I is prime, each I_i is generated by an irreducible polynomial $f_i(\mathbf{z}, x_i)$.

For each i let $h_i(\mathbf{z})$ be a nonzero polynomial obtained by eliminating x_i from $f_i(\mathbf{z}, x_i)$ and the derivative $\partial f_i(\mathbf{z}, x_i)/\partial x_i$, for instance the resultant of $f_i(\mathbf{z}, x_i)$ with respect to x_i, or more generally any nonzero element in

$$(f_i(\mathbf{z}, x_i), \partial f_i(\mathbf{z}, x_i)/\partial x_i) \bigcap k[\mathbf{z}].$$

Proposition 3.68. *Let* $a = (a_1, \dots, a_d) \in k^d$ *be such that* $\forall i$, $h_i(a) \neq 0$. *Then the degree of* A *over* B *is*

$$\ell = \dim_k(k[\mathbf{z}, \mathbf{x}]/(I, z_1 - a_1, \dots, z_d - a_d)).$$

Proof. The choice of a implies that each of the extensions A_i is unramified at the maximal ideal $\mathfrak{p} = (z_1 - a_1, \dots, z_d - a_d)$. Since $k[\mathbf{z}]_\mathfrak{p}$ is regular, it follows from [AB59, Propositions 4.6 and A.1] that their composite $A_\mathfrak{p}$ is an unramified extension of $k[\mathbf{z}]_\mathfrak{p}$ and is therefore a free $k[\mathbf{z}]_\mathfrak{p}$–module. □

Since such points always exist in profusion, we may also find them by trial and error, without the bother of elimination. For instance, evaluation at points $\mathbf{z} = \mathbf{a}$ followed by computation of gcd works extremely fast in most systems. (For details on the structure of unramified extensions, see [Ray70, Chap. V, Th. 1].)

Once the degree has been found, we have the following general primality test. Observe that we have traded U by one of the x_i–variables.

Proposition 3.69. *Let A be an equidimensional reduced algebra and let*

$$B = k[\mathbf{z}] \hookrightarrow A = k[\mathbf{z}, \mathbf{x}]/I,$$

be a Noether normalization as above. Let ℓ be the degree of A over B. For any variable U, the ideal $I \cap k[\mathbf{z}, U]$ is generated by one polynomial $f(\mathbf{z}, U)$. If $f(\mathbf{z}, T)$ has degree ℓ then A is an integral domain if and only if $f(\mathbf{z}, T)$ is irreducible.

Proof. The proof is essentially the same as that of Proposition 3.66, but let us highlight the help supplied by the assumptions on A.

First, note that for any variable U, the ideal $J = k[\mathbf{z}, U] \cap I$ is principal. Indeed, since A is a torsion–free $k[\mathbf{z}]$–module (the equidimensional hypothesis), the chain

$$B = k[\mathbf{z}] \hookrightarrow S = k[\mathbf{z}, U]/J \hookrightarrow A = k[\mathbf{z}, \mathbf{x}]/I,$$

says S is also torsion free as a B–module, which clearly implies that J is equidimensional. Since J has codimension one it must be principal.

Finally, if $\deg f(\mathbf{z}, U) = \ell$, the total ring of fractions of S and A are equal, proving the assertion. □

Remark 3.70. Observe that if h has degree ℓ, but it is not necessarily irreducible, its prime factors can be used to obtain a primary decomposition of I. This is very close to the classical approach to verify primality, mod the equidimensional and radical reductions that might facilitate the task.

Example 3.71. Craig Huneke considered the ideal

$$I \subset R = k[a, b, c, d, e], \quad \text{char } k \neq 2, 5,$$

generated by the following polynomials:

$$5abcde - a^5 - b^5 - c^5 - d^5 - e^5$$
$$ab^3c + bc^3d + a^3be + cd^3e + ade^3$$
$$a^2bc^2 + b^2cd^2 + a^2d^2e + ab^2e^2 + c^2de^2$$
$$abc^5 - b^4c^2d - 2a^2b^2cde + ac^3d^2e - a^4de^2 + bcd^2e^3 + abe^5$$
$$ab^2c^4 - b^5cd - a^2b^3de + 2abc^2d^2e + ad^4e^2 - a^2bce^3 - cde^5$$
$$a^3b^2cd - bc^2d^4 + ab^2c^3e - b^5de - d^6e + 3abcd^2e^2 - a^2be^4 - de^6$$
$$a^4b^2c - abc^2d^3 - ab^5e - b^3c^2de - ad^5e + 2a^2bcde^2 + cd^2e^4$$
$$b^6c + bc^6 + a^2b^4e - 3ab^2c^2de + c^4d^2e - a^3cde^2 - abd^3e^2 + bce^5$$

This ideal has codimension 2 and projective dimension 3. Because it is homogeneous, the rank of $A = R/I$ (relative to any normalizing subring generated by 1–forms) is determined by the Hilbert polynomial; *Macaulay* showed that it is 15. A test promptly affirmed I to be radical and equidimensional.

A Noether normalization of R/I is $B = k[a, b, f]$, $f = d - c$ (if char $k \neq 2$), while $I \cap k[a, b, c, d]$ is generated by the irreducible polynomial

$$a^{10}b^5 + a^5b^{10} + 5a^4b^8c^2d + 10a^3b^6c^4d^2 + 10a^2b^4c^6d^3 + 5a^6b^2c^3d^4 + c^5d^{10}$$
$$+ 5ab^2c^8d^4 + a^{10}d^5 + 2a^5b^5d^5 + 2a^5c^5d^5 + c^{10}d^5 + 5a^4b^3c^2d^6 + a^5d^{10}.$$

Primality and Connectedness

Let R be a ring of polynomials over a field k of characteristic zero and let I be a homogeneous ideal. To check whether I is a prime ideal, we may first test for the presence of height mixed components as in this chapter and then whether $I = \sqrt{I}$ using the methods of Chapter 5.

Proposition 3.72. *Let I be a height unmixed homogeneous radical ideal, and let J be its Jacobian ideal. Then I is prime in the following two cases:*

(a) height$(I + J) > 2 \cdot$ height (I).

(b) *For each $f \in J \setminus I$,*

$$I : f = I, \text{ and}$$
$$(I, f) : J = (I, f).$$

Proof. (a) Let $\mathfrak{p}$ and $\mathfrak{q}$ be two distinct minimal primes of I. From a classical formula (see Theorem A.25), $\mathfrak{p} + \mathfrak{q}$ is a (proper) ideal of height at most $2 \cdot$ height I. Localizing at a minimal prime $\mathfrak{m}$ of this sum, we get a regular local ring $R_{\mathfrak{m}}$ such that $R_{\mathfrak{m}}/IR_{\mathfrak{m}}$ is also a regular local ring and therefore $I_{\mathfrak{m}}$ is a prime ideal, against the assumption.

(b) This is just the criterion for normality (see Theorem 6.40). It follows that R/I decomposes into a product of graded, affine algebras over k. Since 1 is the only idempotent element of degree 0, R/I must be a domain. $\qquad \square$

The preceding indicates the difficulty in controlling the height of a sum of ideals away from Krull's principal ideal theorem. A simple tool that is useful to have at hand is the following.

Proposition 3.73 (Abhyankar–Hartshorne). *Let R be a Noetherian ring without nontrivial idempotents and let I and J be nonzero ideals such that $I \cdot J = 0$. Then* grade$(I+J) \le 1$. *In particular, if R satisfies the condition S_2 of Serre then* height$(I + J) \le 1$.

Proof. We may assume that $I \cap J \subset (0 : J) \cap (0 : I) = 0$, as otherwise for $0 \ne x \in (0 : J) \cap (0 : I)$, $x(I+J) = 0$. Now by the connectedness of Spec(R),

$$I + J \subset (0 : I) + (0 : J) \ne R.$$

Hence localizing at a prime containing $(0 : J) + (0 : I)$, we retain our assumptions $I \ne 0, J \ne 0, I \cap J = 0$, and $I + J \ne R$. Thus we may assume that R is a local ring.

Suppose grade$(I+J) > 0$. Let $x = a + b$, $a \in I$, $b \in J$, be a nonzero divisor; it is clear that $a \ne 0$, $b \ne 0$. Moreover, $a \notin R(a+b)$ for an equation $a = r(a+b)$ yields $(1 - r)a = rb$, which is a contradiction, whether r is a unit or not. Since $(I+J)a \subset R(a+b)$ it follows that grade$(I+J) = 1$. $\qquad \square$

A rephrasing shows that Cohen–Macaulay ideals have primary decompositions that are extremely packed in the following sense.

Proposition 3.74. *Let R be a Noetherian local ring and suppose I is an ideal such that R/I has the property S_2. If $I = J \cap K$ is a proper decomposition then* height $(J + K) \leq$ height $I + 1$.

Proof. To the decomposition $I = J \cap K$, there corresponds an exact sequence

$$0 \to R/I \longrightarrow R/J \oplus R/K \longrightarrow R/(J+K) \longrightarrow 0.$$

Suppose P is an associated prime ideal of $I + K$ of height greater than height $I + 1$. Applying the functor $\mathrm{Hom}_R(R/P, \cdot)$ to the sequence, we obtain the exact sequence

$$\mathrm{Hom}_R(R/P, R/J \oplus R/K) \longrightarrow \mathrm{Hom}_R(R/P, R/(J+K)) \longrightarrow \mathrm{Ext}^1_R(R/P, R/I),$$

in which the first module vanishes because P cannot be contained in an associated of either J or K, and the third module vanishes since R/I satisfies the condition S_2 of Serre. $\qquad\square$

Symbolic Powers

Outside the defining ideals of (known) irreducible varieties, there are not many methods to obtain prime ideals. On the other hand, there are various ways to generate many primary ideals with a given radical.

A fertile source of primary ideals arises from the notion of the symbolic powers of a prime ideal $\mathfrak{P}$. We describe both the absolute and relative versions of the construction. Let $I \subset \mathfrak{P}$ and consider the diagram of natural maps

$$
\begin{array}{ccc}
\mathfrak{P}_I^{(n)} & \longrightarrow & R \\
\downarrow & & \downarrow \\
(\mathfrak{P}^n, I)(R/I)_{\mathfrak{P}} & \longrightarrow & (R/I)_{\mathfrak{P}}
\end{array}
$$

where $\mathfrak{P}_I^{(n)}$ is the inverse image of the nth power of the maximal ideal of the local ring $(R/I)_{\mathfrak{P}}$. It is a $\mathfrak{P}$–primary ideal and it is called the nth–symbolic power of $\mathfrak{P}$ above I. (One tends to drop the subscript when the ideal I is understood.)

If $\mathfrak{P}$ is a radical ideal,

$$\mathfrak{P} = \mathfrak{P}_1 \cap \cdots \cap \mathfrak{P}_r,$$

$\mathfrak{P}_i$ prime, the symbolic powers of $\mathfrak{P}$ are defined by

$$\mathfrak{P}^{(n)} = \mathfrak{P}_1^{(n)} \cap \cdots \cap \mathfrak{P}_r^{(n)}.$$

A more general notion of symbolic power that fits many needs is the following. Let R be a Noetherian ring and I one of its ideals. Suppose

$$I = J \cap L$$

is a decomposition where J is the intersection of all the primary components of I associated to the minimal primes of I, and L is the intersection of primary ideals corresponding to embedded primes of I. The mapping

$$I \mapsto I' = J$$

is well-defined.

Definition 3.75. *Let I be an ideal of the Noetherian ring R. For a positive integer n the nth symbolic power of I is the ideal*

$$I^{(n)} = (I^n)'.$$

In general the symbolic powers of a prime ideal $\mathfrak{P}$ differ significantly from its ordinary powers. One case where they agree is given by:

Proposition 3.76. *Let R be a Cohen–Macaulay ring and let $\mathfrak{P}$ be a prime ideal locally generated by a regular sequence of g elements. Then*

$$\mathfrak{P}^{(n)} = \mathfrak{P}^n, \; \forall n \geq 1.$$

This follows by examining the associated primes of $R/\mathfrak{P}^n$: It follows easy that $\mathfrak{P}$ is the only such ideal. We shall use this feature to give two methods to compute symbolic powers in rings of polynomials.

Proposition 3.77. *Let $\mathfrak{P}$ be a prime ideal of a ring of polynomials. If height $\mathfrak{P} = g$, the ideal $\mathfrak{P}^{(n)}$ can be determined in one of the two following ways. Suppose $f_1, \ldots, f_m$ is a generating set for $\mathfrak{P}$.*

(a) *Let φ be an $r \times m$ matrix of presentation of $\mathfrak{P}$, that is there is an exact sequence*

$$R^r \xrightarrow{\;\varphi\;} R^m \longrightarrow \mathfrak{P} \to 0.$$

Denote by f one of the minors of size $m - g$ of φ which are not contained in $\mathfrak{P}$.
(b) *If k is a field of characteristic zero, let f be an element of the Jacobian ideal of $\mathfrak{P}$ which is not contained in $\mathfrak{P}$.*

(Such elements exist because $\mathfrak{P}_\mathfrak{P}$ is generated by g elements.) Then

$$\mathfrak{P}^{(n)} = \mathfrak{P}^n : f^\infty. \tag{3.17}$$

Proof. In each of these cases, in the localization R_f the ideal $\mathfrak{P}_f$ is locally generated by a regular sequence of g elements. But in such case, the symbolic and ordinary powers of $\mathfrak{P}_f$ coincide, so that $\mathfrak{P}_f^n = \mathfrak{P}_f^{(n)}$, from which the assertion follows. $\qquad\square$

Symbolic Powers by Iteration

Let $R = k[x_1, \ldots, x_n]$ be a polynomial ring over a field k and let $I \subset R$ be an ideal. There is a classical interpretation of the symbolic powers of a radical ideal due to Zariski and Nagata using the notion of differential operators.

Theorem 3.78. *If* char $k = 0$ *and* I *is a radical ideal then for each positive integer* r *we have*

$$I^{(r)} = \{g \in R \mid \frac{\partial^\alpha f}{\partial \mathbf{x}^\alpha} \in I, \forall \alpha, |\alpha| \leq r - 1\}.$$

For an indication of the proof, see [Ei95, Theorem 3.14], where other references are given. We describe one of the methods of Aron Simis ([Sim96]) to compute these powers iteratively based on this result. One of its advantages is that it does not require elimination orders.

Let $\mathbf{f} = \{f_1, \ldots, f_m\} \subset R$ be a set of generators of $I^{(r)}$. Denote by $\Theta(\mathbf{f})$ the Jacobian matrix of $\mathbf{f}$. For an integer $r \geq 0$, $\Psi^{(r)}(\mathbf{f})$ will denote a lift to R of the relation matrix of $\Theta(\mathbf{f})$ over the ring $\overline{R} = R/I^{(r)}$: That is consider the commutative diagram,

$$
\begin{array}{ccccc}
R^p & \xrightarrow{\Psi^{(r)}(\mathbf{f})} & R^m & & \\
\downarrow & & \downarrow & & \\
\overline{R}^p & \longrightarrow & \overline{R}^m & \xrightarrow{\Theta(\mathbf{f})} & \overline{R}^n
\end{array}
$$

where the long row is exact and the vertical maps are the natural ones. $((\mathbf{f}) \cdot \Psi(\mathbf{f}))R$ will denote the ideal generated by the entries of the product matrix $(\mathbf{f}) \cdot \Psi^{(r)}(\mathbf{f})$. If r is fixed throughout the discussion, one sets $\Psi(\mathbf{f}) := \Psi^{(r)}(\mathbf{f})$.

Proposition 3.79. *If* $\mathbf{f} = \{f_1, \ldots, f_m\}$ *is a set of generators of* $I^{(r)}$ *then*

$$I^{(r+1)} = ((\mathbf{f}) \cdot \Psi(\mathbf{f}))R + II^{(r)}.$$

Proof. We use Theorem 3.78. For simplicity, set $\Theta = \Theta(\mathbf{f})$ and $\Psi = \Psi(\mathbf{f}) = (\psi_{jl})$. First, let $g = \sum_j \psi_{jl} f_j \in ((\mathbf{f}) \cdot \Psi)R$. Then, for each $1 \leq i \leq n$, one has

$$\frac{\partial g}{\partial x_i} = \sum_j \psi_{jl} \frac{\partial f_j}{\partial x_i} + \sum_j \frac{\partial \psi_{jl}}{\partial x_i} f_j \in I^{(r)},$$

by definition of Ψ. This means that

$$\frac{\partial^\beta}{\partial \mathbf{x}^\beta}\left(\frac{\partial g}{\partial x_i}\right) \in I, \forall \beta, |\beta| \leq r - 1, \forall i.$$

Since any differential operator $\partial^\alpha g/\partial \mathbf{x}^\alpha$, with $|\alpha| \leq r$, is a k-linear combination of ones of the above form, it follows that $g \in I^{(r+1)}$, as required.

Conversely, let $g \in I^{(r+1)}$. Since $I^{(r+1)} \subset I^{(r)}$, one has $g = \sum_j h_j f_j$, for suitable $h_j \in R$. Applying Θ to the vector $v = \sum_j h_j e_j \in R^m$ yields the vector of $\sum_i R dx_i$ whose ith coordinate is $\sum_j h_j (\partial f_j / \partial x_i)$. We claim that each of these polynomials belongs to $I^{(r)}$. Indeed, $\sum_j h_j (\partial f_j / \partial x_i) = \partial g / \partial x_i - \sum_j (\partial h_j / \partial x_i) f_j$ and $\partial g / \partial x_i \in I^{(r)}$ since $g \in I^{(r+1)}$. We have thus shown that $\bar{v} = \sum_j \bar{h}_j e_j \in \ker \overline{\Theta} = \Im \overline{\Psi}$, where $\bar{}$ denotes taking residues modulo $I^{(r)}$. By a standard calculation,

$$h_j = \sum_l b_l \psi_{jl} + d_j,$$

for certain $b_l \in R, d_j \in I^{(r)}$. Substituting these expressions into $g = \sum_j h_j f_j$ yields the required result. $\square$

Remark 3.80. This result was used by David Eisenbud in writing two *Macaulay* scripts, called *next_symbolic_power* and *symbolic_power*, respectively. The iterative computation of symbolic powers, using these scripts, is fairly fast in many cases.

Problem 3.81. Let R be a Cohen–Macaulay integral domain and let u be an element of the field of fractions of R. Give conditions for $R[u]$ to be Cohen–Macaulay in at least the following cases: u is integral over R, or (ii) u is arbitrary but R is integrally closed.

4

Computing in Artin Algebras

Let k be a field and let A be a finite dimensional k–algebra. The fundamental features of A are its Jacobson radical

$$J = \sqrt{(0)},$$

and the decomposition

$$A/J = K_1 \times \cdots \times K_r, \quad K_i \text{ is a field extension of } k,$$

afforded by its finite set of primitive idempotent elements. Their study is significant for the role in the interplay between computation in Artin algebras and finding primary decomposition.

In this chapter, to emphasize these views we carry out a discussion of the following themes:

- The nilradical of Artin algebras
- Finding idempotents in Artin algebras
- Effecting primary decomposition of zero-dimensional ideals
- Probabilistic methods
- Root finders
- Solving systems of polynomial equations

The first two problems are enhancing faces of the same question, each with its own natural approaches, with the former being a shade of a difference the more amenable of the two. They affect directly the other topics. We collect elements of the structure of the idempotents, various means and tricks to get them. Actually, a more general approach, directed to not necessarily commutative algebras, might have been advisable here.

There is a brief interlude with probabilistic methods, some dependent on the characteristic of the base field and others resorting to generic change of variables or generators.

With regard the other topics, we outline some of the more popular root finders. One of the concerns is how best to deliver, to a numerical equation solver, a set of polynomial equations. The focus will lie on forwarding sets without excess multiplicities and in number equal to the number of variables. A naive presentation of iterative schemes is attempted as a bridge to other realms.

4.1 Structure of Artin Algebras

In this section we study affine Artin rings with the view towards the determination of their Jacobson radical.

Representing an Algebra

There are two ways of describing a commutative, affine algebra of Krull dimension zero:

- $R = k[y_1, \ldots, y_m]/J$, where $J = (f_1, \ldots, f_r)$ is an ideal of codimension m;
- as a k–vector space basis

$$e_1, \ldots, e_n$$

with attached *structure constants* $\{c_{ij\ell}, \ 1 \le i, j, \ell \le n\}$, that is

$$e_i \cdot e_j = \sum_{\ell=1}^{n} c_{ij\ell} e_\ell.$$

The first representation is more packed and it is the usual form the algebra is given; it requires considerably less bookkeeping. Furthermore, a Gröbner basis computation provides the means to find the set of structure constants attached to the basis given by its standard monomials: If $\{m_1, \ldots, m_q\}$ are the standard monomials of some Gröbner basis, then

$$\boxed{\text{NormalForm}(m_i \cdot m_j) = \sum_{\ell=1}^{q} c_{ij\ell} m_\ell.} \qquad (4.1)$$

The second presentation meanwhile permits an easy conversion of questions into linear algebra problems. It is also suitable for computation on non-commutative algebras. If the $c_{ij\ell}$'s are mostly non-trivial, carrying them around is cumbersome for large dimensions and it might be preferable to produce them on-the-fly from the f_j.

Generic Algebras

The relations that the structure constants c_{ijk} of an n-dimensional k-algebra A must satisfy are:

$$c_{ij\ell} = c_{ji\ell} \qquad \text{for the commutative law}$$
$$\sum_\ell c_{ij\ell} c_{\ell kp} = \sum_\ell c_{jk\ell} c_{i\ell p} \qquad \text{for the associative law.}$$

In addition, there must be provisions to give an identity to the algebra, for example

$$c_{1j\ell} = \delta_\ell^j.$$

Let A_n denote the algebra defined by the n^3 symbols $c_{ij\ell}$,

$$k[c_{ij\ell}, \ 1 \le i, j, \ell \le n]/I,$$

with I generated by the three sets of equations above. The k–rational points of the variety $V(I)$ are the n–dimensional commutative k-algebras.

Remark 4.1. Among the questions that can be raised here:

- What is the Krull dimension of A_n?
- Is A_n Cohen–Macaulay?

Jacobson Radical

For a given k–algebra A, an element $x \in A$ defines a k-linear transformation

$$\lambda(x) : A \longrightarrow A, \quad a \mapsto ax.$$

The determinant and trace of $\lambda(x)$ are called the *norm* and the *trace* of x; they are denoted $\text{norm}_{A/k}(x)$ and $\text{trace}_{A/k}(x)$ (or simply by $\text{norm}(x)$ and $\text{trace}(x)$ if no confusion arises).

These two functions of x can be employed, under most circumstances, to give a description of the nil radical J of the algebra A. Let $e_1, \ldots, e_n$ be a k-basis of A, with structure constants $\{c_{ij\ell}\}$. Suppose throughout that char k, the characteristic of k, is larger than n. The condition that $\lambda(x)$ is nilpotent can be expressed in the usual manner:

Proposition 4.2. *Let A be an Artin algebra over the field k whose characteristic is larger than $\dim_k A$. Then $\lambda(x)$ is nilpotent if and only if $\text{trace}(x^i) = 0$, for $i = 1, \ldots, n$.*

Let us we pick $x \in A$,

$$x = x_1 e_1 + \cdots + x_n e_n,$$

with unspecified coefficients and seek the conditions on the x_i's that place x in J. One way is to solve the n equations prescribed by the previous proposition. Instead we "linearize" these equations.

Proposition 4.3. *The element x lies in J if and only if $(x_1, \ldots, x_n)$ is a solution of the set of n linear equations*

$$\sum_i \left(\sum_{p,\ell} c_{ik\ell} c_{\ell pp} \right) x_i = 0, \qquad 1 \le k \le n. \tag{4.2}$$

Proof. To check that $\text{trace}(x^i) = 0$ for all $i \geq 1$, we may as well ask that $\text{trace}(x \cdot y) = 0$ for every element $y \in A$. Indeed, if $J^{r+1} = 0$ and we choose a basis of A put together from bases of the factors J^i/J^{i+1}, it is clear that the matrix for any $\lambda(x)$, $x \in J$, is necessarily lower triangular. But it suffices to have this equation for $y = e_1, \ldots, e_n$. Expanding $x \cdot e_k \cdot e_p$ and adding up the coefficients of e_p in the expansion, we obtain the left hand side of (4.2). $\qquad\square$

Socle of an Algebra

Let A be an Artin ring and let J be its radical. In the lattice of ideals of A, sitting at the opposite end of J there lies another distinguished ideal, the *socle* of A:

Definition 4.4. *The* socle *of the Artin algebra A of Jacobson radical J is the sum of all (nonzero) minimal ideals of A. Equivalently, the socle of A is the annihilator of J. It will be denoted* socle(A).

For example, if k is a field and $A = k[x]/(x^n)$, then x^{n-1} generates the socle of A. In general however it is not very clear how to find the socle from any of the two ways one tends to represent the algebra A.

For a more general polynomial $f(x)$, the algebra $A = k[x]/(f(x))$ decomposes into a finite direct product of local algebras

$$A = A_1 \times \cdots \times A_r,$$

one factor for each irreducible polynomial that divides $f(x)$, and the image of $f'(x)$, the derivative of $f(x)$, in each A_i generates its socle.

The point will be that we shall want to interpret this formula as a statement about the structure of the socle of certain Gorenstein algebras. For more general algebras, it is not well known how to predict, from the generators and relations of the algebra, which elements will generate its socle. An important exception is Theorem 4.8 below.

Northcott Ideals

One of the earliest examples of socle formulas is provided by the following theorem of Northcott [Nor63]. It will also play a role in our discussion of the integral closure in Chapter 6.

Let $\mathbf{u} = u_1, \ldots, u_n$ be a sequence of elements in the ring R, and let $\varphi = (a_{ij})$ be a $n \times n$ matrix with entries in R. Denote by $\mathbf{v} = v_1, \ldots, v_n$ the sequence defined by the product of matrices

$$[\mathbf{v}]^t = \varphi \cdot [\mathbf{u}]^t.$$

The *Northcott ideal* associated to the sequence $\mathbf{u}$ and the matrix φ is $(\mathbf{v}, \det \varphi)$.

Theorem 4.5. *If* grade $(\mathbf{v}) = n$, *then*

$$(\mathbf{v}) : \det \varphi = (\mathbf{u}). \tag{4.3}$$

Proof. It is not difficult to see that we may assume that R is a local ring, and the sequences **u** and **v** are (proper) regular sequences.

From Cramer's rule, the right–hand side of the equation is contained in $(\mathbf{v}):(\mathbf{u})$. We have to show that any element r that is conducted by $D = \det\varphi$ into $(\mathbf{v})$ must be an element of $(\mathbf{u})$.

To prove the assertion we are going to induct on n, the statement being clear for $n = 1$. Using prime avoidance, we can replace the generators $u_1,\ldots,u_n$ of $(\mathbf{u})$ by another sequence $\mathbf{w} = w_1,\ldots,w_n$, such that $v_1,\ldots,v_{n-1},w_n$ is a regular sequence. Furthermore, the change of generators matrix ψ,

$$[\mathbf{u}]^t = \psi \cdot [\mathbf{w}]^t,$$

by Nakayama's Lemma, is invertible over R. We may then use **w** instead of **u**.
Let then

$$D \cdot z = \sum_{i=1}^{n} c_i v_i.$$

Since we also have

$$D \cdot w_n = \sum_{i=1}^{n} A_{in} v_i,$$

where the A_{ij} are the entries of the adjoint of φ, we can write

$$(zA_{1n} - w_n c_1)v_1 + \cdots + (zA_{nn} - w_n c_n)v_n = 0.$$

Because the v_i's form a regular sequence, we must have

$$zA_{nn} - w_n c_n \in (v_1,\ldots,v_{n-1}).$$

Passing to the ring $\overline{R} = R/(w_n)$, the conditions of the theorem remain and the proof is easily completed by the induction hypothesis. $\square$

Corollary 4.6. *For* **v**, **u** *and* $\det\varphi$ *as in the theorem, it holds that*

$$(\mathbf{v}):(\mathbf{u}) = (\mathbf{v},\det\varphi).$$

In particular, let **u** *be a regular system of parameters for a n–dimensional regular local ring and let* **v** *be a regular sequence of length n. Then the socle of the Artin ring $R/(\mathbf{v})$ is generated by* $\det\varphi$.

Proof. This is a particular rephrasing of Corollary A.140. $\square$

Traces of Gorenstein Algebras

Suppose that the algebra A is a Gorenstein ring. This means that $A^* = \mathrm{Hom}_k(A,k)$, with the structure of A–module given in the usual manner (for $r,x \in A, \psi \in A^*$ then $(r \cdot \psi)(x) = \psi(rx)$) is isomorphic to A.

Suppose
$$A = A_1 \times \cdots \times A_r$$
is the decomposition of A into a product of local rings. If $\mathfrak{m}_i$ is the maximal ideal of A_i, then
$$\sqrt{(0)} = J = \mathfrak{m}_1 \times \cdots \times \mathfrak{m}_r$$
$$\mathrm{socle}(A) = (0 : \mathfrak{m}_1) \times \cdots \times (0 : \mathfrak{m}_r).$$

Each module $0 : \mathfrak{m}_i$ is generated by one element s_i. A generator of A^* is any homomorphism φ with the property that $\varphi(s_i) \neq 0 \;\; \forall i$.

Proposition 4.7. *Let A be a Gorenstein algebra and let* $\mathrm{trace}_{A/k}$ *be the ordinary trace of A. Let φ_0 be a generator of A^* and write* $\mathrm{trace}_{A/k} = \alpha \cdot \varphi_0$. *Then α is a generator of the socle of A.*

Proof. Since $\mathrm{trace}(J) = 0$, $\varphi(\alpha J) = 0$ means that for each index i the ideal $L_i = \alpha J \cap A_i$ does not contain the element s_i, and therefore $L_i = 0$ so that $\alpha J = 0$ as well. This means that $\alpha \in \mathrm{socle}(A)$. Finally, if $\alpha A_i = 0$, α would be annihilated by one idempotent element ε, but since $\mathrm{trace}(\varepsilon) \neq 0$ (since the characteristic of k is larger than n), we would have a contradiction. $\qquad\square$

The element α, which is defined up to a unit of A, is called a *different* of A.

Complete Intersections

Let us assume now that $A = k[x_1, \ldots, x_m]/(f_1, \ldots, f_m)$ is a complete intersection. Let $D = \left(\frac{\partial f_i}{\partial x_j} \right)$ be the Jacobian matrix of the polynomials $f_1, \ldots, f_m$.

We now give a result of Tate [Tat70, Theorem A.3(4)] (see generalizations in [ScS75] and [Kun85]) that gives a concrete description of the different of A.

Theorem 4.8. *There exists one generator φ_0 of A^* such that* $\mathrm{trace}_{A/k} = \alpha \cdot \varphi_0$, *where α is the image of D in A.*

Problem 4.9. Describe a process to obtain the socle of an algebra A in terms of the structure constants of A.

Problem 4.10 (Isomorphism Problem). Let A be a finite dimensional algebra over a field k, but not necessarily graded.

- If B is another k–algebra, how to decide that $A \simeq B$?
- If E and F are finitely generated A–modules, how to decide that $E \simeq F$? This is a special case of the first problem as we can consider their *idealizations* $A \oplus E$ and $A \oplus F$.

Problem 4.11. Let A be a finite dimensional algebra over the field k, with structure constants $\{c_{ij\ell}\}$. Describe a method to determine the index of nilpotency of its Jacobson radical J, that is the smallest integer s such that $J^s = 0$.

Problem 4.12. Let $A = k[x_1, \ldots, x_n]/I$ be a zero–dimensional algebra over the field k, let $>$ be a term ordering and set $B = k[x_1, \ldots, x_n]/in_>(I)$. Experiments have shown that the index of nilpotency of the Jacobson radical of A is never larger than that of the radical of B. Is there an explanation (if true)?

4.2 Zero-Dimensional Ideals

In this section we study Artin algebras, $A = k[x_1, \ldots, x_n]/I$, by emphasizing their ideals of definition to check for the existence of zeros and compute the nilradical of A.

Finitely Many Zeros

We begin with the view from the Nullstellensatz. Given an ideal I of the ring of polynomials $k[\mathbf{x}] = k[x_1, \ldots, x_n]$, we denote

$$V(I) = \{ \mathbf{c} = (c_1, \ldots, c_n) \in \overline{k}^n \mid f(\mathbf{c}) = 0 \; \forall \; f(\mathbf{x}) \in I \}, \qquad (4.4)$$

where $\overline{k}$ is the algebraic closure of the field k.

Theorem 4.13. *Let $I \subset R = k[x_1, \ldots, x_n]$ be an ideal. Then*

$$\begin{aligned} V(I) = \emptyset &\Leftrightarrow I = (1) \\ V(I) \text{ is finite} &\Leftrightarrow \dim_k R/I < \infty. \end{aligned} \qquad (4.5)$$

A criterion for detecting this condition is:

Proposition 4.14. *Let $I \subset R = k[x_1, \ldots, x_n]$ be an ideal and G one of its Gröbner bases. I is zero–dimensional if and only if the set $in(G)$ contains a power of each indeterminate x_i.*

Proof. By Macaulay's Theorem, there is an isomorphism of k–vector spaces

$$R/I \simeq R/in(I) = R/(in(G)).$$

The finite dimensionality of $R/in(G))$ is just the condition on the presence of a power of each x_i in $in(G)$. $\qquad \square$

Let I be a zero-dimensional ideal of a polynomial ring $k[x_1, \ldots, x_n]$. We now give the classical construction of the radical of I. In this section, we assume that k is a perfect field.

A Venerable Radical Formula

Consider the case of one polynomial $f(x)$ in one variable. Let $f(x) = \prod_i f_i(x)^{e_i}$ be the primary decomposition of $f(x)$ (the $f_i(x)$ are invisible to us). Since k is perfect, the $f_i(x)$'s have simple roots in the algebraic closure $\bar{k}$.

Theorem 4.15. *If char k does not divide any e_i, then*

$$\sqrt{(f(x))} = (f(x)/\gcd(f(x), f'(x))). \tag{4.6}$$

Proof. From

$$f'(x) = \sum_i e_i \frac{f(x)}{f_i(x)} f_i'(x),$$

and the hypotheses on k and e_i, each prime factor $f_i(x)$ occurs precisely with multiplicity $e_i - 1$ in $f'(x)$, and therefore

$$\gcd(f(x), f'(x)) = \prod_i f_i(x)^{e_i - 1},$$

which establishes the formula.   □

One refers to the generator of $\sqrt{(f(x))}$ as the square-free part of $f(x)$. Note that the formula above falls short of covering all cases.

Radicals

The following is the standard description of the radical of a zero-dimensional ideal.

Theorem 4.16. *Let k be a perfect field and let I be a zero-dimensional ideal of $k[x_1, \ldots, x_n]$. For each $i = 1, \ldots, n$, let*

$$0 \neq f_i(x_i) \in I \cap k[x_i],$$

and denote by $g_i = g_i(x_i)$ the square-free part of $f_i(x_i)$. Then

$$\sqrt{I} = (I, g_1, \ldots, g_n). \tag{4.7}$$

Proof. Clearly the right-hand side of (4.7) is contained in $\sqrt{I}$. On the other hand,

$$A = k[x_1]/(g_1(x_1)) \otimes_k \cdots \otimes_k k[x_n]/(g_n(x_n)) = k[x_1, \ldots, x_n]/(g_1, \ldots, g_n)$$

is a semisimple algebra (see Theorem A.9), and therefore its homomorphic image

$$k[x_1, \ldots, x_n]/(I, g_1, \ldots, g_n)$$

will also be semisimple, which shows that $(I, g_1, \ldots, g_n)$ is a radical ideal.   □

Remark 4.17. It is not necessary to compute $I \cap k[x_i]$, just to find a nonzero element in this intersection. Note also that once $g_1, \ldots, g_{i-1}$ have been constructed, any nonzero element in $(I, g_1, \ldots, g_{i-1}) \cap k[x_i]$ will serve as well.

Radical Zero-Dimensional Ideals

Zero–dimensional ideals exhibit all sorts of structure that gets narrow downed in critical cases such as the following.

Theorem 4.18. *Let I be a radical, zero–dimensional ideal of the polynomial ring $R = k[x_1, \ldots, x_n]$ over the field k. Then I is generated by n elements.*

Proof. Set $(f(x_1)) = I \cap k[x_1]$, $R = k[x_1, \ldots, x_n]$ and let

$$f(x_1) = f_1(x_1) \cdots f_s(x_1)$$

be a prime decomposition of $f(x_1)$ (which is a squarefree polynomial). By the Chinese Remainder Theorem, we have that $I/f(x_1)R$ can be expressed as

$$
\begin{aligned}
& I \cdot R/f(x_1)R \\
&= I \cdot (k[x_1]/(f_1(x_1))[x_2, \ldots, x_n] \times \cdots \times k[x_1]/(f_s(x_1))[x_2, \ldots, x_n]) \\
&= I \cdot k[x_1]/(f_1(x_1))[x_2, \ldots, x_n] \times \cdots \times I \cdot k[x_1]/(f_s(x_1))[x_2, \ldots, x_n]
\end{aligned}
$$

where the summand $I \cdot k[x_1]/(f_i(x_1))[x_2, \ldots, x_n]$ is a zero–dimensional radical ideal of a polynomial ring in $n - 1$ indeterminates over the field $k[x_1]/(f_i(x_1))$. By induction it can be generated by a set $f_{i,1}, \ldots, f_{i,n-1}$ of elements (lifted to) in I.

To assemble a set of generators of n elements for I, let

$$F_i(x_1) = \prod_{j \neq i} f_j(x_1),$$

and for $1 \leq k \leq n-1$ set

$$G_k = \sum_{1 \leq j \leq s} F_j(x_1) f_{j,k}.$$

We claim that

$$I = (f(x_1), G_1, \ldots, G_{n-1}).$$

It suffices to prove that the two ideals agree in each localization at prime ideals containing the polynomial $f(x_1)$. Suppose that $\mathfrak{P}$ is a prime ideal that contains $f_1(x_1)$. This means that $F_1(x_1)$ is a unit modulo $\mathfrak{P}$ but $F_j(x_1) \in \mathfrak{P}$ for $j > 1$. By Nakayama lemma one has

$$
\begin{aligned}
(f(x_1), G_1, \ldots, G_{n-1})_{\mathfrak{P}} &= (f_1(x_1), f_{1,1}, \ldots, f_{1,n-1})_{\mathfrak{P}} \\
&= I_{\mathfrak{P}},
\end{aligned}
$$

which establishes the claim. $\qquad\square$

Massaging the Output

Here we are concerned in interfacing the output of a Gröbner basis computation of a zero–dimensional ideal with the needs of a numerical solver.

For a zero–dimensional ideal I of the polynomial ring $R = k[x_1, \ldots, x_n]$, we have discussed the reason for the need and means to produce the radical of I. Now we assume that I is radical and seek to provide an efficient system of generators for it.

One door to an economical representation of the ideal I is the structure of the conormal module I/I^2 as a module over the semi–simple ring R/I. Let us sketch the situation. We recall that every maximal ideal of R has height n and is generated by n elements, I/I^2 is a free R/I–module of rank n. This implies that we can find an ideal $J \subset I$, generated by n elements $a_1, \ldots, a_n$, that generate I mod I^2, from which it follows that the ideal I/J of the ring R/J is idempotent, $(I/J)^2 = I/J$, and therefore is generated by one element. This arises very concretely in the following manner.

Let $f_1, \ldots, f_m$ be a generating set of I and express each in the form

$$f_1 = g_1 + \sum c_{1j} f_j$$

$$\vdots$$

$$f_m = g_m + \sum c_{mj} f_j,$$

with $g_i \in (a_1, \ldots, a_n)$ and $c_{ij} \in I$. Setting it up in the format of the characteristic polynomial of a matrix, we have that

$$\det \begin{bmatrix} 1 - c_{1,1} & \cdots & -c_{1,m} \\ \vdots & \ddots & \vdots \\ -c_{m,1} & \cdots & 1 - c_{m,m} \end{bmatrix} = 1 - b, \ b \in I.$$

This gives the following representation for I:

$$I = (a_1, \ldots, a_n, b) \tag{4.8}$$
$$b - b^2 \in (a_1, \ldots, a_n). \tag{4.9}$$

As we have seen, I can be generated by n elements, which would be interesting to turn this result into an efficient procedure. We take another approach which by innocuously adding one new variable and a superfluous generator makes the problem more amenable from a numerical perspective.

Proposition 4.19. *Let I be a zero–dimensional radical ideal of R. Then, either we can find a set of $n+1$ generators of I such that the ideal $(I, 1 - t) \subset R[t]$ is generated by $n+1$ elements or else, we can find a proper decomposition $I = I_1 \cap I_2$.*

Proof. We begin by showing how a generating set as in (4.8) leads to our assertion, and then how they are to be found.

We claim the following equality holds:

$$(a_1,\ldots,a_n,b,1-t)=(a_1,\ldots,a_n,1-(1-b)t).$$

It is enough to verify that the element b is contained in the right hand side, and this follows from

$$b(1-(1-b)t)=b-(b-b^2)t,$$

while by (4.9)

$$b-b^2\in(a_1,\ldots,a_n).$$

To find the a_i, proceed as follows. For $a\in I\setminus I^2$, if $I^2{:}a=I$, we have found the first basis element of the module I/I^2. Otherwise we have a proper decomposition

$$I=(I^2{:}a)\cap(I{:}(I^2{:}a)). \qquad (4.10)$$

If $a_1,\ldots,a_r\in I$ have been found such that

$$(I^2,a_1,\ldots,a_{j-1}){:}a_j=I,\ 1\leq j\leq r,$$

and $r<n$, we repeat the previous step where I^2 is replaced by $(I^2,a_1,\ldots,a_r)$.

Suppose no decomposition of I arises and we have found $a_1,\ldots,a_n$. (With extra care the a_i's could be chosen to generate a zero–dimensional ideal using the argument of Proposition 5.35. Let us assume this was done.) To determine b using the determinantal argument might not always be advisable. One can also build b by pieces as follows. Denote $J=(a_1,\ldots,a_n)$. For $c\in I\setminus J$ the sequence

$$(J,c^j),\ j\geq 1,$$

stabilizes, say at $j=s$. Then

$$c^s=rc^{2s}+d,\ d\in J$$

and rc^s is an idempotent mod J. If $(J,c^s)\neq I$, one looks for another element $d\in I$ to add to (J,c^s), where we ensure that $dc^s\in J$ using again the trick used in Proposition 5.35. $\qquad\square$

Remark 4.20. The reader is encouraged to package this in a better format.

4.3 Idempotents versus Primary Decomposition

It comes about in the following manner. Given an equidimensional ideal with a primary decomposition

$$I=Q_1\cap\cdots\cap Q_r,$$

let

$$\sqrt{I}=\sqrt{Q_1}\cap\cdots\cap\sqrt{Q_r}$$

be the corresponding decomposition of its radical. Because the prime ideals are not comparable, determining the $\sqrt{Q_i}$ leads, by residuation, to the Q_i. This means that we may assume that I is an equidimensional radical ideal.

Let us assume that I is a radical, equidimensional and homogeneous. (The addition of a fresh homogeneous variable and saturation makes possible the last assumption.) Let A denote the integral closure; A is a direct product of affine domains. Let A_0 be its component of degree zero. A_0 is a direct product of fields, and there is a correspondence between its set of idempotents and the minimal primes of I. The knowledge of these elements would permit the recovery of each $\sqrt{Q_i}$ (see [EHV92]).

Idempotents in Artin Algebras

Let R be a finite dimensional algebra over a field k, $n = \dim_k R$. Let $\mathbf{e} = \{e_1, \ldots, e_n\}$ be a basis of R over k, and the $\{c_{ijk}\}$ the corresponding structure constants of the basis $\mathbf{e}$. Suppose $\mathbf{x} = x_1, \ldots, x_n$ is a set of indeterminates. Set $E = \sum_{i=1}^n x_i e_i$ and consider the polynomial equations derived from

$$E^2 = E.$$

The solutions in k^n of the n quadratic equations

$$f_\ell = \sum_{i,j} c_{ij\ell} x_i x_j - x_\ell, \text{ for } 1 \leq \ell \leq n, \tag{4.11}$$

is the variety of *idempotents* or *projectors* of R. When k is algebraically closed we shall refer to the solutions as *idempoints*.

The structure of this ideal is easy to describe:

Proposition 4.21. *Set* $I = (f_1, \ldots, f_n) \subset S = k[x_1, \ldots, x_n]$. I *is a zero–dimensional radical ideal of* S.

Proof. The ideal I is independent of the chosen basis. It is also clear that we may extend the field k, so we shall assume that k is algebraically closed.

The algebra R breaks up into local Artin algebras

$$R = R_1 \times \cdots \times R_m,$$

where each R_i can be written as $R_i = k + M_i$, M_i is its radical. It suffices now to prove the assertion for the algebras R_i.

Changing notation, assume $R = k + M$. Pick a basis of R over k, with $e_1 = 1$ and $e_i \in M$, for $i \geq 2$. The system of equations (4.11) has only two roots, corresponding to the trivial idempotents, 0 and 1. To prove the assertion it suffices to note that in the Jacobian matrix

$$\frac{\partial(f_1, \ldots, f_n)}{\partial(x_1, \ldots, x_n)}$$

the diagonal elements equal $1 - 2x_1$, and the off–diagonal entries lie in $(x_2, \ldots, x_n)$. Its determinant cannot vanish at either of those roots.   □

The problem is, how to find the roots of these equations? For although the structure of its roots is rather simple, the number of variables may be very large. We are now going to consider another method to find idempotents, involving the same number of variables above, but which after a preprocessing step, converts into a problem of linear equations. As of now, it will only work in finite fields.

From Primary Decomposition to Factorization

We consider a very concrete problem: Let k be a perfect field and let $f(x)$ and $g(y)$ be irreducible polynomials over k; we seek the prime factorization of $g(y)$ in the ring $S = k[x]/(f(x))[y]$. The latter has a natural structure of an Euclidean domain.

We outline the steps of an algorithm that starting from a primary decomposition of $I = (f(x), g(y))$ produces a factorization of $g(y)$. Since I is a radical ideal, its primary decomposition

$$I = P_1 \cap \cdots \cap P_r$$

contains only maximal ideals. Furthermore, each P_i can be generated by two elements one of which can be chosen to be $f(x)$, $P_i = (f(x), G_i)$, $G_i \in k[x,y]$. G_i can be determined, using the Euclidean algorithm of S, from a set of generators of P_i; furthermore it can be chosen to be monic in y. Altogether:

Proposition 4.22. *The product* $g(y) = G_1 \cdots G_r$ *is a prime decomposition of* $g(y)$ *mod* $f(x)$.

Factoring the polynomial $g(y_1, \ldots, y_m)$ over a field $S = k[x_1, \ldots, x_n]/M$ has an entirely different character. In the case of a single variable, $m = 1$, one can proceed as above; the main difficulty lies in doing divisions in S (see [Buc85]). Of quite different nature is factorization of polynomials in several variables.

4.4 Decomposition via Sampling

This section deals, with various means, with isolating the prime ideals of a zero-dimensional ideal I. There is some emphasis on probabilistic approaches. Unless said otherwise, I is assumed to be a radical ideal.

Primitive Elements

The difficulty of finding the primary decomposition of a zero dimensional ideal $I \subset R = k[x_1, \ldots, x_n]$ is simplified if the algebra $A = R/I$ is actually defined by fewer than n generators. An extreme case of this is the following assertion.

Proposition 4.23. *Suppose the algebra A is reduced and the residue fields A/P_i are separable over k. Let $(a_{i1}, \ldots, a_{in})$, $1 \leq i \leq m$, be the zeros of I in the algebraic closure of k. If the a_{in}'s are distinct, then the natural inclusion*

$$\varphi : k[x_n]/(I \cap k[x_n]) \to k[x_1, \ldots, x_n]/I$$

is an isomorphism.

Proof. To prove that φ is an isomorphism, it suffices to show that after changing the scalars from k to its algebraic closure $\bar{k}$ the mapping φ is surjective. But the assumptions are preserved by this change, and the conclusion is clear.   □

This immediately allows another description of the ideal I: Since the image of each x_i in A is the image of a polynomial $g_i(x_n)$, it follows that

$$I = (x_1 - g_1(x_n), \dots, x_{n-1} - g_{n-1}(x_n), g_n(x_n)),$$

and in particular the primary components of I are determined entirely by the prime factors of $g_n(x_n)$.

These conditions can be realized by a sufficiently generic linear change of variables:

Proposition 4.24. *Let k be a sufficiently large field and let v_1, v_2, ..., v_m be a finite set of distinct points of k^n. For almost all invertible $n \times n$ matrices A, the nth coordinates of the points $A(v_1), \dots, A(v_m)$ are distinct.*

In [GTZ88], and particularly [GM89], several strategies are discussed to effect such changes of variables.

The Odds

Let A be a finite dimensional reduced algebra over the field k. We seek means to break up A into its summands—or ascertain that it is already a field.

A device is found in Emmy Noether's argument. If $A = k[x_1, \dots, x_n]/I$, for any element

$$f \in k[x_1, \dots, x_n],$$

then

$$I = (I, f) \cap (I : f). \tag{4.12}$$

This follows because for any element f of A, the ideal Af is generated by one idempotent. The issue is to find f such that this intersection is proper. If such elements occur in profusion, one could try for a 'sparse' element. The difficulty lies in the following fact. The algebra A decomposes into a direct product of field extensions

$$A = K_1 \times \cdots \times K_m, \tag{4.13}$$

so that the sought after element f has a component which is zero. If k is a small field and the K_i have small dimension over k, there are plenty of such elements. If however k is large, say has characteristic zero, then almost all the elements of A are units, so that the random picking of f leads nowhere.

Reversing the Odds

The drawback here is: If k is a large enough field, for almost all u, $(I, u) = R$. What is needed is a mechanism that reverses these odds, that is, a process

$$u \mapsto P(u)$$

that for almost all u's produces an element $P(u)$ which is a zero divisor of A or reports that A is a field. The key point is to apply P to sparse elements u.

Let us describe one such process. View the algebra

$$A = k[x_1, \ldots, x_n]/I$$

as a finite dimensional vector space over k. For each element $u \in k[x_1, \ldots, x_n]$, let L_u be the linear transformation induced by multiplication by u. Denote by $P_u(t)$ its characteristic polynomial.

Proposition 4.25. *If the cardinality of k is sufficiently large, then for almost all $u \in A$, $P_u(t)$ is the minimal polynomial of L_u. In particular, if for some u the polynomial $P_u(t)$ is irreducible then A is a field. If*

$$P_u(t) = F(t) \cdot G(t)$$

is a nontrivial decomposition then $F(u)$ is a nontrivial zero divisor of A.

Proof. The first assertion is true since in (4.13) almost all elements whose components are all nonzero are generators of the algebra A over the field k.

If $A = k[u]$, for some u, the minimal polynomial of u must equal its characteristic polynomial, from which the second assertion follows. □

Characteristic Polynomial

To obtain polynomials such as $P_u(t)$ the following device may be employed.

Proposition 4.26. *Let $u \in k[x_1, \ldots, x_n]$ and let t be a fresh variable. Set $J = (t - u, I)$ and choose a lexicographic ordering for which t has lowest weight. Let $(f_u(t)) = J \cap k[t]$. If the field k is large enough then for almost all u,*

$$\deg f_u(t) = \dim_k A.$$

Proof. This follows simply from the inclusion

$$k[t]/(f_u(t)) \hookrightarrow A,$$

and the fact that $f_u(t)$ vanishes on the class of u in A. □

Note that if a Gröbner basis for I has been computed, we also have a Gröbner basis for J, and the additional cost of conversion to the new elimination order has a polynomial complexity ([FGLM]).

Actually, the access to the characteristic polynomial $P_u(t)$ gives a more direct route to the decomposition of I than that afforded by equation (4.12). We employ an argument of [GMT89].

Proposition 4.27. *If $P_u(t) = g(t) \cdot h(t)$ is a non–trivial decomposition then*

$$\gcd(g(t), h(t)) = 1.$$

Writing

$$w(t) \cdot g(t) + v(t) \cdot h(t) = 1, \tag{4.14}$$

$w(u) \cdot g(u)$ and $v(u) \cdot h(u)$ are non–trivial, orthogonal idempotents of A.

Proof. We have already remarked that the polynomial $P_u(t)$ is square-free, so that if $P_u(t)$ is not irreducible, the polynomials $g(t)$ and $h(t)$ exist as stated. Furthermore, $w(t) \cdot g(t)$ and $v(t) \cdot h(t)$ share each a proper subset of the roots of $P_u(t)$, and the substitution $t \mapsto u$ evaluates to idempotents of A, necessarily non–trivial ones.   □

Berlekamp Method

The following approach, called to our attention by L. Robbiano, to find equations for the idempotents of R was suggested by Berlekamp's factorization method [Ber70]. It is extensively used in [GMT89] to break up finite dimensional algebras, first over finite fields and then, using Hensel's methods, algebras over the rational numbers.

Let R be a finite dimensional algebra (which for the moment we do not assume to be reduced) over the finite field $k = \mathbb{F}_q$, $q = p^r$. Let e_i, $1 \leq i \leq n$, be a basis of R over k, and set

$$E = \sum_{i=1}^{n} x_i e_i$$

the generic element of R. We want to make use of the fact that all idempotents will satisfy the equation

$$E = E^q,$$

that is

$$\sum_{i=1}^{n} x_i e_i = \sum_{i=1}^{n} x_i^q e_i^q = \sum_{i=1}^{n} x_i e_i^q.$$

The solutions of this system of equations form a subalgebra A of R, which is used in [GMT89] as a vehicle to get the idempotents of R. Let us follow their argument.

Proposition 4.28. *A is a semisimple k-algebra, and $m = \dim {}_k A$ is the number of distinct prime ideals of R.*

Proof. The first assertion is clear since the elements of A are invariant under taking qth powers, and therefore do not contain nilpotent elements. For the same reason, in the decomposition of A into local algebras, each factor must be equal to k. $\square$

Suppose $\{e_1, \ldots, e_m\}$ are the primitive idempotents of R; they form a basis of the algebra A. Let

$$u = \sum_{i=1}^{m} a_i e_i,$$

be an element of A, in which all the a_i's are nonzero. These are the regular elements of A. To produce an idempotent out of u, we distinguish two cases:

(a) If $q = 2^r$, set $e = \sum_{i=0}^{r-1} u^{2^i}$.

(b) If characteristic $k = p \neq 2$, set $t = u^{\frac{q-1}{2}}$ and consider the elements $e_{\pm} = \dfrac{t^2 \pm t}{2}$.

Since $u^q = u$, it follows easily that e and $e_{\pm}$ are always idempotents. Furthermore, if $t \neq 0, \pm 1$, then either e_+ or e_- is non-trivial.

In case (a), we can also write

$$e = \sum_{i=1}^{m} \text{Trace}(a_i) e_i,$$

for the trace function of the Galois extension $\mathbb{F}_q/\mathbb{F}_2$, a non-trivial linear mapping. Note that the values of $\text{Trace}(a_i)$ are 0, 1.

In case (b),

$$e_{\pm} = \sum_{i=1}^{m} \frac{(1 \pm a_i^{\frac{q-1}{2}})}{2} e_i,$$

where $a_i^{\frac{q-1}{2}} = \pm 1$.

Proposition 4.29. *In each of these two cases, the idempotent is non-trivial with probability at least $\frac{1}{2}$.*

Proof. In case (a), to fail all the $\text{Trace}(a_i)$ must be 0, or all must be 1. In case (b), to fail all a_i's are either quadratic non-residues, or all are nonzero quadratic residues. A straightforward count leads to the assertion. $\square$

Sampling Zeros Through Eigenvalues

The following scheme has been used in methods to sample the zeros of systems of equations ([ASt88], [Laz81]).

Theorem 4.30. *Let $A = R/I$ be a zero–dimensional affine algebra over a field k, and let $P_1, \ldots, P_s$ be the zeros of I in an algebraic closure of k. For $f \in R$, let L_f be the k–linear transformation of A induced by multiplication by f. Then the eigenvalues of L_f are $f(P_1), \ldots, f(P_s)$.*

Proof. We may assume that k is algebraically closed. Let

$$0 \subset A_r \subset \cdots \subset A_1 \subset A_0 = A,$$

be a composition series of A as a module. Note that each A_i is invariant under L_f, so that the eigenvalues of L_f are those obtained by the action of L_f on the factors A_i/A_{i+1}. Each of these is a module isomorphic to R/M_i, where M_i is the maximal ideal of one of the zeros of I. If v_i is a generator of A_i/A_{i+1}, $fv_i = a_iv_i, a_i \in k$. But a_i is just f evaluated at P_i. □

Letting f run through the set of variables, $x_1, \ldots, x_n$, generates the finite pools of values of the coordinates of the zeros of the system of equations,

$$V(I) \subset Z_1 \times \cdots \times Z_n.$$

A key issue is how to further refine this set.

4.5 Root Finders

This section has a very informal character, being intended as a bridge—or, a mere plank—to the world of numerical computation of systems of equations.

Let $f_1, \ldots, f_m$ be polynomials in n variables over a field k, $f_j \in k[x_1, \ldots, x_n]$. The problem of finding the solutions of the equations

$$f_1 = f_2 = \cdots = f_m = 0, \tag{4.15}$$

in some extension field of k, occurs very frequently in science and technology.

It is a common view to consider the set of equations $\{f_1 = \cdots = f_m = 0\}$ as being equivalent to the set of zeros of the polynomial ideal generated by the f_j's. While strictly speaking this is correct, it overlooks an important fact. When the equations are derived from a model, it carries information about how the algebraic variety is built down, from successive intersections, in a process akin to sculpting. Such structures, e.g. flags, may be lost when other generators for the ideal are considered.

It is important, to the extent possible, not to disturb the original set of equations. Doing so, may cause symmetry, sparseness, to disappear. This is one of least understood issues with the subject.

Root finders, or *solvers*[1], are algorithms or numerical schemes to find the zeros of such sets of equations. They tend to be broadly defined, being amenable to all kinds of equations, transcendental or algebraic. They come in all colors & shapes and there is a vast literature regarding them.

Our interest here is on solvers that can benefit by a preconditioning that may be provided by symbolic computation packages. They tend to have a dynamical character, being driven either by iteration or by differential equations. One of their characteristics is the fact that they work more naturally on systems of equations where the number of equations equals the number of variables.

Secant and Newton Dynamics

Iterative schemes for solving systems of polynomial (or more general equations) are based on converting the ingredients of the system

$$\mathbf{f} = 0: \qquad f_1 = f_2 = \cdots = f_m$$

into a mapping

$$\psi : \mathbb{R}^n \mapsto \mathbb{R}^n,$$

so that the zeros of $\mathbf{f} = 0$ correspond to the fixed points of ψ.

This is an active arena of mathematical activity for which the author lacks the expertise to discuss in detail. We shall make however some comments on the kinds of data such methods require to act properly (see the volume [AG90] for several expert reviews).

Let us begin by recalling two of the oldest such schemes. Let $f(x)$ be a polynomial in one variable.

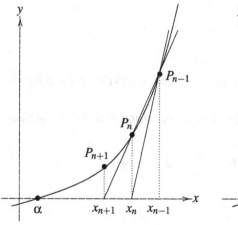

The secant method.

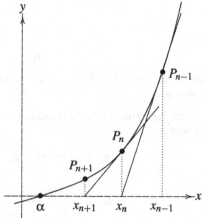

Newton's method.

---

[1] or *Zerors?*

The *secant* method, from an initial choice of two points x_0 and x_1, generates a sequence

$$x_{n+1} = S_f(x_{n-1}, x_n) = x_n - \frac{x_n - x_{n-1}}{f(x_n) - f(x_{n-1})} f(x_n). \tag{4.16}$$

In *Newton*'s method, from an initial guess x_0, one has iterates

$$x_{n+1} = N_f(x_n) = x_n - \frac{f(x_n)}{f'(x_n)}. \tag{4.17}$$

The latter can be rephrased and generalized as

$$\psi : \mathbb{R} \mapsto \mathbb{R},$$

$$\psi(x_n) = x_n - \lambda(x_n) \cdot \frac{f(x_n)}{f'(x_n)},$$

where $\lambda(\cdot)$ is chosen to try to contain the values of ψ to a small neighborhood of the initial guess and to speed up the convergence to its fixed point α (if any).

These methods require the nonsingularity of $f(x)$ at $x = \alpha$. They are broadly classified by the rates at which the sequences approach α. Thus Newton method, $\lambda(x) = 1$, exhibits quadratic convergence, under broad conditions, $|x_{n+1} - \alpha| \leq C|x_n - \alpha|^2$, for some constant C. Choosing $\lambda(x)$ so that it amounts to applying Newton's method to

$$g = \frac{f}{\sqrt{f'}},$$

that is

$$H_f(x) = N_g(x),$$

gives the method of Halley, which is a cubic scheme (see [ST95] for a delightful discussion).

There are immediate generalizations to systems of equations: In the case of Newton's method,

$$x_{n+1} = N_f(x) = x_n - D(f)(x_n)^{-1} \cdot f(x_n),$$

$$D(f) = \frac{\partial(f_1, \ldots, f_n)}{\partial(x_1, \ldots, x_n)},$$

provided the set of equations is a complete intersection without singular points. This gives the incentive for computing the radical of a zero-dimensional ideal since it satisfies those two requirements.

Method of the Steepest Descent

Another scheme, which does not require that the number of equations and of variables be the same, uses the *gradient*. Suppose

$$f_1, \ldots, f_m \in \mathbb{R}[x_1, \ldots, x_n]$$

are real polynomials (or complex polynomials that are converted to real polynomials by the doubling of the x_i's and of the f_j's) and set

$$F = \sum_{j=1}^{m} f_j^2 \quad \text{or} \quad F = \sum_{j=1}^{m} f_j \overline{f_j}.$$

Then for some step size $h > 0$,

$$x_{n+1} = x_n - h \cdot \frac{\operatorname{grad} F(x_n)}{|\operatorname{grad} F(x_n)|}$$

defines a process to drive an initial choice to a minimum of F often enough.

Path Continuation

We describe briefly a numerical solver and the kind of data it requires to run smoothly.

Let

$$V(I) : f_1 = \cdots = f_m = 0$$

be a set of equations, $f_j \in k[x_1, \ldots, x_n]$. One seeks another set, closely related to $V(I)$, but with more accessible solutions. Suppose that $m = n$, $k = \mathbb{C}$. The *path following* approach to the solution of the system of equations consists in studying homotopies linking this set, $V(I)$, to the solutions of another, better known, system

$$g_1 = \cdots = g_n = 0.$$

The simplest of these are linear homotopies that have the form

$$H(x,t) = (1-t) \cdot G(x) \cdot r + t \cdot F(x) \tag{4.18}$$

where $F(x)$ and $G(x)$ are the polynomial maps from k^n into itself, defined by the f_i's and g_i's. In a reversal of the usual practice, one applies techniques from the theory of numerical solution of differential equations to solve algebraic equations.

The graph of $H(x,t) = 0$ consists of paths joining of the kind sketched in the figure:

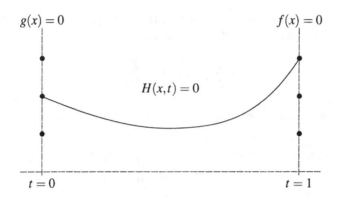

The differential equation permits tracing out the paths that connect the known points (the solutions of $g = 0$) to the desired points (the solutions of $f = 0$). There are several strategies for picking the g's and modifying the homotopy (e.g. by rotating the vector g).

This works best when there are natural choices for the g's and the f's are stripped of excess multiplicities: It is critical that $H(x,t)$ defines smooth curves in the interval $[0,1]$.

These requirements again highlight the need to solve the following problem: Given a zero–dimensional ideal I of $k[x_1,\dots,x_n]$, find another ideal, with the same zeros but generated by n elements and without excess multiplicities.

Boxing Zeros

The numerical schemes mentioned above, for their well functioning, require start up positions near a root. Getting there is however usually very expensive except for simple cases.

Sturm Sequences

Let $f(x) \in k[x]$ be a squarefree polynomial and consider the sequence of polynomials that arises from applying the Euclidean algorithm to $f(x)$ and its derivative $f'(x)$ (with a sign switch):

$$f_0 = f(x)$$
$$f_1 = f'(x)$$
$$f_2 = -f_0 + q_2 f_1$$
$$\vdots$$
$$f_{r-1} = -f_{r-3} + q_{r-1} f_{r-2}$$
$$f_r = -f_{r-2} + q_r f_{r-1}$$

where f_r is a nonzero element of k. If k is a subfield of $\mathbb{R}$, the sequence

$$S(f) = \{f_0, f_1, \ldots, f_r\}$$

has two important properties: (i) for $0 \leq j < r$, f_j and f_{j+1} do not have a common zero; (ii) for $0 < j < r$, if $f_j(\alpha) = 0$ for some $\alpha \in k$ then $f_{j-1}(\alpha)$ and $f_{j+1}(\alpha)$ have opposite signs. $S(f)$ is called *the Sturm sequence* of $f(x)$ (see [Mil92] for a modern discussion and recent extensions).

For any α which is not a root of any of the f_j's, one defines (the *winding number of the sequence*) $W(S(f))(\alpha)$ as the number of sign changes in the sequence

$$\{f_0(\alpha), f_1(\alpha), \ldots, f_r(\alpha)\}.$$

Theorem 4.31 (Sturm). *Let $\alpha < \beta$ be elements of k neither of which is a root of any of the $f_j \in S(f)$. Then*

$$\left| \{a \mid f(a) = 0\} \cap [\alpha, \beta] \right| = W(S(f))(\beta) - W(S(f))(\alpha). \qquad (4.19)$$

Proof. Partition the interval $[\alpha, \beta]$, $\alpha = a_0 < a_1 < \cdots < a_n = \beta$, in such a manner that none of the f_j's has a root at a_i and no more than one root in the interval $[a_{i-1}, a_i]$. From

$$W(S(f))(\beta) - W(S(f))(\alpha) = \sum_{i=1}^{n} (W(S(f))(a_i) - W(S(f))(a_{i-1})),$$

the property (ii) above implies that the summands here are either 1 or 0, according to whether $f(x)$ has a root in $[a_{i-1}, a_i]$ or not. $\qquad \square$

Remark 4.32. Sequences with these properties may be used on general continuous functions but the means to engender the sequence from a given function are spotty. There are also generalizations to systems of n functions in n variables in the literature (see [Mil92], [GTr94], [Swe95]).

The Degree of a Map

The proper setting to consider extensions of the preceding is through degree theory of continuous maps. Let $f(\mathbf{x}) = (f_1, \ldots, f_n)$ be a set of polynomials in n indeterminates, and denote by $D(f)$ its Jacobian determinant. Consider a nice region $\Omega \subset \mathbb{R}^n$ with a boundary which is topologically a sphere (typically a polyhedron) and suppose that $f(\mathbf{x})$ does not vanish on the boundary of Ω. This permits defining a *degree* $\deg(f, \Omega)$: There is an associated continuous function on the sphere S^{n-1} into itself, which at the homology level maps the generator of $H_{n-1}(S^{n-1}, \mathbb{Z}) = \mathbb{Z}\varepsilon$ into $\deg(f, \Omega)\varepsilon$.

A classical result here is:

Theorem 4.33 (Kronecker). *If $\{\alpha \mid f(\alpha) = 0\}$ consists of regular points of $f(\mathbf{x})$ then $\deg(f, \Omega)$ counts the (finite) number of points in $V(f) \cap \Omega$ where $D(f) > 0$ minus the points where $D(f) < 0$.*

As a rule computing this number is very hard and the methods tend not to benefit from the fact that the functions are polynomials. (See [GTr94] for pointers to the literature.)

5

Nullstellensätze

A *Nullstellensatz* is the description of the nil radicals of the ideals in a given class, the quintessential example of which is Hilbert's Nullstellensatz ([Hil93]). For ideals in more narrowly defined classes, it is be desirable to have sharper descriptions of their radicals.

The aim of this chapter is to set up an environment to develop methods—when not closed formulas—for computing the radical of an ideal of the ring R of polynomials over a field k. Typical are results in the form of closed formulas for the radical of an unmixed ideal of R—an ideal all of whose primary components have the same codimension. They are generalizations, to polynomials in several variables, of the elementary:

$$\sqrt{(f)} = (f):f' = (f/\gcd(f,f')). \tag{5.1}$$

This formula is a unique gem of efficiency with signposts to other opportunities.

Efficient methods to compute radicals are of fundamental interest, for both theoretical and practical reasons. As demonstrated by Brownawell [Br87] and Kollár [Ko88], there are fundamental gains to be obtained if in the membership problem one seeks only radical membership.

We want to convert these embedded efficiencies as pathways to enhance discussion of primary decomposition in Chapter 3 with the closely allied problem of finding zeros of polynomials equations dealt with in Chapter 4, and various approaches to the determination of the integral closure of a ring in Chapter 6.

A mix of differential and homological approaches are used to produce explicit formulas and algorithms that compute radicals. When they fail, other methods with good programmability can often be brought in.

This chapter gives a fairly complete discussion of the Jacobian criteria that will be used in all constructions. For brevity, it will be limited to the characteristic zero case, with proofs influenced by the spirit of some old approaches of Zariski [Zar65].

The radical formulas themselves depend on some arcane of the theory of the Jacobian ideals following [EHV92]. Some of these methods are highly dependent on the ability to generate regular sequences fairly efficiently: this turns out to a bottleneck in computational algebra, with many challenges awaiting the interested reader. We view several techniques to meet the difficulties, a method of Eisenbud and Sturmfels [EiS94] to produce regular sequences from "sparse generic combinations," a mélange of recursive uses of the explicit radical formulas with standard tricks involving the Chinese Remainder Theorem, and even venture into the characteristic sets of Ritt and Wu to look for regular sequences in triangular sets of polynomials.

Among applications we consider the retrieval of the isolated zeros of systems of equations. Finally, there is a brief discussion of how to find pieces of the radical of an ideal among the coefficients of its syzygies.

5.1 Radicals via Elimination

We begin with the higher dimensional analog of the method used in Chapter 4 to find radicals of zero–dimensional ideals (Theorem 4.16). It relies on rounds of eliminations. We will be following [KLo91], which contains additional details; see also [GTZ88]. Some constructions, such as finding the square free part of a polynomial, may require the field to have characteristic zero, or at least be perfect.

Independent Sets of Variables and Radicals

Let I be an ideal of $R = k[\mathbf{x}] = k[x_1, \ldots, x_n]$ and suppose height $(I) = m$. Set $d = n - m$, the dimension of R/I and denote by

$$\sqrt{I} = I_m \cap I_{m+1} \cap \cdots \cap I_n$$

a decomposition of the radical of I, where I_j is an equidimensional radical ideal of height j (we set $I_j = R$ if such component is not present). In this representation the associated primes belonging to different components are not comparable.

By assumption, there are subsets of $\mathbf{x} = \{x_1, \ldots, x_n\}$ consisting of d variables (after suitable relabeling) with the following property: The subring $S = k[x_1, \ldots, x_d]$ satisfies $I \cap S = (0)$ and any larger subset of $\mathbf{x}$ leads to a non-trivial intersection. Thus for each x_i, $i > d$, there exists a polynomial $0 \neq f_i \in I \cap k[x_1, \ldots, x_d, x_i]$. Furthermore, the embedding

$$k[x_1, \ldots, x_d] \hookrightarrow k[x_1, \ldots, x_n]/I \qquad (5.2)$$

guarantees that some minimal prime of I_m contracts to 0 in the subring. The converse also holds, any minimal prime of I_m will contract to 0 in some subring of polynomials $k[x_{i_1}, \ldots, x_{i_d}]$.

Write $f_i = g_i \cdot h_i$, where g_i is the gcd of the coefficients of f_i viewed as a polynomial of $S[x_i]$. If K is the field of fractions of S, by Gauss Lemma, g_i is still the content of $f_i \in K[x_i]$. If h_i is not square free, let b_i be its square free part. One has $p_i = g_i b_i \in \sqrt{I}$. Obviously the difficulty here lies mostly in obtaining the elements $f_1, \ldots, f_m$, but they can be derived from Gröbner basis computations associated to elimination orders. For one of these, let q be the least common multiple of the leading coefficients of the leading terms of G viewed as polynomials of $K[x_{d+1}, \ldots, x_n]$.

Radical Formulas

The polynomials $p_1, \ldots, p_d$ and q then provide for the following description of a radical component of $\sqrt{I}$.

Theorem 5.1. *The equality*

$$\sqrt{I} K[x_{d+1}, \ldots, x_n] \cap k[x_1, \ldots, x_n] = (I, p_1, \ldots, p_d) : q^\infty$$

holds and this radical ideal J is the intersection of all the minimal primes $\mathfrak{P}$ of I_m which have the property $\mathfrak{P} \cap k[x_{d+1}, \ldots, x_n] = 0$.

The proof is straightforward from the various constructions (see [KLo91]). A point worth making is that the element q does not lie in any minimal prime of J and therefore from $q^\infty \cdot J \subset \sqrt{I}$, it must be contained in all the other minimal primes of $\sqrt{I}$.

Corollary 5.2. *Let KL be the family of all subsets $\gamma = \{x_{i_1}, \ldots, x_{i_d}\}$ of indeterminates, with the property of (5.2). Each γ gives rise to a radical ideal J_γ and*

$$I_m = \bigcap \{J_\gamma, \quad \gamma \in KL\}.$$

The situation is more amenable when $k[x_1, \ldots, x_d]$ is a Noether normalization of R/I. Indeed in this case

Corollary 5.3. *If in addition to the conditions above the f_i's are monic then*

$$I_m = (I, p_1, \ldots, p_m) : q^\infty.$$

To get the lower dimensional components of $\sqrt{I}$ one can proceed in several ways.

Proposition 5.4. *Denote by P the ideal generated by all the p_i's for all choices of γ, and similarly denote by Q the ideal generated by all q's. If I is in Noether position, choose simply from the p_i's and q in Corollary 5.3. Then*

$$\sqrt{I} = I_m \cap \sqrt{(I, P, Q)}.$$

Observe that the ideal (I, P, Q) has codimension at least $m+1$ and its minimal primes may contain ideals which are not minimal over I, but such primes would contain I_m and therefore would be filtered out in the intersection.

Exercise 5.5 (Logar). Let I be an ideal generated by binomials of the ring of polynomials $k[x_1, \ldots, x_n]$, that is elements of the form $ax^\alpha - bx^\beta$, $a, b \in k$ but possibly zero, and let $f = x_1 \cdots x_n$. Prove that the ideal $I : f^\infty$ is unmixed and radical.

Exercise 5.6. Let I be an ideal generated by monomials and at most two binomials. Describe how to find the radical of I without looking up [EiS96].

5.2 Modules of Differentials and Jacobian Ideals

Let $R = k[x_1, \ldots, x_n]$, k is a field, and let $I = (f_1, \ldots, f_p)$ be an ideal of R. We shall recall some basic properties of the module of Kähler differentials of the algebra $A = R/I$, its Fitting invariants and their connection to regularity criteria for A. The full references will be [Kun86] or [Mat80].

Kähler Differentials

Definition 5.7. *Let k be a commutative ring and let A be a k–algebra. A derivation of A over k with values in an A–module E is a k–linear mapping $d\colon A \to E$ that satisfies Leibniz rule,*
$$d(a \cdot b) = a \cdot d(b) + b \cdot d(a).$$

The set of all such mappings define a module $\mathrm{Der}_k(A, E)$, more precisely, a functor from the category of A–modules into itself. If A is the polynomial ring $R = k[x_1, \ldots, x_n]$, the assignment

$$f \in A \rightsquigarrow df = \left(\frac{\partial f}{\partial x_1}, \cdots, \frac{\partial f}{\partial x_n} \right) \in A^n \tag{5.3}$$

defines a derivation, and any derivation of A factors through it. In other words, we may identify $\mathrm{Der}_k(A, E) = \mathrm{Hom}_A(A^n, E)$. It is common to denote the images of the variables x_i by dx_i.

Given an arbitrary k–algebra A, any module with the similar property will be called the universal module of k–derivations, or the module of Kähler differentials. It will be denoted by $\Omega_{A/k}$, and the corresponding mapping

$$d\colon A \to \Omega_{A/k}$$

will be called the *universal derivation* of A.

The relationship between the modules of Kähler differentials of R and R/I is simply expressed by the exact sequence

$$I/I^2 \xrightarrow{\ d\ } \Omega_{R/k} \otimes_R A \longrightarrow \Omega_{A/k} \to 0,$$

where d is the universal derivation: $df = \sum_{i=1}^n \frac{\partial f}{\partial x_i} dx_i$. Since $\Omega_{R/k} \otimes_R A = A^n$, the exact sequence gives rise to the presentation

$$A^p \xrightarrow{\ \varphi\ } A^n \longrightarrow \Omega_{A/k} \to 0, \tag{5.4}$$

where

$$\varphi = \left(\frac{\partial(f_1, \ldots, f_p)}{\partial(x_1, \ldots, x_n)} \right) \bmod I.$$

Jacobian Ideals of an Algebra

The matrix φ is called the Jacobian matrix of the system of generators f_j's. Since it is taken modulo I, it has a measure of invariance. The *Jacobian ideals* of I, or R/I, are the Fitting ideals of $\Omega_{A/k}$. They are denoted

$$J_a(I) := I_{n-a}(\varphi).$$

If $m = \text{codim } I$, the *Jacobian ideal* proper is the ideal of $m \times m$ minors of φ; it shall be denoted by J.

Since the $J_a(I)$ are the Fitting ideals of the module $\Omega_{A/k}$, they are independent of the generating set of I, and behave well with regard to many processes such as localization and completion. Repeatedly we abuse the notation, and take $J_a(I) + I$ for the Jacobian ideals of A, in other words, we view $\Omega_{A/k}$ as an R–module.

Remark 5.8. (a) One of the most direct consequences of this definition is that the module of Kähler differentials behave well relative to *base change*: If $k \to K$ is a homomorphism of commutative rings, then $\Omega_{A \otimes_k K/K} = \Omega_{A/k} \otimes_k K$.

(b) It follows also that if k is a field and $A = R/I$ is a reduced affine k–algebra and $\mathfrak{p}$ is a minimal prime ideal whose residue field $k(\mathfrak{p})$ is separably generated over k, then $\Omega_{A/k} \otimes_A A_{\mathfrak{p}}$ is a vector space over $A_{\mathfrak{p}}$ of dimension equal to tr deg $k(\mathfrak{p})/k$. This will allow us to determine the generic rank of some modules of differentials.

Partial Derivatives and Variables

We give a brief proof of the Jacobian regularity criterion using ideas of [Sei66] and [Zar65]. The rings here all contain a field of characteristic zero. For terminology, if D is a derivation of R and I is an ideal such that $D(I) \subset I$, then we say that I is a differential ideal (for D).

Proposition 5.9. *Let D be a derivation of a commutative Noetherian ring R with $D(x) = 1$. Then x is a nonzero divisor.*

Proof. Let $a \in I =$ annihilator of x. From $ax = 0$ one gets $a + xD(a) = 0$, that is $a \in Rx$. Suppose it has been proved already that $I \subset Rx^{n-1}$. From $ax = rx^n = 0$ one has $x(D(r)x^{n-1} + nrx^{n-2}) = 0$ and thus $D(r)x^{n-1} + nrx^{n-2} \in I \subset Rx^{n-1}$. Applying D to this expression $n - 2$ times, we get $n!r \in Rx$. This shows $a \in \bigcap Rx^n$, which vanishes by Krull intersection theorem. $\qquad\square$

The derivation D of the pair (D, x), by abuse of terminology, is referred as the *partial derivative* with respect to x.

Proposition 5.10. *Let $(R, \mathfrak{p})$ be a reduced Noetherian local ring of Krull dimension 1 and let D be a derivation of R such that $Dx = 1$ for some $x \in \mathfrak{p}$. Then R is an integral domain.*

Proof. Let P be a minimal prime of R and denote by I its annihilator. Note that $P \cap I = 0$ since R is reduced. For $a \in I$, $b \in P$, we have $D(ab) = aD(b) + bD(a) = 0$ so that $D(b) \in P$ and $D(a) \in I$ (so both I and P are differential ideals). By the previous result, x is a regular element and therefore Rx is $\mathfrak{p}$–primary. In particular $I^n \subset Rx$ for some n. Let $b \in I^n$, $b = rx$. Thus $D(b) = r + xD(r)$ and $r \in Rx$ as $D(I^n) \subset I^n$. An easy induction and another use of the intersection theorem leads to $I^n = (0)$ and thus $I = (0)$. $\square$

The following result of Seidenberg ([Sei66]) is one of our tools for deriving Jacobian criteria.

Theorem 5.11. *Let R be a reduced Noetherian ring with finite integral closure R', and let D be a derivation of R. Then D has an extension to a derivation of R'. Furthermore, the conductor J of R' into R is a differential ideal.*

Proof. If we use the quotient rule for differentiation, D extends to a derivation of the total ring of fractions Q of R. The issue is whether R' is invariant under this extension (still noted by D), for which purpose we employ a pretty construction of Seidenberg [Sei66].

Let z be an element of R'; it is clear that there exists a regular element d of R such that $d \cdot z^n \in R$ for all n; it suffices to consider n bounded by the degree of an integral equation of z with respect to R. Conversely, any element of Q with this property is integral over R (for non-Noetherian rings the notions differ).

Denote by $R[[t]]$ the ring of formal power series in the variable t over R. Since R' is Noetherian, $R'[[t]]$ is also Noetherian and is the integral closure of $R[[t]]$. Set

$$\varphi(a) = \sum_{k \geq 0} \frac{D^k(a)}{k!} t^k, \quad \text{for } a \in Q$$
$$\varphi(t) = t.$$

φ extends to an isomorphism of $Q[[t]]$ (still denoted by φ), restricting to an isomorphism of $R[[t]]$. Applying it to an equation $d \cdot z^n = r \in R$, we get

$$\varphi(d) \cdot (\varphi(z))^n = \varphi(r),$$

which shows that $\varphi(z)$ is integral over $R[[t]]$, and therefore has all of its coefficients in R', proving the first assertion.

For the remainder, it suffices to notice that from $J \cdot R' \subset R$, we have $D(J) \cdot R' + J \cdot D(R') \subset R$, so $D(J)$ also conducts R' into R. $\square$

Remark 5.12. If one does not want appeal to the description of the integral closure of $R[[t]]$, another argument goes as follows. Note that from $\varphi(d)(\varphi(z))^n \in R[[t]]$, we have

$$d\varphi(d)(\varphi(z) - z)^n \in R[[t]], \quad \forall n \geq 0,$$

from which it follows easily that the coefficient of t^n in the expansion is $d^2 D(z)^n$.

Theorem 5.13. *Let* $(R, \mathfrak{p})$ *be a local ring of dimension n that is the localization of an affine k–algebra. Let* $D_1, \ldots, D_n$ *be a family of derivations of R and* $x_1, \ldots, x_n$ *a family of elements of* $\mathfrak{p}$ *such that the determinant of the matrix* $(D_i(x_j))$ *is a unit of R. Then* $x_1, \ldots, x_n$ *is a regular sequence in R. If R is reduced, then R is a regular local ring.*

Proof. Taking convenient linear combinations of the derivations we can assume that the Jacobian matrix is the identity. This means that Rx_i is a differential ideal with respect to D_j, for $j \neq i$. By Proposition 5.9, the x_i will form a regular sequence, in particular R is a Cohen–Macaulay ring.

To prove the second assertion, consider the ideal $I = (x_1, \ldots, x_{n-1})$. For each minimal prime ideal P of I, the derivations $D_1, \ldots, D_{n-1}$ induce derivations of R_P and by induction R_P is regular. This implies that R/I is a reduced ring.

Finally we use Proposition 5.10 on the reduced ring $A = R/I$ and the derivation induced on it by D_n. By Theorem 5.11, the conductor J of A' into A is a differential ideal. Since it must be primary to the maximal ideal of A, it contains a power of x_n, which by repeated differentiation proves that $J = A$. Thus A is a reduced, integrally closed local ring of dimension 1 so must be a domain and therefore a discrete valuation domain with (x_n) its maximal ideal. □

Regularity Criteria

The main use of the Jacobian ideals is as a carrier of the regularity of the algebra A. We begin by listing the Jacobian criteria to be used later (see [Kun86, Theorem 7.2]).

Theorem 5.14. *Let A be an affine algebra over a field k of characteristic zero. For any prime ideal* $\mathfrak{p}$ *the following are equivalent*

(a) $A_\mathfrak{p}$ *is a regular local ring.*
(b) $\Omega_{A_\mathfrak{p}/k}$ *is a free* $A_\mathfrak{p}$*–module.*

In this case the rank of $\Omega_{A_\mathfrak{p}/k}$ *is* $\dim A_\mathfrak{p} + \operatorname{tr} \deg k(\mathfrak{p})/k.$ *(*$k(\mathfrak{p})$ *is the residue field of* $A_\mathfrak{p}$*.)*

Of these conditions, (b) has a wider validity being independent of the characteristic. To ascertain it, one uses the Jacobian ideals. There are versions of this result for finite characteristics; see [Kun86] and [Mat80, Theorem 64].

We shall want to use these criteria in the following form:

Corollary 5.15. *Let k be a field of characteristic zero and let* $A = R/I$ *be an affine algebra with Jacobian ideal J. For any prime ideal* $I \subset \mathfrak{p} \subset R$, *the localization* $A_\mathfrak{p}$ *is a regular local ring if and only if* $J \not\subset \mathfrak{p}$. *In particular,* $\mathfrak{p}$ *contains a single primary component of I which is a prime ideal.*

Of particular significance to our discussion of the integral closure in Chapter 6 is the following criterion ([Lip69]):

Theorem 5.16 (Lipman). *Let A be a reduced, equidimensional algebra over a field k of characteristic zero. If the Jacobian ideal J of A is invertible then A is a regular ring.*

Rather than establishing the criteria individually, we show how they can be derived from our previous discussion. Let A be an affine algebra over a field of characteristic zero, and let $\mathfrak{p}$ be a maximal ideal. For simplicity, we denote still by $\mathfrak{p}$ the maximal ideal of the localization $A_\mathfrak{p}$. We begin by observing that $\Omega_{A_\mathfrak{p}/k}$ is generated by $d(f)$, $f \in \mathfrak{p}$ (cf. (5.3)). Indeed, for $u \in A \setminus \mathfrak{p}$, its image in $A/\mathfrak{p}$ has a separable minimal polynomial $f(x)$. From $f(u) \in \mathfrak{p}$, we get $f'(u)d(u) \in d(\mathfrak{p})$, which proves our assertion since $f'(u) \notin \mathfrak{p}$.

The argument also shows that any minimal generator of $\Omega_{A_\mathfrak{p}/k}$ can be chosen as $\sum_j a_j d(y_j)$, with $y_j \in \mathfrak{p}$, an expression that can be rewritten as

$$d(\sum_j a_j y_j) - \sum_j y_j d(a_j),$$

which from Nakayama lemma shows that every minimal generator of $\Omega_{A_\mathfrak{p}/k}$ can be taken of the form $d(y)$, $y \in \mathfrak{p}$.

If therefore $\Omega_{A_\mathfrak{p}/k}$ has a free summand, it may be chosen in the form $A_\mathfrak{p} d(y)$, $y \in \mathfrak{p}$. The decomposition allows the definition of a derivation $D : A_\mathfrak{p} \to A_\mathfrak{p}$ such that $D(y) = 1$. If $\Omega_{A_\mathfrak{p}/k}$ has a free summand of rank equal to $\dim A_\mathfrak{p}$, such as in the case of Theorem 5.16, we obtain from Theorem 5.13 that $A_\mathfrak{p}$ is a regular local ring.

Exercise 5.17 (B. Kostant). Let k be an algebraically closed field of characteristic zero and let $= (f_1, \ldots, f_m) \subset k[x_1, \ldots, x_n]$. Suppose that the radical of I is a prime ideal and that for some point $p \in V(I)$ the rank of the Jacobian matrix

$$\frac{\partial(f_1, \ldots, f_m)}{\partial(x_1, \ldots, x_n)}$$

at p is m. Prove that I is a prime ideal.

Exercise 5.18 (Zariski–Nagata). Prove Theorem 3.78.

5.3 Generic Socles

We shall define here a generalization of the socle of an Artin algebra. Its framework is that of the primary decomposition of ideals, of which we recall some specialized concepts.

Generic Socle Formulas

The following technical notion will play an important role in our approach to the radical ([EHV92]).

Definition 5.19. *Let I be an unmixed ideal of codimension m, and let* $P_1,\ldots,P_s$ *be its associated prime ideals. Denote* $B_t = (R/I)_{P_t}$. *By a generic socle formula we mean an ideal* $G \subset R$ *such that for each* $1 \leq t \leq s$

$$G \cdot B_t = \begin{cases} B_t, & \text{if } B_t \text{ is a domain} \\ \text{otherwise a nonzero submodule of the socle of } B_t. \end{cases}$$

If G is generated by a single element f, we shall call it a generic socle generator.

It is clear that generic socle generators always exist: If R and I are as above, the total ring of fractions of R/I is the semi–local algebra

$$B_1 \times \cdots \times B_s,$$

and the existence of the socle generator follows from the Chinese Remainder Theorem. The point is to try and identify them in the data associated with the ideal I.

Tate Formula

Socle formulas have been very difficult to identify. One that plays a key role in the exposition here is the following formulation of Theorem 4.8:

Theorem 5.20. *Let k be a field and let* $A = k[[x_1,\ldots,x_n]]/I$ *be a finite dimensional k–algebra. Assume* $\dim_k A$ *is not divisible by the characteristic of k. Denote by J the Jacobian ideal of A. If A is a complete intersection then J generates the socle of A. Conversely, if k has characteristic zero and A is not a complete intersection then* $J = 0$.

We only comment on the proof of the last assertion. We may assume that I has a generating system $\{f_1,\ldots,f_m\}$ for which each subset of n elements, $\{f_{i_1},\ldots,f_{i_n}\}$, is a regular sequence. From the first part of the theorem, the corresponding Jacobian determinant will generate the socle of

$$B = k[[x_1,\ldots,x_n]]/(f_{i_1},\ldots,f_{i_n}),$$

and therefore is contained in I, since the image of I is a proper ideal in B.

Socle Formulas and Radicals

The reason for the usefulness of generic socle formulas is:

Proposition 5.21. *Let I be a height unmixed ideal and let G be a generic socle formula for I. Then*

$$\boxed{\sqrt{I} = I : G.} \tag{5.5}$$

Proof. Since I is height unmixed, both sides of the equation are unmixed ideals of the same codimension, so it suffices to prove equality at localizations $R_\mathfrak{p}$, where $\mathfrak{p}$ is a minimal prime of I. But then it is clear from the definition of G. $\square$

Remark 5.22. The need will arise in Chapter 6 of the *full* generic socle of the ideal I. It is defined by the formula

$$L = I : \sqrt{I}.$$

5.4 Explicit Nullstellensätze

In this section we describe variations on a closed formula for the radical of a complete intersection, and its use to find the radical of an ideal. It requires that certain elements be in place. To make up for this, we provide other Jacobian methods to compute radicals.

The following result provides the tools to seek some generic socle formulas, and through them compute radicals.

Theorem 5.23. *Let* $I = (f_1, \dots, f_p) \subset k[x_1, \dots, x_n]$ *be an unmixed ideal of codimension* m, *which is generically a complete intersection, and let*

$$k[\mathbf{z}] = k[z_1, \dots, z_d] \hookrightarrow A = R/I$$

be a Noether normalization of the k–algebra A. *Let* J *be the ideal generated by the* $n \times n$ *minors of the Jacobian matrix*

$$\varphi = \frac{\partial(z_1, \dots, z_d, f_1, \dots, f_p)}{\partial(x_1, \dots, x_n)}.$$

Then J *is a generic socle formula for* I.

The assertion on J means that at each minimal prime P of A, $J \cdot A_P$ is a nonzero ideal that is contained in the socle of A_P. Observe that if I is a complete intersection, then J is generated by a single element. Furthermore, if the z_i's are linear forms on the x_j's—as it will be usually the case—then a large block of the matrix φ is scalar.

Another point that will be made in the proof will be a restriction on the characteristic of the field k. Loosely speaking, characteristics below the degree of A should be avoided.

Proof. There will be a sequence of changes of base rings each accompanied by an examination of the Jacobian ideal.

The degree of A over $k[\mathbf{z}]$ is the dimension of the vector space

$$r = \dim_{k(\mathbf{z})} A \otimes_{k[\mathbf{z}]} k(\mathbf{z}).$$

If A is a graded ring and the $z_i's$ are forms of degree 1, then r is the usual degree of the ring, or of the ideal I. In general the z_i are linear forms in the x_j, and we may then assume, without changing the Jacobian ideal, that the z_i's form a subset of the x_i's.

We shall assume that the characteristic of k is larger than r. The condition on the characteristic implies that for any minimal prime P of A the field of fractions of A/P is separably generated over k. This will permit several changes of base fields in such way that the nil radical will change simply by extension.

Consider the module $\Omega_{A/k[\mathbf{z}]}$ of $k[\mathbf{z}]$–differentials of the algebra A. From the Jacobi–Zariski sequence of $k \to k[\mathbf{z}] \to A$, it follows that $\Omega_{A/k[\mathbf{z}]}$ is precisely defined by the matrix φ taken mod I.

Since I is an unmixed ideal, A is a torsion–free $k[\mathbf{z}]$–module, and we pass from $k[\mathbf{z}]$ to its field of fractions K. The algebra $B = A \otimes_{k[\mathbf{z}]} K$ is a direct product

$$B = B_1 \times \cdots \times B_s$$

of local Artinian algebras whose residue fields K_i are separable over K. Each B_t is by hypothesis a complete intersection. Finally we pass from K to a joint Galois closure L of all the K_t/K. This will imply that $C = A \otimes_{k[z]} L$ is a product of local Artinian algebras

$$C = C_1 \times \cdots \times C_q$$

with residue fields isomorphic to L. By Cohen's Theorem, each C_i is an algebra of the form

$$L[[y_1,\ldots,y_m]]/(g_1,\ldots,g_m),$$

where $(g_1,\ldots,g_m) = IC_i$.

Note that the Jacobian ideal of the g_i's with respect to the y_j is given by the image of J. This implies that

$$J \cdot C_i = \det\left(\frac{\partial(g_1,\ldots,g_m)}{\partial(y_1,\ldots,y_m)}\right) \cdot C_i.$$

Since the dimension of each C_i relative to L is at most r, which is less than the characteristic of L, we apply Theorem 5.20 (see [ScS75, Korollar 4.7] and also [Kun86, Exercise 3, p. 382]) to conclude that J is as stated. $\square$

We give two immediate consequences of Theorem 5.23.

Theorem 5.24. *Let $I = (f_1,\ldots,f_m)$ be an ideal of codimension m, and let J be its Jacobian ideal. Then*

$$\boxed{\sqrt{I} = I{:}J.} \tag{5.6}$$

This formula is the analog of (5.1), with J playing the role of the derivative.

Proof. It will be enough to show that for each minimal prime P of I, if $(R/I)_P$ is not a domain then the image of J lies in the socle of $(R/I)_P$.

Suppose $I \subset k[x_1, \ldots, x_n]$ has codimension m. We make a change of base field k to another large enough to allow for a change of variables such that the subring $k[x_{j_1}, \ldots, x_{j_{n-m}}]$ is a Noether normalization of R/I for any subset of $n - m$ variables. For instance, pass from k to the rational function field $k(u_{ij})$ in n^2 new indeterminates u_{ij} and make a linear change to the original variables, as specified by the matrix (u_{ij}).

Now we compute the Jacobian ideal with respect to the new variables. Each $m \times m$ minor of the Jacobian matrix, by the argument of the Theorem 5.23, ends up in the socle of the localization.   □

Theorem 5.25. *Let I be an ideal whose primary components all have codimension m. Let $(f_1, \ldots, f_m)$ be an ideal of codimension m contained in I, and let J_0 be its Jacobian ideal. Then*

$$\sqrt{I} = (I_0 : J_0) : ((I_0 : J_0) : I). \tag{5.7}$$

Proof. By Theorem 5.24, J_0 is a generic socle formula of I_0, and therefore by Proposition 5.21 $\sqrt{I_0} = I_0 : J_0$. Every minimal prime of I is a minimal prime of I_0, so that $\sqrt{I_0} = \sqrt{I} \cap L$, where L, if not equal to (1), is the intersection of prime ideals of height m, none of which is an associated prime of I. L is therefore given by $\sqrt{I_0} : I$, and I is obtained by $\sqrt{I_0} : L$.   □

There will be 'dynamic' constraints on the characteristic of the ground field, connected to the choice of the complete intersection I_0.

Example 5.26. Let k be a field of characteristic zero and let $f \in k[x_1, \ldots, x_n]$ be a polynomial such that the ideal I generated by its partial derivatives is zero–dimensional. Then one has the pleasant formula

$$\sqrt{I} = I : H(f),$$

where $H(f)$ is the Hessian of f.

Remark 5.27. We use this opportunity to correct an error in [EHV92] that was pointed out to its authors by Julie Rehmeyer. Its algorithm 2.2 is misstated, the Jacobian ideal considered should be that defined by the $n \times n$ minors as in Theorem 5.23, not that of $c \times c$ minors.

Radicals of General Ideals

Let now I be an ideal, written as $I_1 \cap I_2 \cap I_3$. I_1 is the intersection of the primary components of codimension $m = $ height I, while I_2 is the intersection of the primary components of codimension at least $m+1$ but still associated to irreducible compo-nents. The ideal I_3 is the joint contribution of the embedded primary components. One has:

$$\sqrt{I} = \sqrt{I_1} \cap \sqrt{I_2}.$$

The previous method to find radicals of equidimensional ideals together with the techniques we employed in Chapter 3 in detaching equidimensional components yields the following algorithm to get the radical of a general ideal.

Algorithm 5.28 *Let I be an ideal of codimension m, with a representation*

$$I = I_1 \cap I_2 \cap I_3,$$

as above and let $I_0 = (f_1, \ldots, f_m)$ be a regular sequence in I. Let J be the Jacobian ideal of I_0.

(a) $\sqrt{I_1}$ *can be determined as follows:*

$$
\begin{aligned}
\sqrt{I_0} &= I_0 : J \\
L &= \sqrt{I_0} : I \\
\sqrt{I_1} &= \sqrt{I_0} : L
\end{aligned}
\tag{5.8}
$$

(b) *The stable value of*

$$
\begin{aligned}
I : (\sqrt{I_1})^j &= (I : (\sqrt{I_1})^{j-1}) : \sqrt{I_1} \\
&= I_2 \cap I_4,
\end{aligned}
\tag{5.9}
$$

where I_4 is the intersection of some of the primary components of I_3.

In the ideal $I_2 \cap I_4$, some of the primary components of I_4 may no longer be embed-ded. What this amounts to is that a certain amount of unnecessary computation may just creep in.

Example 5.29. As a simple illustration, using the *Macaulay* system, choose $f = x^3y^2 + y^3z^2 + x^2z^3$ and let

$$I = (3x^2y^2 + 2xz^3, 2x^3y + 3y^2z^2, 2y^3z + 3x^2z^2)$$

be the ideal generated by the partial derivatives of f. I has codimension 2 and projec-tive dimension 2, so has (x, y, z) as its only embedded prime, $I = I_1 \cap I_3$. It is enough to determine the radical of I_1.

Let $I_0 = (3x^2y^2 + 2xz^3, 2x^3y + 3y^2z^2)$; I_0 is a complete intersection and denote by J its Jacobian ideal. Carrying out the computation as in (a), we get

$$\sqrt{I} = \sqrt{I_1} = (xy, yz, xz).$$

A more realistic example is the following. Let I be the ideal generated by the 2×2 minors of

$$\varphi = \begin{bmatrix} x^2 & -y^2 + zw & z^2 - zw \\ -y^2 + zw & w^2 & x^2 - y^2 \\ z^2 - zw & x^2 - y^2 & z^2 \end{bmatrix},$$

and let J be the ideal generated by the 3 minors along the diagonal. Both ideals have codimension 3, so that we can apply the procedure above. $\sqrt{J}$ has a rather long list of generators but

$$\sqrt{I} = (I, x^3z + xz^3 + x^3w - xy^2w - 2xz^2w - xzw^2 - xw^3).$$

A Break

The following observation is extremely useful as it leads to efficiencies in assisting to compute the radical of I by steps.

Remark 5.30. The sheer sizes of determinantal ideals is a deterrent to using Jacobian methods. We note however the following fact, which follows from the previous discussion. If I is a complete intersection and $f \in J$ is an element of its Jacobian ideal, then

$$I{:}f = \begin{cases} R \text{ if } f \in I \\ I \text{ in which case } I = \sqrt{I} \\ L \text{ is the intersection of some of the minimal primes of } I. \end{cases}$$

The Method of the Lower Jacobians

In the same framework of Jacobian ideals, there is another method to compute radicals in [EHV92]. It makes extensive use of the lower Jacobian ideals and often offers significant performance gains. It addition, it is more programmable.

Theorem 5.31. *Let k be a perfect field, and let $R = k[x_1, \ldots, x_n]$ be a polynomial ring. Assume $I = (f_1, \ldots, f_m) \subset R$ is an ideal of codimension g. Denote by $d = n - g$ the dimension of I. If the characteristic of k is not zero, suppose that the nil radical of R/I is generated by elements whose index of nilpotency is less than the characteristic of k. If for some integer $a \geq d$ we have*

$$\dim J_{a+1}(I) < d$$

then

$$L = I : J_a(I)$$

has the same equidimensional radical as I. Further, if $a = d$ then L is radical in dimension d; that is, the primary components of L having dimension d are prime.

We only consider the characteristic zero case, and refer to [EHV92] for the full proof.

Proof. Suppose that $\dim J_{a+1}(I) < d$. To check that L has the same equidimensional radical as I, it is enough to verify that for each prime $\mathfrak{p} \supset I$ of dimension d we have

$$I : J_a(I) \subset \mathfrak{p},$$

or equivalently $J_a(I)_{\mathfrak{p}} \not\subset I_{\mathfrak{p}}$. If it were otherwise the case, for $A = R/I$, the module $(\Omega_{A/k})_{\mathfrak{p}}$ has all of its Fitting ideals equal to $A_{\mathfrak{p}}$. It is therefore a free $A_{\mathfrak{p}}$–module, and by the ordinary Jacobian criterion, $A_{\mathfrak{p}}$ is a field of transcendence degree $a+1$ over k. Since $a+1 > d$, we have a contradiction.

For the remainder of the assertion, the condition implies that I is a complete intersection at each prime $\mathfrak{p}$ of dimension d, and Theorem 5.25 will apply. $\square$

5.5 Finding Regular Sequences

The methods based on Theorem 5.25 to compute radicals require an efficient mechanism to select regular sequences. Furthermore, to be truly effective, the chosen sequence must reflect the naturality of the given generators of the ideal, which may not be a simple task. In contrast, Theorem 5.31 has broader validity and the added feature of simpler programmability.

Generic Regular Sequences

Regular sequences can be found in generic combinations of the generators of an ideal. More precisely, let $R = k[x_1, \ldots, x_n]$ be the ring of polynomials over a sufficiently large field k. If

$$I = (\mathbf{f}) = (f_1, \ldots, f_r)$$

is an ideal of codimension c and $\varphi = (a_{ij})$ is a sufficiently generic $r \times c$ matrix with entries in k, then the entries of $\mathbf{f} \cdot \varphi$ generate an ideal of codimension g.

Although straightforward, this method has several drawbacks: (i) there is the loss of whatever sparseness is present in the data, which is a resource which must be preserved; (ii) if the f_i's are homogeneous of different degrees and a homogeneous regular sequence is required I must be replaced by a subideal of the same codimension but generated in equal degrees, e.g.

$$\sum_{i=1}^{r} (x_1, \ldots, x_n)^{m-d_i} f_i, \ m \geq d_i = \deg(f_i);$$

(iii) the 'regular sequence' comes without a certificate. On the plus side, if the f_i's are forms of the same degree and the ideal I has an 'uncomplicated' primary decomposition the method is surprisingly effective.

A related approach is the following. Consider the sequence of integers

$$\text{height}(f_1) \leq \text{height}(f_1, f_2) \leq \cdots \leq \text{height}(f_1, \ldots, f_r) = g.$$

By Krull's principal ideal theorem, there is an associated sequence

$$1 = i_1 < i_2 < \cdots < i_g \leq r,$$

such that

$$\text{height}(f_1, \ldots, f_{i_j}) = j, \quad 1 \leq j \leq g.$$

The i_j's are the places where the heights of the ideals in the chain defined by the $(f_1, \ldots, f_{i_j})$'s change value.

Proposition 5.32. *For each $1 \leq j \leq g$, let*

$$h_j = \sum_{\ell = i_{j-1}+1}^{i_j} c_\ell f_\ell$$

be a generic linear combination of the elements $f_{i_{j-1}+1}, \ldots, f_{i_j}$. Then $h_1, \ldots, h_g$ is an ideal of codimension g.

Proof. This is an immediate consequence of Theorem A.19. Here generic means that the element h_j must avoid the associated primes of the ideal $(f_1, \ldots, f_{i_{j-1}})$, of codimension $j - 1$.   □

Setwise Regular Sequences

To deal with some of these difficulties, there is a method of [EiS94] that is an elegant variation of the generic approach but that seeks to contain the loss of sparseness. It is based on:

Theorem 5.33. *If F is a set of polynomials whose initial terms, with respect to some term order on $R = k[x_1, \ldots, x_n]$, generate an ideal of codimension c, then F can be partitioned into disjoint subsets $F_1, \ldots, F_c$ such that for each i, the initial terms of the polynomials in F_i have a nontrivial common factor. For any such partition, and for almost every choice of coefficients $r_{i,f} \in k$, the linear combinations*

$$f_1 = \sum_{f \in F_1} r_{1,f} f$$

$$\vdots \tag{5.10}$$

$$f_c = \sum_{f \in F_c} r_{c,f} f$$

generate an ideal of codimension c. Further, each polynomial in F may be multiplied by any product of the factors of its initial term without spoiling this property.

The corresponding algorithm is (cf. [EiS94]):

Algorithm 5.34 *Let F be a set of generators of the ideal I, and pick a term order.*

- *Enlarge F toward a Gröbner basis until $in(F)$ generates an ideal of codimension c. Next replace this partial Gröbner basis by a minimal subset which has codimension c.*
- *To partition F as in the Theorem, do as follows: Choose a prime $(x_{i_1}, \ldots, x_{i_c})$ containing $in(F)$. For $p = 1, 2, \ldots, c$ define F_p inductively to be the set of all elements of*

$$F \setminus \bigcup_{j<p} F_j$$

  whose initial terms are divisible by x_{i_p}.
- *Choose random elements $r_{i,f}$ in k and verify that the polynomials f_i in (5.10) generate an ideal of codimension c. If they do not, try a new random choice.*

Partition of the Unity

We discuss next a mechanism that calls on Theorem 5.24 itself to generate its own regular sequences. Actually it only produces an ideal which is locally a complete intersection; but this is obviously all that is required to use Theorem 5.25. It is more of an approach than an algorithm proper. It permits however a great deal of *manual* control. Its shape and usage are based on the following elementary observations (cf. [Vas92]).

Proposition 5.35. *Let R be a reduced Noetherian ring with n minimal prime ideals. Let I be an ideal of codimension at least one. Then*

(a) *For $a \in I$ there exists $b \in I \cap (0:a)$ such that $a + b$ is a regular element.*
(b) *If $a \neq 0$ for each $0 \neq b \in I \cap (0:a)$, $(0:a) \neq (0:(a+b))$.*
(c) *Iteration leads to a regular element contained in I using at most n steps.*

Proof. (a): Any prime ideal $\mathfrak{p} \supseteq (a, I \cap (0:a))$ either contains I, which has codimension at least one, or will contain $(a, (0:a))$. Since R is reduced the latter has also codimension at least 1. This means that there exists a regular element of the form $ra + b$, with $b \in I \cap (0:a)$. Again using that R is reduced, $a + b$ is regular as well.

(b): This is immediate since any annihilator of $a + b$ must annihilate a and b.

(c): Let $a_1, a_2, \ldots, a_r$ be obtained by this process: in the notation above, we repeatedly set $a = a_1 + \cdots + a_{j-1}$, $b = a_j$. The descending chain of ideals

$$0 : a_1 \supset 0 : (a_1 + a_2) \supset \cdots \supset 0 : (a_1 + a_2 + \cdots + a_r)$$

gives rise to a similar sequence in the total ring of fractions S of R. Since S has a composition series of length n, the last ideal vanishes if $r \geq n$. □

Remark 5.36. In practice this runs as follows. Suppose $f_1, \ldots, f_s \in I$ have been chosen so that they generate an ideal of codimension $s <$ height I. Let

$$L = \sqrt{(f_1, \ldots, f_s)}.$$

We follow the scheme above, but with colon ideals computed relative to L. We note by f_{s+1} the element of I obtained from the lift. The degree of manual control over the sparseness comes in because in selecting the a_j's we may also use reduction modulo the previously chosen a_i for $i < j$.

Set–Theoretic Generation of Ideals

Related to this scheme to find regular sequences is the following well-known result ([EE73], [Stc72]):

Theorem 5.37. *Let S be a Noetherian ring of dimension $d - 1$. Then the radical of any ideal I of $R = S[x]$ is the radical of a d–generated ideal.*

Their proof is fairly constructive already. It is based on pseudo–division of polynomials and an appropriate induction argument on the dimension of S. We will just rewrite it in sketching out some moves to make it more amenable for a Gröbner basis computation.

We assume that S is a reduced ring. Let $I = (f_1, \ldots, f_m)$. In their proof it is argued that there exists a regular element u of S and an element h_1 of I such that

$$u \cdot I \subseteq (h_1).$$

The issue is how u is to be found. We employ the argument of Proposition 5.35. Given two elements f, g of I, with $\deg f(x) \geq \deg g(x)$ (as polynomials in x), let α be the leading coefficient of $g(x)$. For some power of α we have

$$\alpha^s \cdot f = q \cdot g + r, \; \deg r < \deg g.$$

Replace then f by r and keep processing the list of generators of I until we have a nonzero element $a_1 \in S$ such that $a_1 \cdot I \subseteq (g_1)$, with $g_1 \in I$.

Repeat this step on the generators of $I \cap (0\!:\!a_1)$, if the latter is nonzero. This produces a sequence $a_1, \ldots, a_r \in S$, with corresponding elements $g_i \in I$ such that

$$\begin{cases} a_i \cdot a_j = 0 \text{ if } i \neq j \\ a_1 + a_2 + \cdots + a_r \text{ is regular on } S \\ a_i \cdot I \subseteq (g_i) \end{cases}$$

Observe that when the process is applied to an ideal such as $I \cap (0\!:\!a_1)$, it returns an element $b_1 \in (0\!:\!a_1)$ such that

$$b_1 \cdot (I \cdot (0\!:\!a_1)) \subset b_1 \cdot (I \cap (0\!:\!a_1)) \subset (g_2), g_2 \in I.$$

From this equation we then select some $0 \neq a_2 \in b_1 \cdot (0\!:\!a_1)$.

Set now

$$u = a_1 + a_2 + \cdots + a_r$$
$$h_1 = a_1 g_1 + a_2 g_2 + \cdots + a_r g_r,$$

and consider the image I^* of I in $(S/\sqrt{(u)})[x]$. By induction select $h_2, \ldots, h_d \in I$ whose images generate an ideal with the same radical as I^*.

It suffices to verify

$$\sqrt{I} = \sqrt{(h_1, h_2, \ldots, h_d)}.$$

Let $\mathfrak{p}$ be a prime ideal with $(h_1, \ldots, h_d) \subseteq \mathfrak{p}$. If $u \in \mathfrak{p}$, by hypothesis $I \subseteq \mathfrak{p}$. Assume otherwise; then $a_i \notin \mathfrak{p}$ for some i which implies that $a_j \in \mathfrak{p}$ for $j \neq i$. Since $h_1 \in \mathfrak{p}$, this means that $g_i \in \mathfrak{p}$ which from the equation $a_i \cdot I \subseteq (g_i)$ finally implies $I \subseteq \mathfrak{p}$, as desired.

Remark 5.38. This arrangement leaves much to be desired: There is a great deal less of control here than that afforded by Proposition 5.35.

Wu–Ritt Characteristic Sets

Another place to look for regular sequences is among *triangular sets* of polynomials, particularly among some of the characteristic sets of Wu and Ritt (see [GM91]).

Definition 5.39. *Let $R = k[y_1, \ldots, y_n]$ be a polynomial ring and let $\mathbf{f} = f_1, \ldots, f_m$ be a set of polynomials. $\mathbf{f}$ is a triangular set if there is a permutation $x_1, \ldots, x_n$ of the y_i such that*

$$f_i \in k[x_1, \ldots, x_i, x_{m+1}, \ldots, x_n] \setminus k[x_1, \ldots, x_{i-1}, x_{m+1}, \ldots, x_n].$$

Each such polynomial is written

$$f_i = a_{0,i} x_i^{r_i} + \cdots + a_{r_i,i}, \; a_{p,i} \in k[x_1, \ldots, x_{i-1}, x_{m+1}, \ldots, x_n], \; a_{0,i} \neq 0.$$

Clearly, if these polynomials are all monic ($a_{0,i} = 1$), then $(\mathbf{f})$ is an ideal of codimension m. In the Wu–Ritt process, one uses pseudo-division on each f_i by the earlier f_j, $j < i$. Since there are algorithms to produce such sets, it is of interest to find more general conditions that ensure that codim $(\mathbf{f}) = m$. A step towards this goal is to find criteria for the equality

$$\text{codim } (\mathbf{f}) R_h = m$$

to hold, where $h = \prod_{i=1}^{m} a_{0,i}$.

5.6 Top Radical and Upper Jacobians

The methods we have discussed thus far to find the radical of the equidimensional components of an ideal all require that we process the higher dimensional components first. It would be desirable to target any dimension independently. Through the upper Jacobian ideals it is possible to access the light component of the radical of an ideal ([Vas92]).

Let I be an ideal of codimension m, minimally generated by $m + r$ elements; r is the *deviation* of I. According to Krull principal ideal theorem (see Theorem A.20), any minimal prime of I has codimension at most $m + r$. We are going to key on those primes.

Let I be an ideal and let

$$I = Q_1 \cap \cdots \cap Q_s$$

be one of its irreducible primary decompositions. The ideals Q_i can be collected by codimension: let I_i be the intersection of those components of codimension $m + i$. The decomposition can be refined by breaking up each I_i into two pieces: I_i', corresponding to minimal primes of I, and I_i'' obtained from the embedded primes of I. If one of these is not present, such as I_0'' or I_i', $i > r$, by abuse of notation we set it equal to R.

The radical of I is then given by the expression

$$\sqrt{I} = \sqrt{I_0'} \cap \sqrt{I_1'} \cap \cdots \cap \sqrt{I_r'}.$$

It suggests that one focus on obtaining formulas for the $\sqrt{I_i'}$'s.

Definition 5.40. *Given an ideal I of deviation r, $\sqrt{I_r'}$ is the* top radical *of I. It is noted* topradical(I).

Conjecture 5.41. Let I be an ideal of codimension m and deviation r, and denote by L the ideal generated by the minors of size $m + r$ of the Jacobian matrix of I. Then

$$\text{topradical}(I) = I : L.$$

The following provides a measure of support (assume from now on that k is a field of characteristic zero):

Theorem 5.42. *This conjecture holds if every embedded prime of I has codimension at most $m + r$. In particular it holds if $v(I) \geq \text{proj dim }_R R/I$.*

Proof. We have

$$I : L = (I_0 : L) \cap (I_1' : L) \cap (I_1'' : L) \cap \cdots \cap (I_r' : L) \cap (I_r'' : L),$$

since by assumption $I_j'' = R$ for $j > r$. We claim that all the quotients on the right side, with the possible exception of $I_r' : L$, are equal to R. This will suffice to establish the claim since at each minimal prime $\mathfrak{p}$ of I_r', we have $I_\mathfrak{p} = (I_r')_\mathfrak{p}$ and $L_\mathfrak{p}$ is nothing but the Jacobian ideal of $(I_r')_\mathfrak{p}$; we may then apply Theorem 5.25.

We begin by showing that $I_r'':L = R$. If this is not so, any associated prime of the left–hand side has codimension $m + r$. Let $\mathfrak{p}$ be then one of its primes and localize R at $\mathfrak{p}$ (but keep the notation); denote also $A = R/I$. (We warn the reader about a possible confusion: Sometimes we shall say that an ideal is zero when it would be more appropriate to say it is zero mod I.)

As before we may assume that $A = k[[x_1,\ldots,x_n]]/I$, and $\mathfrak{p}$ is the maximal ideal of A. We consider the Jacobian ideals of the Artin algebras $B_s = A/\mathfrak{p}^s$. Because A has positive dimension, L will map into the Jacobian ideal of B_s, for each s. If B_s is not a complete intersection, by Theorem 5.20, its Jacobian ideal vanishes and thus $L \subset \mathfrak{p}^s$. On the other hand, if B_s is a complete intersection its socle must be $\mathfrak{p}^{s-1}B_s$ and $L \subset \mathfrak{p}^{s-1}$. This means that

$$L \subset \bigcap_{s \geq 0} \mathfrak{p}^s,$$

which by Krull intersection theorem (see Theorem A.12) implies $L = 0$.

The case of a component such as $I_j':L$, for $j < r$, or $I_j'':L$, $1 \leq j < r$, is easier to deal with. In fact, if $\mathfrak{p}$ is a minimal prime of say I_j' and $(I_j')_\mathfrak{p}$ is not a complete intersection, then already the $m + j$–sized minors of the Jacobian matrix of $(I_j')_\mathfrak{p}$ vanish by Theorem 5.20. On the other hand, if $(I_j')_\mathfrak{p}$ is a complete intersection, the rank of its Jacobian matrix is $m + j < m + r$, so L vanishes at $(I_j')_\mathfrak{p}$ anyway.

The last assertion follows from the Auslander–Buchsbaum formula (A.23) and standard facts on depth.   □

The Top of a System of Equations

A case of significance is that of a set of equations with isolated zeros. Suppose

$$V(I) = V(\mathfrak{p}_1) \cup \cdots \cup V(\mathfrak{p}_r) \cup V(\mathfrak{Q}_1) \cup \cdots \cup V(\mathfrak{Q}_s),$$

where the $\mathfrak{p}_i$ have positive dimension and the $\mathfrak{Q}_j$ are maximal ideals. The issue is how to recover the latter. The following zeroes in on the isolated zeros of special sets of polynomials (cf. [Vas92]).

Corollary 5.43. *Let $f_1,\ldots,f_n \in k[x_1,\ldots,x_n]$ be a set of n polynomials, and let Δ be its Jacobian determinant. Then*

$$\boxed{(f_1,\ldots,f_n):\Delta}$$

is the radical of the minimal primary components of dimension 0. In particular this ideal is either (1) or a complete intersection.

To illustrate, suppose $I = (f(x,y)g(x,y), f(x,y)h(x,y))$, where $f(x,y)$ is the gcd of I. Denote by Δ the Jacobian determinant of these two generators of I, and denote by Δ_0 the Jacobian of g, h. A simple calculation shows that

$$I{:}\Delta = ((g,h){:}\Delta_0){:}f.$$

By Theorem 5.24

$$I{:}\Delta = \sqrt{(g,h)}{:}f,$$

in agreement with the assertion of the theorem.

Theorem 5.44. *Let I be an ideal of $R = k[x_1, \ldots, x_n]$. We can find n polynomials $f_1, \ldots, f_n$ generating an ideal with the same radical as I. If Δ is the Jacobian determinant of the f_j's, then $(f_1, \ldots, f_n){:}\Delta$ is the top radical of I.*

A Homological Nullstellensatz

It is not likely, through strictly homological means, that one should be able to produce the radical of an ideal or even large chunks of it. Here we make some observations on this, and begin quoting a result of Gerson Levin [Lev90].

Theorem 5.45. *Let R be a local ring, I an ideal in R of finite projective dimension, which is not a complete intersection. Let J be the content of I, i.e. the ideal of all coefficients of relations on a minimal set of generators of I. Let $r = 1 + \operatorname{proj} \dim_R R/I$; then $(I : J)^r \subset I$.*

The harder part of the proof is the assertion on the index of nilpotency. That $I{:}J$ is contained in the radical of I can be shown as follows. Localize at a minimal prime of I, so that we may assume that I is primary for the maximal ideal of the local ring $(R, \mathfrak{p})$. If $\nu(I_{\mathfrak{p}}) < \nu(I)$, $J_{\mathfrak{p}} = R$ and there is nothing to prove. If $\nu(I_{\mathfrak{p}}) = \nu(I)$, $I_{\mathfrak{p}}$ is not a complete intersection and $J_{\mathfrak{p}}$ is a primary ideal properly containing $I_{\mathfrak{p}}$, and consequently $I_{\mathfrak{p}}{:}J_{\mathfrak{p}}$ is also primary. Of interest is the fact that the quotient is then larger than I and therefore is a better approximation of $\sqrt{I}$ than I itself.

Here is another link between the content of an ideal and radicals. We leave its proof as an exercise for the reader.

Proposition 5.46. *Let I and J be ideals, $I \subset J$. Denote by L the content of I. If $I{:}J \not\subset \sqrt{L}$, then $\nu(I) \geq \nu(J)$.*

Exercise 5.47. Let I be an ideal of codimension two given as the maximal minors of a $n \times n + 1$ matrix φ, $I = I_n(\varphi)$. Prove that

- If $I_n(\varphi){:}I_{n-1}(\varphi) = I_n(\varphi)$, then I is generically a complete intersection.
- It always holds $\sqrt{I_n(\varphi){:}I_{n-1}(\varphi)} = \sqrt{I_n(\varphi)}$.

Exercise 5.48. Let I be an ideal and let H be the 1–dimensional homology module of a Koszul complex built on a set of generators of I. Prove that the annihilator of H is contained in $\sqrt{I}$.

6

Integral Closure

Let A be an integral domain with field of quotients K. An element $x \in K$ is integral over A when there exists an equation

$$x^n + a_{n-1}x^{n-1} + \cdots + a_0 = 0, \qquad a_i \in A. \tag{6.1}$$

All these elements form a new ring, $A \subset B \subset K$, the *integral closure* of A. The ring A is *integrally closed*, or *normal*, if $A = B$.

There are slight variations of this notion which arise by placing restriction on the coefficients a_i's or by taking x in some other extension of A. In all we shall we treat the following:

- Integral closure of a ring
- Integral closure of an ideal
- Integral closure of a morphism

Our major aim is to consider methods for the determination of B from some specification of A. One of the major instances of this question is that of finding the integers of a number field, where progress has been considerable, and fundamental efficiencies have been achieved (see the beautiful exposition by Lenstra [Len92]). Here, we discuss affine rings, in which the general question lies on the brink of intractability but with many cases deemed of sufficient interest to warrant the effort. To complicate the issue, except perhaps in the case of curves singularities, it has not been possible to convert the number theoretic techniques into corresponding efficiencies. On the other hand, the geometric framework of affine domains allows the use of differential means without equivalence in the number theory setting.

Classically the question was dealt with in the following manner. Let S be a polynomial ring over a computable field k, and let $A = S/P$ be an affine domain of dimension d. The problem of describing, in a reasonably constructive fashion, the integral closure B of A has been repeatedly visited since it was dealt with originally by Dedekind [D1882] and Kronecker [Kro82]. At their best, one would seek to express the integral closure of A as an affine domain $k[y_1, \ldots, y_m]/Q$. Their marvelous argument, whose outline bears recalling, considered a Noether normalization

$$k[\mathbf{z}] = k[z_1, \ldots, z_d] \hookrightarrow A$$

whose field of quotients K defines a separable extension of $k(\mathbf{z})$ (e.g. [Nag62, (10.15)]). The discriminant D of any basis of $K/k(\mathbf{z})$ is shown to conduct B into A,

$$B \hookrightarrow D^{-1}A.$$

Identifying the numerators of these fractions constitute the main task in completing the description of B. More recent approaches to this step have been introduced in [Sei75] and [Sto68], with a fuller analysis of them being given in [Tra86].

A difficulty lies in the fact that we are asked to solve equations which are not detailed! The availability of Gröbner basis algorithms enables treatments of the question by looking for batches of integral elements. Instead of taking a step back, such as represented by a Noether normalization, often very onerous, one looks for various ring–theoretic constructions which augment A into its integral closure. Our exposition follows the method that is proposed in [Va91a], along with other enhancements (see [Va94a]). It relies on using the more bountiful elements in the Jacobian ideal of A for some of the same purposes as the discriminants.

There are two basic ways the ring A can be specified:

- As an affine algebra $A = k[x_1, \ldots, x_n]/P$
- As a subalgebra $k[f_1, \ldots, f_m] \subset k[x_1, \ldots, x_n]$.

The second is typical of problems in the theory of invariants when m may be large. The description of B may be sought out in the same form, or, as in the classical case, through a set of generators in the field of fractions of A.

In the first case n is called the embedding dimension of A and the integral closure $\overline{A}$ is expected to be described in similar form. This presentation can be used to test whether A is normal. In the other case, A may be given by a large number of generators which are polynomials in the variables x_i. It may have a description in the first format at the possible cost of using a large number of new variables. It also lacks general normality criteria.

In this chapter we deal mostly with the first representation, leaving the other problem to a skimpy treatment in the next chapter. Nothing specific will be said on what remains an unexplored territory, the integral closure of algebras over $\mathbb{Z}$.

The more general problem of finding the integral closure of a ring A in one of its overrings B is also treated here following [BV93]. The motivation for such methods comes from the presence of such constructions in the Zariski's Main Theorem, to create means to effect Stein factorizations (see [Har77]), and in the computation of rings of invariants.

The other problem we discuss is that of the integral closure of an ideal I: it is defined by the solutions in A' of equations such as (6.1) in which $a_i \in I^i$. The more interesting cases are those in which A is already integrally closed. There is a costly conversion to a problem of finding the integral closure of another algebra–the Rees algebra of the ideal I–but we shall look for more direct approaches.

6.1 Integrally Closed Rings

The framework we use is the normality criterion of Krull–Serre ([Mat80, Theorem 39]). For ease of reference we recall its assertion.

Theorem 6.1. *Let A be a Noetherian integral domain. A is integrally closed if and only if the following conditions hold:*

(a) *(Condition R_1): For each prime $\mathfrak{p}$ of codimension one, $A_\mathfrak{p}$ is a discrete valuation domain.*

(b) *(Condition S_2): Every ideal I of codimension two contains a regular sequence on A with two elements.*

The explanation for this criterion is elementary. Given any integral domain A, one has:

Proposition 6.2. *Let A be a Noetherian integral domain. Then*

$$A = \bigcap A_\mathfrak{p},$$

where $\mathfrak{p}$ runs over the associated primes of principal ideals.

The conditions ensure that each such localization is a discrete valuation domain.

There are two ingredients that go into the formation of the integral closure of a Noetherian domain, and whose effect is to enlarge the ring into the integral closure. The task is: given an affine domain A, find an integral rational extension B that satisfies both (S_2) and (R_1). The integral closure B will be determined by two types of extensions. First, through a preprocessing, that enables (S_2); this is the so–called S_2–ification of A. It will be discussed in the next two sections. It is followed by a sequence of extensions of a second kind, each a step in the desingularization of A in codimension one; it could be labeled the R_1–ification of A.

> $A \longrightarrow S_2$-ification $\longrightarrow$ desingularization in codimension one

Actually, the two processes can be executed in the reverse order without changing the outcome. Thus, for example, if A already satisfies (R_1) then B can be computed in a single step. In some cases, for instance if $\dim A = 2$, it may be advisable to use the sequence above (see Remark 6.5).

Rings of Endomorphisms

It is useful to express issues of integral closure in terms of rings of endomorphisms. Let A be an integral domain and assume x is an element of its field of fractions that is integral over A. From an equation

$$x^n + c_{n-1}x^{n-1} + \cdots + c_0 = 0,$$

setting

$$I = (1, x, \ldots, x^{n-1})$$

we obtain a finitely generated submodule of the field of fractions of A (i.e. a fractional ideal of A) which has the property

$$x \in \mathrm{Hom}_A(I, I).$$

Conversely, for any such submodule I, by the Cayley–Hamilton theorem, every element of $\mathrm{Hom}_A(I, I)$ is naturally identified to an element of the field of fractions of A integral over this ring.

An important task is to identify ideals I which are guaranteed to test whether the ring is integrally closed and/or to provide fresh elements of its integral closure.

The criterion given in Theorem 6.1 being non-constructive, it will be allied to the following reformulation of normality:

Theorem 6.3. *The Noetherian integral domain A is normal if and only if for every nonzero ideal I,*

$$\mathrm{Hom}_A(I, I) = A.$$

This actually converts into the following description of the integral closure B of the ring A:

$$B = \bigcup_{I \subset A} \mathrm{Hom}_A(I, I).$$

It will only be required the testing of a few such rings of endomorphisms. They will be derived from ideals whose definition arises from a representation

$$A = k[x_1, \ldots, x_n]/P.$$

If A is an affine domain over a field of characteristic zero, let ω_A denote its canonical module and J its Jacobian ideal. Then we will show that A is normal if and only if:

- $A = \mathrm{Hom}_A(\omega_A, \omega_A)$;
- $A = \mathrm{Hom}_A(J^{-1}, J^{-1})$.

These two modules, ω_A and J^{-1}, are very natural, as will be explained later. Basically,

$$\widetilde{A} = \mathrm{Hom}_A(\omega_A, \omega_A)$$

turns on the property S_2. The other ring,

$$C = \mathrm{Hom}_A(J^{-1}, J^{-1})$$

is guaranteed to be larger than A if the ring is not already normal. Moreover it offers a more direct approach to the computation of its ring of endomorphisms than $\mathrm{Hom}_A(J, J)$.

If the characteristic of k blocks the use of Jacobian ideals, a more systematic use of the canonical module produces other classes of appropriate rings of endomorphisms (see Theorem 6.37).

Embedding Dimension

The process of repeatedly adding new variables in the drive towards the integral closure must be looked into. We recall that if A is an affine ring over a field k, its *embedding dimension* (relative to k) is the smallest integer n such that $A = k[x_1, \ldots, x_n]/I$.

The question is, how are the embedding dimensions of A and of $\overline{A}$ related? If the latter is much larger than the former, $\overline{A}$ can be hard to get. Suppose that A is an integral domain and consider a Noether normalization

$$R = k[x_1, \ldots, x_d] \hookrightarrow A \hookrightarrow \overline{A}.$$

Denote by e the torsionfree rank of A over R. If A is a standard graded algebra then e is the multiplicity $\deg(A)$ of A.

Proposition 6.4. *If the ring $\overline{A}$ is Cohen–Macaulay then its embedding dimension is at most $d + e - 1$. In particular this is the case if $\dim A \leq 2$.*

Proof. $\overline{A}$ is Cohen–Macaulay precisely when $\overline{A}$ is a free R–module, necessarily of rank e. In addition to the identity, it requires $e - 1$ module generators over R, hence the estimate. $\square$

Remark 6.5. (a) This argument suggests that
  if $\dim A = 2$, to reach $\overline{A}$ through a sequence of extensions

$$A \longrightarrow A_1 \longrightarrow \cdots \longrightarrow A_n = \overline{A},$$

taking the A_i's with the S_2 condition assures that each such extension is a free R–module of rank e. This ensures a measure of control over the number of variables required throughout the process.

  (b) A side effect is that the embedding dimension of $\overline{A}$ may often be smaller than that of A. The practice says that the process amounts to replacing one variable by another. On the other hand, if $\dim A > 2$, when $\overline{A}$ is not always Cohen–Macaulay, one can expect the embedding dimension to become much higher. If A is a graded domain over a field of characteristic zero and dimension 3, it can be proved that $\overline{A}$ can be generated by about e^2 module generators over R. More precisely, one has ([UVp]):

Theorem 6.6. *Let A be a standard graded domain over a field k of characteristic zero, of dimension d and multiplicity $\deg(A) > 1$. If the integral closure B of A has depth $d - 1$ then*

$$\mathrm{emb}(B) \leq d + \deg(A)(\deg(A) - 1).$$

Another element of control of the integral closure is given in the following application of Proposition 2.71:

Theorem 6.7. *Let A be a standard graded domain and let R be any of its standard Noether normalizations. Suppose that both A and its integral closure B are Cohen–Macaulay. If A is generated over R by elements of degree at most $\mathrm{r}(A)$ then B will be generated over R by elements of degree at most $\mathrm{r}(A)$.*

Note that $\mathrm{r}(A)$ is the so-called reduction number of A (see Chapter 9). In the case of a Cohen–Macaulay algebra, it is read off the Hilbert series of A, no Noether normalization being required. One natural application occurs when $\dim A = 2$. First, we replace A by its S_2–ification $\tilde{A}$, which is Cohen–Macaulay. This ring (and B as well) is not necessarily a standard algebra, which really does not matter once we have the Noether normalization of A.

6.2 Multiplication Rings

We now discuss ways to enlarge an integral domain A into its integral closure B and the roles the conditions S_2 and R_1 play in the process. The simplest way to achieve is through the notion of *multiplication rings* (or *idealizers*).

Definition 6.8. *Let A be a commutative ring and let I be an ideal of A. The ring of endomorphisms $C = \mathrm{Hom}_A(I, I)$ is the multiplication ring of I.*

Somewhat more generally, one can take for I any finitely generated submodule L of the total ring of fractions of A. Such modules are called fractionary ideals. By abuse of terminology, we refer to these rings simply as *multipliers*. If I is finitely generated and contains regular elements, C is contained in the integral closure of A. The basic theme is to identify ideals which, in the event that A is not normal, guarantee that their multipliers are indeed larger than A.

Example 6.9. The multiplier of a principal ideal (from now on A is an integral domain), $I = Ax$, is obviously A. If $A = k[x^3, x^4, x^5] \subset k[x]$ and $I = (x^3, x^4)$, since $x^5 \notin I$ then a polynomial $f = a + bx + cx^2 + \cdots$ lies in $\mathrm{Hom}_A(I, I)$ if and only if $b = c = 0$.

Multipliers and Duals

From now on, A is an affine ring given by a presentation $A = k[x_1, \ldots, x_n]/P$. If I is an ideal containing regular elements, $\mathrm{Hom}_A(I, I)$ consists of the elements $x \in K$, the total ring of fractions of A, such that

$$x \cdot I \subset I.$$

Thus

$$x \cdot I^{-1} \subset I^{-1}$$

also holds, and we have

$$\mathrm{Hom}_A(I, I) \subset \mathrm{Hom}_A(I^{-1}, I^{-1}) \subset B.$$

The second multiplier can be written in another form. Let

$$I^{-1} = \mathrm{Hom}_A(I, A),$$
$$C = (I \cdot I^{-1})^{-1};$$

both I^{-1} and C are fractionary ideals of A.

Proposition 6.10. *The ring of endomorphisms of I^{-1} and of $(I^{-1})^{-1}$ is*

$$\boxed{C = (I \cdot I^{-1})^{-1}.} \tag{6.2}$$

Proof. If $x \in C$, the equation

$$x \cdot I \cdot I^{-1} = x \cdot I^{-1} \cdot I \subset A$$

implies that $x \cdot I^{-1} \subset I^{-1}$, and therefore x belongs to $\mathrm{Hom}_A(I^{-1}, I^{-1})$. On the other hand, every element of this endomorphism ring is realized by multiplication by an element in the field of fractions of A, so that if $z \cdot I^{-1} \subset I^{-1}$ we have $z \cdot I^{-1} \cdot I \subset I^{-1} \cdot I \subset A$, and thus $z \in C$.

For the other assertion, from the argument above we already know that the multiplier of $(I^{-1})^{-1}$ contains the multiplier of I^{-1}. But the former is the inverse of $I^{-1} \cdot (I^{-1})^{-1}$; since this ideal contains $I \cdot I^{-1}$, its inverse will be correspondingly smaller. □

Finally, for completeness, the following criterion for integrally closed domains provides the necessary setting.

Proposition 6.11. *Let A be a Noetherian integral domain. Then A is integrally closed if and only if for each nonzero ideal I*

$$(I \cdot I^{-1})^{-1} = A.$$

Proof. We have seen that $(I \cdot I^{-1})^{-1}$ is the endomorphism ring C of the ideal I^{-1}. Therefore if A is integrally closed, $A = C$.

For the converse, first observe that the condition is retained by any localization of A. Here it suffices to show that for any prime ideal $\mathfrak{p}$ such that depth $A_\mathfrak{p} \leq 1$, the ring $A_\mathfrak{p}$ is integrally closed. Indeed A is always the intersection of all such localizations.

Let I be a nonzero ideal of $A_\mathfrak{p}$. If I is not principal, $L = I \cdot I^{-1}$ is a proper ideal. On the other hand, if x is a nonzero element of $\mathfrak{P} = \mathfrak{p} A_\mathfrak{p}$, $\mathfrak{P}$ is associated to (x) so there exists $y \notin (x)$ such that $y\mathfrak{P} \subset (x)$. This obviously implies that $\dfrac{y}{x} \in L^{-1}$, against the assumption. $\square$

Trace Rings

The ideal $\text{trace}(I) = I \cdot I^{-1}$ is the so–called *trace* ideal of I. We want to determine it and its dual by means other than multiplying the divisorial ideals I and I^{-1}. Let us discuss a general method first.

The relation

$$\text{Hom}_A(I,I) \subset \text{Hom}_A(I^{-1},I^{-1}) = \text{Hom}_A((I^{-1})^{-1},(I^{-1})^{-1})$$

shows that the multiplication rings of I and I^{-1} coincide if I is a reflexive ideal.

Endomorphism Rings of Duals

Let us give the required steps to the determination of the multiplication ring of I^{-1}. It contains several embedded efficiencies when compared to the determination of $\text{Hom}_A(I,I)$.

Algorithm 6.12 *Let $J \subset R = k[x_1,\ldots,x_n]$ and $I = J/\mathfrak{p} \subset A = R/\mathfrak{p}$. Assume that J contains regular elements modulo $\mathfrak{p}$. The ring of endomorphisms*

$$C = (I \cdot I^{-1})^{-1}$$

is determined as follows:

- *Let $a \in J$ be regular modulo $\mathfrak{p}$, and let α denote its residue class.*

$$I^{-1} = ((\alpha) :_A I)\alpha^{-1} = (((a,\mathfrak{p}) :_R J)/\mathfrak{p})\alpha^{-1},$$

and therefore

$$\begin{aligned} I \cdot I^{-1} &= (((a,\mathfrak{p}) :_R J)J/\mathfrak{p})\alpha^{-1} = ((La,\mathfrak{p})/\mathfrak{p})\alpha^{-1} \\ &= L/\mathfrak{p}, \end{aligned}$$

for some ideal $\mathfrak{p} \subset L$.

- *Since a is regular modulo $\mathfrak{p}$, L can be recovered as*

$$L = (La, \mathfrak{p}) :_R a.$$

- *Finally,*

$$C = (I \cdot I^{-1})^{-1} = (((a, \mathfrak{p}) :_R L)/\mathfrak{p})\alpha^{-1}.$$

- *To get the equations for C use (2.17): If*

$$((a, \mathfrak{p}) : L) = (b_1, \dots, b_m), \quad then$$
$$C = R[y_1, \dots, y_m]/P, \quad where$$
$$P = (\mathfrak{p}, ay_1 - b_1, \dots, ay_m - b_m) : a^\infty.$$

When the computation of syzygies is possible over the ring A the ideal $I \cdot I^{-1}$ can be found as follows.

Proposition 6.13. *Let*

$$A^m \xrightarrow{\varphi} A^n \longrightarrow I \to 0$$

be a presentation of I. Let ψ be the $n \times p$ matrix, whose columns generate the relations of φ^* (the transpose of φ). Then

$$\boxed{I \cdot I^{-1} = I_1(\psi).} \tag{6.3}$$

Proof. The module I^{-1} is naturally isomorphic to $\text{Hom}(I, A)$ and

$$I \cdot I^{-1} = \sum \alpha(I), \quad \alpha \in \text{Hom}(I, A).$$

Any such map α has a lift $\beta: A^n \longrightarrow A$, $\beta = \alpha \cdot \pi$, so that $\beta \cdot \varphi = 0$. This means that $\varphi^t \cdot \beta^t = 0$, and β corresponds to a syzygy of φ^t. But the image of β is determined by what it does to the basis of A^n, in other words, to the coordinates of β^t. $\qquad\square$

Hypersurface Rings

We illustrate the previous situation with the important case of a hypersurface. Let

$$f \in R = k[x_1, \dots, x_n]$$

be a polynomial. A *hypersurface ring* is a ring of the form $B = R/(f)$.

These rings have an important role in the construction of the integral closure according to the following elementary observation.

Proposition 6.14. *For any affine integral domain A over a large field k, there exists a hypersurface integral domain B such that $\overline{A} = \overline{B}$.*

Proof. Let $R \hookrightarrow A$ be a Noether normalization and let u be a generator of the field of fractions of A over the field of fractions of R. Since u can be taken in A, the subextension $B = R[u] \hookrightarrow A$ gives the required hypersurface ring. □

Set $L = (f, \frac{\partial f}{\partial x_1}, \ldots, \frac{\partial f}{\partial x_n})$; then $J = L/(f) \subset A = R/(f)$ is the Jacobian ideal of A. We assume that A is not normal but that L is a Cohen–Macaulay ideal, necessarily of codimension two. (This will be the case for a plane curve.)

Let

$$0 \to R^n \xrightarrow{\alpha} R^{n+1} \longrightarrow L \to 0,$$

be a resolution of L. The matrix β obtained from α by deleting the first row (the row that corresponds to the generator f of L) yields a projective resolution of J as an R–module:

$$0 \to R^n \xrightarrow{\beta} R^n \longrightarrow J \to 0.$$

Tensoring with A we get the exact sequence of Cohen–Macaulay A–modules

$$0 \to J \longrightarrow A^n \xrightarrow{\varphi} A^n \longrightarrow J \to 0.$$

Because A is a Gorenstein ring, dualizing the sequence we get another exact sequence

$$0 \to \mathrm{Hom}(J, A) \longrightarrow A^n \xrightarrow{\varphi^*} A^n \longrightarrow \mathrm{Hom}(J, A) \to 0.$$

If we now use the result of Eisenbud [Ei80] on the relationships between maximal Cohen–Macaulay modules over hypersurface rings and matrix decompositions (the kernel of φ^* is given by the image of its matrix adjoint) we obtain:

Proposition 6.15. *In the situation above,*

$$J \cdot J^{-1} = \Delta/(f), \tag{6.4}$$

where Δ is the ideal generated by the minors of size $n-1$ of the matrix β. In addition, if $g \in \Delta$ is not divisible by f, then

$$C = \mathrm{Hom}(J \cdot J^{-1}, A) = (((f,g):\Delta)/(f))g^{-1}. \tag{6.5}$$

6.3 S_2–ification of an Affine Ring

In this section we consider enlarging an affine ring into an integral extension satisfying the condition (S_2). It is the least expensive step in the construction of the integral closure.

The problem consists in for a given Noetherian ring A with total ring of fractions Q, to find integral extensions

$$A \hookrightarrow B \hookrightarrow Q,$$

with the property that if $\mathfrak{p} \subset A$ has height at least 2 then $\mathfrak{p}B$ has grade at least 2 (see [NVa93]).

Definition 6.16. *The S_2–ification of A is a finite, minimal extension $A \subset B \subset Q$ satisfying Serre condition (S_2).*

If A is an integral domain, there is a natural candidate for B. Let

$$A^{(1)} = \bigcap A_{\mathfrak{p}}, \qquad \text{height } \mathfrak{p} = 1.$$

Since A is Noetherian, it is easy to see that $A^{(1)}$ is an integral extension of A with the property that if x is an element of the field of quotients of A and its conductor into A is an ideal of height at least two then $x \in A^{(1)}$. Unfortunately this representation is non–constructive.

Canonical Module

The approach followed here uses the canonical module of A, an object that is obtained from a presentation of A as an affine ring. We shall need some elementary properties of the canonical module of a ring. See Appendix A and the additional references mentioned there.

Theorem 6.17. *Let $A = S/P$, where S is a Gorenstein ring and let P be an ideal of codimension g. Then $\omega_A = \mathrm{Ext}_S^g(A, S)$ is a canonical module for A.*

This module has the following properties:

(a) ω_A satisfies the (S_2)–condition.
(b) $\mathrm{Hom}_A(*, \omega_A)$ is a self–dualizing functor on the category of A–modules with the condition (S_2).

This leads to the following description of the S_2–ification of A. As a matter of terminology, we use *extension* to denote an injective, finite extension of A.

Theorem 6.18. *Let A be a Noetherian ring with canonical module ω_A, and suppose that A is generically a Gorenstein ring. Then $B = \mathrm{Hom}_A(\omega_A, \omega_A)$ is the minimal extension of A with the property (S_2).*

The condition that A is generically a Gorenstein ring is equivalent to saying that its canonical module may be identified to an ideal J containing regular elements.

Proof. For a given extension $A \subset C$ (C may actually be taken as a A–module) consider the diagram of natural maps

$$
\begin{array}{ccc}
A & \longrightarrow & C \\
\downarrow & & \downarrow \\
\operatorname{Hom}_A(\operatorname{Hom}_A(A, \omega_A), \omega_A) & \longrightarrow & \operatorname{Hom}_A(\operatorname{Hom}_A(C, \omega_A), \omega_A).
\end{array}
$$

If C is an extension, satisfying (S_2), there results an embedding

$$B = \operatorname{Hom}_A(\omega_A, \omega_A) \hookrightarrow C,$$

that proves the assertion. $\qquad\qquad\square$

Another way to get to B is through a subring $C \subset A$, A finite over C, for which the canonical module ω_C is known, as is the case of the Noether normalization of an affine ring A. In this case, $\omega_A = \operatorname{Hom}_C(A, \omega_C)$, see [HK71].

Theorem 6.19. *The S_2–ification of A is given by $\operatorname{Hom}_C(\operatorname{Hom}_C(A, \omega_C), \omega_C)$.*

Proof. From above we obtain the identifications

$$
\begin{aligned}
B = \operatorname{Hom}_A(\omega_A, \omega_A) &= \operatorname{Hom}_A(\operatorname{Hom}_C(A, \omega_C), \operatorname{Hom}_C(A, \omega_C)) \\
&= \operatorname{Hom}_C(\operatorname{Hom}_C(A, \omega_C) \otimes_A A, \omega_C) \\
&= \operatorname{Hom}_C(\operatorname{Hom}_C(A, \omega_C), \omega_C)
\end{aligned}
$$

by adjointness. $\qquad\qquad\square$

Corollary 6.20. *If C has (S_2) and is Gorenstein in codimension 1, then*

$$B = \operatorname{Hom}_C(\operatorname{Hom}_C(A, C), C).$$

Proof. Note that $D = \operatorname{Hom}_C(\operatorname{Hom}_C(A, C), C)$ satisfies (S_2), being a C–dual. By Theorem 6.18 we have an inclusion of extensions of A

$$B \hookrightarrow D,$$

which by assumption is an isomorphism (as C–modules) in codimension at most 1. A depth count shows that its cokernel must be trivial. $\qquad\qquad\square$

In the special case of a rational extension we have:

Corollary 6.21. *Suppose further that A is contained in the total ring of fractions of C. If F is the conductor of A with respect to C, then $B = F^{-1}$.*

Remark 6.22. It would be interesting to have estimates of the number of generators of the ring B. In Chapter 9, using the notion of extended degrees, we mention the case where $\dim A \leq 4$ (see Exercise 9.101).

Computation of the Canonical Ideal

We describe a procedure to represent the canonical module of an affine domain A as one of its ideals. This facilitates the computation of the S_2–ification of A.

Let

$$A = R/P = k[x_1,\ldots,x_n]/(f_1,\ldots,f_m) \qquad (6.6)$$

where P is a prime ideal of codimension g. Pick a regular sequence

$$\mathbf{z} = \{z_1,\ldots,z_g\} \subset P.$$

The canonical module ω_A has the following representation

$$\omega_A = \operatorname{Ext}_R^g(R/P,R) \simeq ((\mathbf{z}):P)/(\mathbf{z}).$$

Theorem 6.23. *Let P and $(\mathbf{z})$ be as above. There exists*

$$a \in (\mathbf{z}):P \setminus (\mathbf{z})$$
$$b \in (a,\mathbf{z}):((\mathbf{z}):P) \setminus P. \qquad (6.7)$$

If we set

$$L = (b((\mathbf{z}):P)+(\mathbf{z})):a$$

then

$$((\mathbf{z}):P)/(\mathbf{z}) \simeq L/P \subset R/P.$$

Proof. It is clear that a can be so selected. To show the existence of b, we localize at P. In the regular local ring R_P, PR_P is generated by a regular sequence $\mathbf{y} = \{y_1,\ldots,y_g\}$ so that we can write

$$\mathbf{z} \equiv \mathbf{y} \cdot \varphi$$

for some $g \times g$ matrix φ. By Theorem 4.5,

$$((\mathbf{z}):P)_P = (\mathbf{z},\det \varphi)$$

and the image of $\det \varphi$ in $(R/(\mathbf{z}))_P$ generates its socle.

On the other hand, ω_A is a torsion–free A–module, so that the image of a in $((\mathbf{z}):P/(\mathbf{z}))_P$ does not vanish. This means that $(a,\mathbf{z})_P$ must contain $(\mathbf{z},\det \varphi)$, which establishes (6.7).

Let L be defined as above. For $r \in (\mathbf{z}):P$ we can write

$$rb \equiv ta \bmod (\mathbf{z}).$$

Another such representation

$$rb \equiv sa \bmod (\mathbf{z}),$$

would lead to an equation

$$(t - s)a \in (\mathbf{z}).$$

But a already conducts P into $(\mathbf{x})$, so that $t - s$ must be contained in P, since the associated prime ideals of $(\mathbf{z})$ have codimension g. This defines a mapping

$$(\mathbf{z}): P/(\mathbf{z}) \longrightarrow L/P,$$

in which the (non-trivial) class of a is mapped to the (non-trivial) class of $b \in R/P$. This completes the proof, as both are torsion–free R/P–modules of rank one.   □

The Double Dual

There are other approaches to realize the S_2–ification of the ring A. Our method of choice involves the computation of the canonical ideal as above. There will be times when certain conditions are present that may be worthwhile to take advantage of. We are going to discuss one of them.

For ease of reference we rephrase a well-known description of the bidual of a finitely generated module over a Krull domain ([Sam64]).

Proposition 6.24. *Let R be an integral domain that satisfies (S_2) and which is Gorenstein in codimension* 1. *Let M be a finitely generated R–module. Then*

$$M^{**} = \bigcap M_{\mathfrak{p}}, \text{ height } \mathfrak{p} = 1.$$

The intersection is taken in the vector space $V = M \otimes R_{(0)}$. In particular, if M is a finite algebra over R then M^{**} is also a finite R–algebra.

Proof. The proof is the same as the case of a normal domain in [Sam64].   □

We assume that we have a Noether normalization of A

$$R = k[\mathbf{z}] = k[z_1, \ldots, z_d] \hookrightarrow A.$$

A is a torsion free R–module, and the process that gives the Noether normalization also yields a presentation of A,

$$R^p \xrightarrow{\varphi} R^q \longrightarrow A \to 0.$$

The double dual of A is the module

$$A^{**} = \operatorname{Hom}_R(\operatorname{Hom}_R(A, R), R).$$

Proposition 6.25. *The module A^{**} is a subring of B.*

Proof. Follows from Proposition 6.24. □

Note that as a consequence of Theorem 6.1, B is a reflexive R–module. We must now find a more explicit description of A^{**}, for which purpose we make use of the presentation of A.

For an ideal I of an integral domain A, $I^{-1} = A\!:\!I$ has the usual meaning:

$$I^{-1} = \{x \in \text{field of fractions of } A \mid x \cdot I \subset A\}.$$

I^{-1} is an A–module, canonically isomorphic to $\text{Hom}_A(I,A)$. As an R–module I^{-1} is reflexive if A is a reflexive R-module. To see this, it is enough to show that I^{-1} is the R–dual of some R–module. From $A = \text{Hom}_R(A^*,R)$, we have by adjointness

$$\text{Hom}_A(I,A) = \text{Hom}_A(I,\text{Hom}_R(A^*,R)) = \text{Hom}_R(I \otimes_A A^*,R).$$

Proposition 6.26. *Let J be the first non–vanishing Fitting ideal of A as an R–module. Let f, g be two relatively prime polynomials in J. (Since A is a torsion–free R–module, height $J \geq 2$.) Then it holds*

$$A^{**} = \bigcup_{t \geq 1} (A\!:\!(f,g)^t). \tag{6.8}$$

Proof. From Proposition 6.24, it follows that the right–hand side of this formula, say C, is contained in A^{**}. (At the same time, this shows that the right–hand side has a stable value, since B is a finitely generated R–module.) Conversely, let $x \in A^{**}$ and let L denote the R–conductor of A^{**} into C, that is

$$L = \{r \in R \mid rx \in C\}.$$

L is an ideal of R whose radical contains J. Indeed, if $\mathfrak{p}$ is a prime containing L but not J, $A_\mathfrak{p}$ is a free $R_\mathfrak{p}$–module and therefore coincides with $A_\mathfrak{p}^{**}$. But $A_\mathfrak{p} \subset C_\mathfrak{p}$ contradicts the definition of L. This means that for some integer t, $(f,g)^t \cdot x \subset A$ and therefore $x \in C$. □

The point of this construction is that A^{**} will satisfy condition (S_2). The other overrings to be constructed will also preserve this property. We have the following:

Theorem 6.27. *If A satisfies the condition (R_1) then $B = C$.*

Proof. C is a finite rational extension of A, and therefore inherits the Serre's condition (R_1) from it. □

In many cases it is unnecessary to find the Noether normalization, since it is only used to provide the elements f and g. It is clearly enough that f and g define part of a normalizing system of parameters and lie in the ideal that defines the Cohen–Macaulay locus of A.

Proposition 6.28. *Let $A = k[\mathbf{x}]/P$. The formula of Proposition 6.26 holds if f and g are chosen such that* height$(P, f, g) = 2 +$ height P *and belong to the ideal*

$$L = \bigcap_{j \geq r+1} \text{annihilator}(\text{Ext}^j_{k[\mathbf{x}]}(A, k[\mathbf{x}])). \tag{6.9}$$

Generators and Relations

There remains to obtain a presentation of $A^{**} = k[y_1, \ldots, y_m]/Q$. Once that is done, we replace A by A^{**}.

We use a method developed in [Vas89], exploiting the fact that the transform is computed relative to a two–generated ideal. There will be two steps to consider:

Algorithm 6.29 *Let A be an affine domain and let f, g be a regular sequence as above.*

(a) *The formula*

$$A(t) := A : (f, g)^t = (\bigcap_{1 \leq i \leq t} (Af^t :_A g^i f^{t-i})) f^{-t} = (\bigcap_{1 \leq i \leq t} (Af^i :_A g^i)) f^{-t}$$

yields $a_1 f^{-r}, \ldots, a_s f^{-r}$, such that $A^{**} = A[a_1 f^{-r}, \ldots, a_s f^{-r}]$, where r is the stable value of t.

(b) *For each $a_i f^{-r}$ consider the linear polynomial $f^r U_i - A_i$ where the U_i are indeterminates and A_i is a representative in $k[x_1, \ldots, x_n]$ of a_i. Let $K = (P, f^r U_1 - A_1, \ldots, f^r U_s - A_s)$. Let $Q = \bigcup_{t \geq 1} (K :_{k[x_1, \ldots, x_n, U_1, \ldots, U_s]} f^t)$ be the "rational closure" of K. Then*

$$A^{**} = k[x_1, \ldots, x_n, U_1, \ldots, U_s]/Q.$$

The proof of correctness is clear from the discussion. An advantage of computing the transforms $A(t)$ lies in the convenience of testing for stabilization.

In the sequel, when we replace A by a larger ring C, the latter will always be a reflexive A–module and therefore (S_2) passes from A to C.

Example 6.30. The affine domain defined by the prime ideal of Example 3.71 fits nicely here. $A = k[a, b, c, d, e]/P$ satisfies the condition of Theorem 6.27. Its integral closure is $B = k[a, b, c, d, e, u, v]/Q$, with Q generated by the polynomials (where P is the radical of the ideal generated by the first three polynomials):

$ab^3c + bc^3d + a^3be + cd^3e + ade^3$

$a^2bc^2 + b^2cd^2 + a^2d^2e + ab^2e^2 + c^2de^2$

$a^5 + b^5 + c^5 + d^5 - 5abcde + e^5$

$a^4bc - ab^4e - 2b^2c^2de + a^2cde^2 + bd^3e^2 + du$

$a^3b^2c - bc^2d^3 - b^5e - d^5e + 2abcde^2 - ev$

$a^2b^2cd + b^2c^3e + a^4de - 2bcd^2e^2 - abe^4 - cu$

$bc^5 - 2ab^2cde + c^3d^2e - a^3de^2 + be^5 - bv$

$b^5c + a^2b^3e - abc^2de - ad^3e^2 + cv$

$ab^2c^3 + abcd^2e - a^2be^3 - de^5 + dv$

$abc^2d^2 - b^3c^2e + ad^4e - a^2bce^2 + b^2d^2e^2 - cde^4 + bu$

$b^3c^2d - cd^2e^3 - av$

$bcd^4 - c^4de - au$

$abc^4 - b^4cd - a^2b^2de + ac^2d^2e + b^2c^2e^2 - bd^2e^3 - eu$

$ab^4c^2e^3 - a^4b^3c^2d + a^2b^3c^4e + a^6bcde + ab^6cde + 3b^4c^3d^2e\backslash$

$\quad -a^3b^4de^2 - 3a^2b^2c^2d^2e^2 - b^3cd^4e^2 + a^2bd^4e^3 - b^5c^5 - a^3b^2ce^4\backslash$

$\quad +bc^2d^3e^4 - abcde^6 - b^2cd^2u + a^2d^2eu + c^2de^2u - abcdev + e^5v - v^2$

$b^4cde^4 + 2a^3d^3e^4 + 4a^2b^2de^5 - ac^2d^2e^5 + 2bd^2e^7 + abcdeu\backslash$

$\quad +ac^2d^2v - b^2c^2ev - 4a^2ce^2v - bd^2e^2v + uv$

$ab^3c^6 + a^5bc^3d - a^2b^4c^2de - 2ab^2c^4d^2e + a^5cd^3e\backslash$

$\quad -a^2b^3d^3e^2 - 2abc^2d^4e^2 - a^2b^2c^3e^3 + ad^6e^3 - a^2bcd^2e^4 - cd^3e^6\backslash$

$\quad +b^2c^2eu - bd^2e^2u - u^2 + bc^3dv - cd^3ev - ade^3v$

($\backslash$ denotes continuation.)

Rees Algebras

Let R be a ring and I one of its ideals. The S_2–ification of Rees algebras has a number of peculiar features. For a given algebra $R[It]$, for simplicity we let $S_2[It]$ denote the S_2–closure of $R[It]$.

One–dimensional Rings

The following elementary calculation provides the full picture of the S_2–ification of a class of Rees algebras.

Theorem 6.31. *Let* $(R, \mathfrak{m})$ *be an* 1*–dimensional Cohen–Macaulay local ring with infinite residue field, and let* I *be an* $\mathfrak{m}$*–primary ideal. Let* $J = (a)$ *be a reduction of* I *such that* $r_J(I) = r$. *Then the* S_2*–ification of the Rees algebra* $R[It]$ *is the ring* $A[at]$, *where* $A = I^r a^{-r}$.

Proof. Since $I^{r+1} = aI^r$, we have

$$A^2 = I^{2r}a^{-2r} = (I^{2r}a^{-r})a^{-r} = I^r a^{-r} = A.$$

Therefore A is a ring such that $R \hookrightarrow A$. On the other hand, for $n \geq r$,

$$I^n t^n = I^n a^{-n} a^n t^n$$
$$= I^{n-1} a^{-n+1} a^n t^n$$
$$\vdots$$
$$= I^{n-(n-r)} a^{-n+(n-r)} a^n t^n$$
$$= (I^r a^{-r}) a^n t^n$$
$$= A a^n t^n.$$

Therefore we have

$$A[at] \hookleftarrow R[It] = R + It + \cdots + I^{r-1} t^{r-1} + A a^r t^r + A a^{r+1} t^{r+1} + \cdots.$$

For $n \geq r$, one also has

$$(I^n, I^n t^n) \cdot A = (I^{n+r} a^{-r}, I^{n+r} a^{-r} t^n) = (I^n, I^n t^n) \subset R[It].$$

This implies $S_2[It] = A[at]$ since $A[at]$ already has (S_2), being isomorphic to a polynomial ring over a one–dimensional Cohen–Macaulay ring. $\square$

Equimultiple Ideals

An ideal I of height g is *equimultiple* if there is a subideal $J \subset I$, generated by g elements, such that $I^{r+1} = JI^r$ for some integer r.

Proposition 6.32. *Let $(R, \mathfrak{m})$ be a Gorenstein local ring and let I be an ideal admitting a minimal reduction J generated by a regular sequence. Assume that $I^{r+1} = JI^r$. Then $S_2[It] = \sum S_n t^n$ can be determined as follows:*

$$S_n = \bigcap_{\ell=0}^{\infty} (J^{n+\ell} : B_\ell),$$

where

$$B_\ell = J^\ell \cap (J^{\ell+1} : I) \cap \cdots \cap (J^{\ell+r} : I^r).$$

Proof. If J is generated by a regular sequence of g elements, the canonical module of $R[Jt]$ is isomorphic to $(1,t)^{g-2}$ (see [HSV87, Corollary 2.95]). Furthermore, as $R[Jt]$–modules, $(1,t) \simeq (JR[Jt])^{-1}$. We now argue that $R[Jt]$ is Gorenstein in codimension one. By the above, it suffices to show that the ideal $JR[Jt]$ is principal in codimension one. Since the ideals $JR[Jt]$ and $JtR[Jt]$ are isomorphic and $R[Jt]/(J,Jt)R[Jt] = R/J$ has dimension $\dim R - g$, the ideal $(J,Jt)R[Jt]$ has codimension $g+1$, and therefore the free locus of $JR[Jt]$ has codimension at least g, which amply establishes the claim. According to Corollary 6.20, the ring $S_2[It]$ is the double dual of $R[It]$ as a $R[Jt]$–module.

The expressions for B_ℓ and S_n are just the formulation of these operations of graded modules over $R[Jt]$ as submodules of $Q[t]$, with Q the total ring of fractions of R. $\square$

A first observation of this construction is that for equimultiple ideals the S_n's define a decreasing filtration, and thus $S_2[It]$ is a *bona fide* Rees algebra.

There is an amusing aspect to the algorithm outlined above: it is less forbidding than it appears, with the computation of the 'infinite' intersection being the more straightforward step of the two. The actual steps are as follows. First observe that from the reduction equation there exists a nonzero divisor $x \in J$ such that $xI^i \subset J^i$, $\forall i$.

Algorithm 6.33 *Let* $J = (a_1, \ldots, a_g)$ *and* $B = R[T_1, \ldots, T_g]$. *Let* H *denote the ideal generated by the Koszul relations* $a_i T_j - a_j T_i$. *(That is* $R[Jt] = B/H$.*)*

- *For* $1 \leq j \leq r$
    - *Denote by* M_j *the ideal of* B *generated by forms of degree* j *that give a lift of the containment* $xI^j \subset J^j$.
    - *Denote* $L_j = (f_j, H) : M_j$, *where* $0 \neq f_j \in I^j$.
- *Set*
$$L = L_1 \cap \cdots \cap L_r.$$
- *Then*
$$S_2[It] = \operatorname{Hom}_B(L/H, B/H) x^{-1}.$$

Note that all we have outlined is how to compute the double dual of $xR[It]$, as an ideal of $R[Jt]$ and removed x when done.

Exercise 6.34. Let $\triangle$ be a simplicial complex and let $A = k[\triangle]$ be the corresponding face ring over the field k. Prove that the S_2–ification of R is a face ring $k[\triangle']$ for some complex $\triangle'$. Is there a nice description of $\triangle'$?

6.4 Desingularization in Codimension One

We must find a way to further enlarge A, if necessary, into its integral closure B. This brings us to the problem of desingularization in codimension one. The approach here will fail to work in many cases of interest as it depends on Jacobian ideals, and we restrict ourselves mostly to characteristic zero, or at least to cases where certain separability conditions are met.

We begin by analyzing the ideal $I \cdot I^{-1}$ and how it is likely to lead to a non-trivial extension of A.

Proposition 6.35. *Let* R *be a Noetherian integral domain and let* I *be a nonzero ideal. Then* $I \cdot I^{-1} = A$ *if and only if for each prime ideal* $\mathfrak{p}$ *of grade 1 the ideal* $I_\mathfrak{p}$ *is principal, in other words,* $I \cdot I^{-1}$ *is not contained in any such prime ideal.*

Proof. This follows from the notion of grade of an ideal. An ideal J of grade 1 has the property that there is a nonzero element x such that J consists of zero divisors of the module $A/(x)$. This means that there is an element $y \in A \setminus (x)$ such that $yJ \subset (x)$, and thus $\dfrac{y}{x} \in J^{-1} \setminus A$.

The assertion now follows since the formation of I^{-1} and taking localization at a prime are commuting operations. $\square$

Proposition 6.36. *Let R be a domain with the property (S_2) and let L be an ideal. Then*

(a) *The dual ideal $\mathrm{Hom}_R(L,R)$ satisfies (S_2);*
(b) *If L is unmixed of codimension 1 then $D = \mathrm{Hom}_R(L,L)$ satisfies (S_2).*

Proof. We only give the proof of part (b), the more interesting of the two assertions.

From the exact sequence

$$0 \to L \longrightarrow A \longrightarrow A/L \to 0,$$

it follows that since any prime ideal P of codimension at least 2 contains regular elements on A/L, the module L satisfies (S_2).

We show that every regular sequence x,y is a regular sequence on D. For completeness we give the standard argument. From the exact sequence

$$0 \to L \xrightarrow{\;x\;} L \longrightarrow L/xL \to 0,$$

we have the derived exact sequence

$$0 \to \mathrm{Hom}_A(L,L) \xrightarrow{\;x\;} \mathrm{Hom}_A(L,L) \to \mathrm{Hom}_A(L,L/xL) = \mathrm{Hom}_A(L/xL,L/xL).$$

Since y is a regular element on the last module, it follows that it is regular on the submodule D/xD. $\square$

Canonical Modules and Multipliers

In the case of finite characteristic, when we may not always depend on the Jacobian ideal, the theory of the canonical module can be depended upon to provide the necessary multipliers.

Suppose first that A is not Gorenstein, with a canonical ideal ω. Then a natural place to start is by considering the multiplier of ω^{-1}:

$$C = (\omega \cdot \omega^{-1})^{-1}.$$

If $A = C$, the ring A is Gorenstein in codimension 1.

If A is known to be a Gorenstein ring in codimension one then almost any ideal can be used instead of J^{-1}. This is useful in some cases of characteristic positive.

Theorem 6.37. *Let A be a Gorenstein ring and let I be an ideal containing regular elements. If I is not always principal in codimension one then the endomorphism ring of I is strictly larger that A.*

Proof. We may localize A at a prime P of codimension 1, $R = A_P$. Suppose $\mathrm{Hom}_R(I, I) = R$. Consider the exact sequence

$$0 \to I \cdot I^{-1} \longrightarrow R \longrightarrow R/L \to 0,$$

where we put $L = I \cdot I^{-1}$. Applying $\mathrm{Hom}_R(\cdot, R)$ we obtain the exact sequence

$$0 \to R \xrightarrow{\varphi} \mathrm{Hom}_R(I \cdot I^{-1}, R) \longrightarrow \mathrm{Ext}^1_R(R/L, R) \to 0.$$

But we have

$$\mathrm{Hom}_R(I \cdot I^{-1}, R) \simeq \mathrm{Hom}_R(I \otimes_R \mathrm{Hom}_R(I, R), R)$$
$$\simeq \mathrm{Hom}(I, \mathrm{Hom}_R(\mathrm{Hom}_R(I, R), R)) \simeq \mathrm{Hom}_R(I, I) = R,$$

since every ideal of R is reflexive and by adjointness. This means that φ is realized by multiplication by an element of R, necessarily the identity since 1 is mapped to itself. This implies that $\mathrm{Ext}^1_R(R/L, R) = 0$, and therefore $L = R$, as claimed. $\square$

Multipliers of Jacobian Ideals

The general principle in using multiplication rings as a foundry to algorithms for the computation of the integral closure is the following: Identify a class of ideals with the property that if $A \neq \overline{A}$ then $\mathrm{Hom}_A(I, I) \neq A$. To be useful the ideal I must be directly derived from one of the presentations of A.

We are going to use (see [Va91a]) an opening provided by the regularity criterion of Theorem 5.16.

Algorithm 6.38 *Let A be an affine domain satisfying Serre's condition (S_2), let J be its Jacobian ideal, and let B be the integral closure of A.*

(a) *If height $J \geq 2$, then $B = A$.*
(b) *If height $J = 1$, then $C = \mathrm{Hom}_A(J^{-1}, J^{-1})$ is a subring of B properly containing A.*
(c) *Replace A by C until (a) is realized.*

Proof. If height $J \geq 2$, A satisfies (R_1) and is therefore normal by Theorem 6.1. If however height $J = 1$, we claim that $A \neq C \subset B$. Indeed, the ideal $J \cdot J^{-1}$ must have grade 1 as otherwise $J_\mathfrak{p}$ would be invertible for each prime $\mathfrak{p}$ of height 1; by Theorem 5.16 $A_\mathfrak{p}$ would be regular. But an ideal I of a Noetherian domain A has grade 1 if and only if $I^{-1} \neq A$ (by Proposition 6.35 or see [Kap74, Exercise 2, p. 102]), and therefore C is a proper extension of A. Moreover, since $C = \mathrm{Hom}_A(J \cdot J^{-1}, A)$, it retains (S_2) from A. $\square$

We rephrase this method in view of our earlier discussion:

Algorithm 6.39 *Let A be an affine domain over a field of large characteristic. Then $\overline{A}$ can be determined as follows:*

- If ω_A is the canonical ideal of A, set

$$A_1 = \mathrm{Hom}_A(\omega_A, \omega_A).$$

- For each integer $n \geq 2$, let

$$A_{n+1} = \mathrm{Hom}_{A_n}(J_n^{-1}, J_n^{-1}) = (J_n \cdot J_n^{-1})^{-1},$$

where J_n is the Jacobian ideal of A_n. The sequence stabilizes at $\overline{A}$.
- If $\dim A \leq 3$, each A_n is guaranteed to be generated by at most

$$\begin{cases} \deg(A)^2 - \deg(A) + 3 \text{ generators if } \dim A = 3 \\ \deg(A) + \dim A - 1 \quad \text{generators if } \dim A \leq 2 \end{cases}$$

Computational Test of Normality

It may be advisable to test for normality, so for completeness we discuss the standard normality test and a variant mechanism.

Theorem 6.40. *Let k be a field of characteristic zero and let A be an k–affine domain of field of fractions Q and let J denote its Jacobian ideal. Then A is normal if and only if*

$$\boxed{A = A :_Q J.} \tag{6.10}$$

Proof. This is just the assertion that J is an ideal of grade at least two, so that at each prime $\mathfrak{p}$ such that depth $A_{\mathfrak{p}} = 1$ the localization is a discrete valuation domain. Since A is the intersection of all such integrally closed domains (see [Kap74, Theorem 53]), it is also normal. □

There will be occasions when it is not necessary to compute the full Jacobian ideal J, as any of its subideals that satisfy the equation above suffices for certification of normality. One of these may be found as follows (see [LS81]):

Theorem 6.41. *Let A be an affine domain and let R be one of its Noether normalizations. Suppose the field of fractions of A is a separable extension of the field of fractions of R. Let $A = R[x_1, \ldots, x_n]/P$ be a presentation of A and let J_0 denote the Jacobian ideal of A relative to the variables $x_1, \ldots, x_n$. If B is the integral closure of A then $J_0 \cdot B \subset A$. In particular if $A = A:J_0$ then A is integrally closed.*

One of its many consequences are:

Corollary 6.42. *Let A be an affine domain over a field of characteristic zero and denote by J its Jacobian ideal. If the height of J is at least 2 then the integral closure of A is given by $B = A : J$.*

Corollary 6.43. *Let A be an affine domain over a field of characteristic zero and denote by J its Jacobian ideal. If B is the integral closure of A, the conductor $c(B/A)$ contains J.*

Let us make explicit the situation of the integral closure of an affine domain A over a perfect infinite field k, when the Jacobian ideal J has height at least two. Let f, g be elements of J such that $\text{height}(f, g) = 2$. Then

Proposition 6.44. *The integral closure B of A is given by*

$$B = A : (f, g).$$

Proof. Whenever $\text{height}(J) \geq 2$, B is simply the S_2–ification of A,

$$B = \bigcap_{\text{height}_{\mathfrak{p}}=1} A_{\mathfrak{p}}.$$

Indeed, since $(f, g) \subset J$ is contained in the conductor of B/A, B is contained in the right hand side of the expression. On the other hand, by the Jacobian criterion A and B agree in codimension one.   □

In other words, the elements of B are obtained from the syzygies $a \cdot f - b \cdot g = 0$ as they give rise to $a/g \in B$. In fact, collecting the coefficient ideal of f in these syzygies, we have

$$B = A[Lf^{-1}].$$

This calculation is realized as follows. Write $A = R/I$, where R is a ring of polynomials in n variables. For a set $f_1, \ldots, f_m$ of generators of I, we consider the syzygies S of $F, G, f_1, \ldots, f_m$, where F and G are lifts of f and g. The desired syzygies over A are the images of S in A.

We make the following observation about the degrees of these generators that is useful in a study of the complexity of these procedures ([UVp]). Let $G = \{g_1, \ldots, g_s\}$ be a Gröbner basis of the ideal $J = \{F, G, f_1, \ldots, f_m\}$, and write $\delta = \max\{\deg(F), \deg(G), \deg(f_1), \ldots, \deg(f_m)\}$. The set S will arise from division relations amongst the g_i's and therefore S will be generated in degrees bounded by $\Delta = \max(\deg(g_i))$. According to [BKW93, Appendix], and especially [Giu84], Δ is bounded by a polynomial in δ of degree a^n (with a of the order of $\sqrt{3}$). For standard graded algebras, much lower bounds are to be found in [UVp], obtained by more geometric arguments.

Unlike the singular locus, through the Jacobian criterion, there is no explicit description of the non–normal locus. This opens the door to *ad hoc* methods, each suited to a limited situation. In the case of Rees algebras, there are a few procedures that have been proved useful. We discuss here two methods. The proof of the first of these will be left as an exercise. Its practice only requires that part of the Jacobian ideal be known with the stated property.

Proposition 6.45. *Let R be a normal affine domain over a field of characteristic zero, and let I be an ideal. Let*

$$0 \to L \longrightarrow R[T_1, \ldots, T_n] \longrightarrow R[It] \to 0$$

be the presentation of the Rees algebra of I. Let J denote the Jacobian ideal of $R[It]$. Then I is a normal ideal if and only if

$$(L, I):J = (L, I).$$

The second method is less general but it is modeled on the situation of a Rees algebra $R[It]$ of a local ring R that is normal on the punctured spectrum of R.

Proposition 6.46. *Let R be a Noetherian domain with a finite integral extension S. Let J be the conductor of the extension. Suppose there is a prime ideal $\mathfrak{p}$ and an element $f \notin \mathfrak{p}$ with the following properties:*

(i) *J contains a power of $\mathfrak{p}$;*
(ii) *$\mathfrak{p}R_f$ is a principal ideal.*

If grade $(\mathfrak{p}, f) \geq 2$ *then* $R = S$.

Proof. Suppose that grade $(\mathfrak{p}, f) \geq 2$, but $R \neq S$. The latter says that J is a proper ideal of grade 1. Let P be one of its associated prime ideal; it follows that the local ring R_P has depth one, and therefore $\mathfrak{p} \subset P$, but $f \notin P$. If $\mathfrak{p} = P$, R_P is a discrete valuation domain and thus $J_P = R_P$, which contradicts the choice of P. So we must have $\mathfrak{p} \neq P$, which in turn implies that PR_P contains a regular sequence with 2 elements since $\mathfrak{p}_P$ is a principal ideal. $\square$

Canonical Module

Let R be a 1–dimensional ring. The method of enlarging R in order to approach its integral closure has consisted in determining the endomorphism ring of ideals, more precisely computing a module of the form I^{-1} for some appropriate ideal.

Let us discuss how I^{-1} and R compare by examining the length of the module I^{-1}/R. (See [BH93, Theorem 3.3.10] for the required tools.)

Proposition 6.47. *Let R be a 1–dimensional Cohen–Macaulay ring with a canonical module ω. For each ideal I containing a regular element*

$$\ell(I^{-1}/R) = \ell(\omega/I \cdot \omega).$$

Proof. We observe that $R = \mathrm{Hom}_R(\omega, \omega)$ and

$$I^{-1} = \mathrm{Hom}_R(I, R) = \mathrm{Hom}_R(I, \mathrm{Hom}_R(\omega, \omega))$$
$$= \mathrm{Hom}_R(I \otimes \omega, \omega) = \mathrm{Hom}_R(I \cdot \omega, \omega)$$

with natural identifications. On the other hand, applying to the exact sequence

$$0 \to I \longrightarrow R \longrightarrow R/I \to 0$$

the functor $\mathrm{Hom}_R(\cdot, \omega)$, and taking into account that $\mathrm{Ext}^1_R(\cdot, \omega)$ is an exact, self–dualizing functor on modules of finite length, we obtain that

$$\ell(\omega/I \cdot \omega) = \ell(\mathrm{Ext}^1_R(\omega/I \cdot \omega, \omega) \simeq \mathrm{Hom}_R(I \cdot \omega, \omega)/\mathrm{Hom}_R(\omega, \omega)).$$

Corollary 6.48. *Let R be a 1–dimensional affine domain and let $\overline{R}$ be the integral closure of R. Then $\ell(\overline{R}/R) \geq$ Cohen–Macaulay type of R.*

Proof. We apply the theorem to the conductor ideal I of $\overline{R}$ with respect to R. If ω is the canonical module of R then the length of the module $\omega/I\omega$ is obviously not smaller than its minimal number of generators. $\qquad\square$

For curves, and some low dimensional rings, this method gives a reasonable performance. The reader will note that we have not tapped the rich vein of methods of desingularization of curves that bring in very early the underlying geometry. The stakes tend to be higher here since "curves" and one–dimensional rings are not equivalent objects. There is a large bibliography on this topic, that is periodically visited and enlarged by geometers, number theorists and code theorists (see [Vaz91]).

Problem 6.49. In the case of a curve, find reasonable conditions for the ring of Proposition 6.15 to be the integral closure of A.

6.5 Discriminants and Multipliers

In this section we discuss some elements of the methods of [Sei75] and [Sto68] in their use of Noether normalizations and discriminants in the construction of the integral closure of an affine domain A. It will turn out that the same benefit can be obtained by using selected elements of the Jacobian ideal of A. The exposition is influenced by the analysis in [Tra86].

Let B denote the integral closure of A, and let L be the conductor of B/A: L is the largest common ideal of the rings A and B. We want to exploit the fact that any nonzero element $d \in L$ provides a representation

$$B = d^{-1}J \subset d^{-1}A, \tag{6.11}$$

where J is an ideal of A. The problem is how to identify this numerator ideal. We are going to make use of some of the techniques to compute radicals developed in Chapter 5.

Discriminant

The representation (6.11) requires the identification of elements d in the conductor of B/A. Earlier, in Theorem 6.41, some mention was already made about this issue. Two ways in which this can be directly approached are:

(a) Let A be an affine domain of field of fractions K, and let

$$k[\mathbf{z}] = k[z_1, \ldots, z_d] \hookrightarrow k[x_1, \ldots, x_n]/\mathfrak{p} = A \qquad (6.12)$$

be a Noether normalization of A. Let $\mathbf{y} = \{y_1, \ldots, y_s\}$ be a basis of K over $k(\mathbf{z})$.

If the extension $K/k(\mathbf{z})$ is separable, and Trace is the ordinary trace function of the field extension, the *discriminant* of the basis $\mathbf{y}$ is the determinant ([Ab90], [Nag62])

$$D = \det(\mathrm{Trace}(y_i y_j)).$$

When the y_i's are chosen to be integral over A, D is a non vanishing element that conducts B into A.

(b) In characteristic zero, Corollary 6.43 provides a more direct access to conductor elements.

The method of choice here will be (b), since we are going to use techniques developed earlier.

Multipliers

We begin by exhibiting some elementary properties of the ideal J.

Proposition 6.50. *Let J be the ideal in* (6.11). *Then*

(a) $B = \mathrm{Hom}_A(J, J)$.
(b) $J^2 = d \cdot J$, *in particular* (d) *and J have the same radical.*
(c) J *is an unmixed ideal of codimension* 1.

Proof. (a) is clear, while (b) follows since $B \cdot B = B$. To verify (c), note that

$$A/J \hookrightarrow B/J = B/dB.$$

The ring B satisfies the condition (S_2) so that B/dB has no embedded primes.   □

Augmenting the Ring

With an element such as d we have access at least to the radical of the numerator ideal J. One of the advantages of the methods here is that they bypass the computation of the S_2–closure of the ring.

Proposition 6.51. *Let A be a Noetherian integral domain and let L be a nonzero radical ideal. If $A = \mathrm{Hom}(L, L)$, then for any prime ideal $\mathfrak{q} \supset L$ with depth $A_\mathfrak{q} = 1$, $A_\mathfrak{q}$ is a discrete valuation ring.*

Proof. Let $L = \mathfrak{p}_1 \cap \cdots \cap \mathfrak{p}_n$ be the primary decomposition of L and let $L \subset \mathfrak{q}$ be as in the assertion. We may assume that A is a local ring and $\mathfrak{q}$ is its maximal ideal without changing any of the assertions.

Assume first that $\mathfrak{q}$ is not any of the $\mathfrak{p}_i$. The ideal $\mathfrak{q}^{-1} \cdot L$ must be contained in each $\mathfrak{p}_i$, as otherwise localizing at this prime would yield a contradiction. Thus $\mathfrak{q}^{-1} \cdot L \subset L$, which shows that $\mathfrak{q}^{-1} \subset A$, against the hypothesis on the depth of A. On the other hand, if we have the equality $\mathfrak{q} = \mathfrak{p}_i$, and A is not a discrete valuation ring, $\mathfrak{q} \cdot \mathfrak{q}^{-1} = \mathfrak{q}$ would again lead to a contradiction. $\qquad\square$

Corollary 6.52. *Let A be a Noetherian domain with a finite integral closure B. Let d be a nonzero element in the radical of the conductor of B/A, and let $L = \sqrt{(d)}$. If $\mathrm{Hom}_A(L,L) = A$, then $A = B$.*

Proof. By Proposition 6.2,

$$A = \bigcap A_\mathfrak{q}, \ \text{depth } A_\mathfrak{q} = 1.$$

We show that each of these localizations is a discrete valuation domain and therefore A will be integrally closed. If $d \notin \mathfrak{q}$, $A_\mathfrak{q} = B_\mathfrak{q}$ and there is nothing to show. In the other case we apply the previous Proposition. $\qquad\square$

An immediate source of elements such as d is the following.

Corollary 6.53. *Let A be an affine domain over a field of characteristic zero and let d be a nonzero element in the Jacobian ideal. If*

$$A = \mathrm{Hom}_A(\sqrt{(d)}, \sqrt{(d)}),$$

then A is integrally closed.

Remark 6.54. It is clear that the same assertion will hold true if we replace $\sqrt{(d)}$ by the radical of the Jacobian ideal. There may be certain efficiencies in picking fewer generators to start with.

Example 6.55. Let $f \in k[x,y]$ be an irreducible polynomial, and consider the ring $A = k[x,y]/(f)$. To apply Corollary 6.53 here would require two steps: the computation of the radical and of the multiplication ring. Here is a simple formula for the first step, that uses Theorem 5.24. Pick say, $d = f_x = \frac{\partial f}{\partial x}$ (assumed not zero). In the ring $k[x,y]$, the ideal (f, f_x) is a regular sequence, so that

$$\sqrt{(f, f_x)} = (f, f_x):J,$$

where J is its Jacobian ideal. In this case $J = (f_x \cdot f_{xy} - f_y \cdot f_{xx})$, and we obtain

$$\sqrt{(f, f_x)} = (f, f_x):f_y f_{xx}.$$

For the computation of the multiplication ring, one could employ Proposition 2.16.

Remark 6.56. One may call an ideal I of a ring A *closed* if $\mathrm{Hom}_A(I,I) = A$. We have based out pursuit of the integral closure on finding non closed ideals. The closed ones are good markers for this landscape and an study is carried out in [BV01]. If A is a 1–dimensional local domain which is not Gorenstein the following are non closed ideals:

- Any power ω^n, $n \geq 2$, where ω is the canonical ideal of A;
- Any irreducible ideal not isomorphic to ω.
- $\omega^{-1} = \mathrm{Hom}_A(\omega, A)$.

An issue is, which of these is such that $\ell(\mathrm{Hom}_A(I,I)/A)$ is largest?

6.6 Integral Closure of an Ideal

In this section we consider a related notion of integral closure that applies to ideals instead of to rings.

Definition 6.57. *Let R be a ring and I one of its ideals. An element $z \in R$ is integral over I if there is an equation*

$$z^n + a_1 z^{n-1} + \cdots + a_n = 0, \qquad a_i \in I^i. \tag{6.13}$$

The set $\bar{I}$ of all such elements is called the integral closure *of I.*

It is obvious that the task of computing the *radical*, $\sqrt{I}$, of the ideal I is much simpler as one is only required to describe the solutions of equations of the form

$$X^n - a = 0, \qquad a \in I.$$

A convenient way to rephrase the notion of the integral closure of I is to consider the Rees algebra of the ideal I, $R[It]$, and observe that $z \in \bar{I}$ if and only if the polynomial $zt \in R[t]$ is integral over the subring $R[It]$. More precisely, the integral closure of the graded subring $R[It]$ in $R[t]$ is the graded subring

$$R + \bar{I}t + \cdots + \overline{I^n}t^n + \cdots.$$

In particular $\bar{I}$ is an ideal. If $\bar{I} = I$ we say that I is *integrally closed*, and if all powers I^n are integrally closed we say that I is a *normal* ideal.

Example 6.58. (a) Any radical ideal is integrally closed.

(b) Let $x^{\nu_1}, \ldots, x^{\nu_m}$ be monomials in the variables $\mathbf{x} = \{x_1, \ldots, x_n\}$. The integral closure of $I = (x^{\nu_1}, \ldots, x^{\nu_m})$ is also generated by monomials. This follows from the general property that the integral closure of a graded subring for any grading (in this case $R[It]$ has the fine grading of $R[t]$) is also graded. If $\mathbf{x}^v \in \bar{I}$, it will satisfy an equation

$$(\mathbf{x}^v)^\ell \in I^\ell,$$

and therefore we have the following equation for the exponent vectors,

$$\ell \cdot v = u + \sum_{i=1}^m r_i \cdot v_i, \quad r_i \geq 0, \quad \sum_{i=1}^m r_i = \ell.$$

This means that $v = \dfrac{u}{\ell} + \alpha$, where α belongs to the convex hull of $v_1, \ldots, v_m$. The vector v can be written as (set $w = \frac{u}{\ell}$)

$$v = \lfloor w \rfloor + (w - \lfloor w \rfloor) + \alpha,$$

and it is clear that the integral vector

$$v_0 = (w - \lfloor w \rfloor) + \alpha$$

also has the property that $\mathbf{x}^{v_0} \in \bar{I}$. Thus $\bar{I}$ is generated by $\mathbf{x}^v$, where v lies either on the integral convex hull of the v_i's, or lies very near the rational convex hull, that is differ from a vector in the convex hull by another vector with positive entries which are less than 1. In symbols, if C denotes the rational convex hull of $\{v_1, \ldots, v_m\}$ and

$$B = [\bar{0}, 1) \times \cdots \times [\bar{0}, 1) = [\bar{0}, 1)^n,$$

then $\bar{I}$ is generated by the monomials x^v with

$$v \in (C + B) \bigcap \mathbb{N}^n.$$

For example, if $I = (x^4, y^3)$, then $\bar{I} = (x^4, y^3, x^2y^2, x^3y)$.

A first bound results directly from this description of the integral closure:

Proposition 6.59. *Let I be a monomial ideal of $k[x_1, \ldots, x_n]$, generated by monomials of degree at most d. Then $\bar{I}$ is generated by monomials of degree at most $d + n - 1$.*

Another connection between the two notions of integral closure is given in the following:

Proposition 6.60. *Let R be an affine domain. Then*

(a) *Let $b \in R$. If $\bar{R}$ is the integral closure of R then $\overline{(b)} = \bar{R}b \cap R$.*

(b) *If R is an affine domain and I is an ideal of R then $\overline{I^{s+1}} = I\overline{I^s}$ for $s \gg 0$.*

Proof. The first assertion is clear, while the second follows from the fact that the integral closure of the Rees algebra $R[It]$ is a finitely generated graded module over $R[It]$. □

Let us exhibit one class of normal ideals.

Proposition 6.61. *Consider the following sets of distinct indeterminates*

$$X = \{x_1,\ldots,x_\ell\}, \qquad Y = \{y_1,\ldots,y_m\}, \qquad Z = \{z_1,\ldots,z_n\},$$

$R = k[X,Y,Z]$ *be a polynomial ring over a field k, and let $I = (x_i y_j z_k \mid x_i \in X, y_j \in Y, z_k \in Z)$. Then I is a normal ideal of R.*

Proof. We will show that I^p is integrally closed for all $p \geq 1$. Let $\overline{I^p}$ be the integral closure of I^p and let $f \in \overline{I^p}$ be a monomial. We write

$$f = x_{i_1}^{a_1} \cdots x_{i_r}^{a_r} y_{j_1}^{b_1} \cdots y_{j_s}^{b_s} z_{k_1}^{c_1} \cdots z_{k_t}^{c_t}.$$

Since $f^m \in I^{mp}$ for some $m > 0$ we can write

$$f^m = x_{q_1}^{d_1} \cdots x_{q_\lambda}^{d_\lambda} M,$$

where M is a monomial whose support does not contain any of the variables in $Y \cup Z$. We obtain $m \sum_{i=1}^{r} a_i = \sum_{i=1}^{\lambda} d_i \geq mp$, which implies $\sum_{i=1}^{r} a_i \geq p$, and a similar argument shows $\sum_{i=1}^{s} b_i \geq p$ and $\sum_{i=1}^{t} c_i \geq p$. Therefore $f \in I^p$. □

Complete Versus Integrally Closed Ideals

There is a beautiful characterization of integral closure of ideals in [ZS60]. Recall that if R is an integral domain of field of quotients K, then a valuation of R is a valuation domain V, $R \subset V \subset K$. The integral closure of R is determined by these valuation domains,

$$\overline{R} = \bigcap_{R \subset V \subset K} V.$$

An ideal I is said to be *complete* if $I = R \cap \bigcap_{R \subset V \subset K} IV$. One then has (see [ZS60, p. 350]):

Theorem 6.62. *An ideal I of an integral domain is complete if and only if it is integrally closed.*

This gives the following interpretation of the integral closure of an ideal:

Theorem 6.63. *Let I be an ideal of an integral domain R. An element $x \in R$ is integral over I if and only if for every homomorphism*

$$\varphi : R \mapsto V,$$

where V is a valuation domain, one has

$$v(x) \geq v(I).$$

(Here v is the valuation function on V and $v(I)$ its minimum on $I \setminus 0$.) In particular, two ideals I, J have the same integral closure if and only if $v(I) = v(J)$ for every valuation.

Equations of Integral Dependence

Where do equations of type (6.13) come from? Part of the answer is provided by the following:

Proposition 6.64. *Let R be an integral domain and I an ideal. An element $z \in R$ is integral over I if and only if there is a faithful, finitely generated R–module M such that $z \cdot M \subset I \cdot M$.*

Proof. If z is as in (6.13), take

$$M = \sum_{i=1}^{\ell-1} R a_i z^{\ell-i}.$$

Conversely, if m_i, $i = 1, \ldots, \ell$, are the generators of M, from the set of equations

$$z m_i = \sum_j r_{ij} m_j, \; r_{ij} \in I,$$

through the usual trick in the Cayley–Hamilton theorem, we get that the determinant of the $\ell \times \ell$ matrix

$$\begin{bmatrix} z - r_{11} & \cdots & -r_{1\ell} \\ \vdots & \ddots & \vdots \\ -r_{\ell 1} & \cdots & z - r_{\ell\ell} \end{bmatrix}$$

annihilates M. This gives the required equation. □

Test Modules

One way to rephrase the notion of integrally closed ideal is:

Proposition 6.65. *Let R be an integral domain. An ideal $I \subset R$ is integrally closed if and only if for each finitely generated faithful R–module M, M/IM is faithful as an R/I–module.*

Observe that if M is a module for which this fails, that is

$$IM :_R M = L \neq I,$$

then we have found new elements of $\bar{I}$, since L is integral over I. This provides the means to construct a bit more of $\bar{I}$. The issue is to find, for a given ideal I, natural modules to check for this property. They will be referred to as *test modules*. The most natural such modules are: I itself or its powers, or modules of syzygies such as those in a resolution or in Koszul complexes defined over I. We will exploit modules related to the radical of I.

Remark 6.66. One way to cast the equation of integrality of an element z with respect to the ideal I is: Setting $L = (I, z)$ one has $L^n = IL^{n-1}$. This means that a test of integrality consists in the comparison of the powers L^n to IL^{n-1}, for all $n \geq 1$. Without estimates for how far to go, this approach is not very useful. It can be phrased instead as a Gröbner basis problem as follows. Let $I = (a_1, \ldots, a_r)$ and denote by $A \subset B$ the Rees algebras of I and L respectively. Write the presentations

$$A = R[T_1, \ldots, T_r]/P, \; T_i \mapsto a_i t,$$
$$B = R[T_1, \ldots, T_r, U]/Q, \; U \mapsto zt.$$

The element z is integral over I if and only if there is an element

$$U^n + h_1(T_1, \ldots, T_r)U^{n-1} + \cdots + h_n(T_1, \ldots, T_r) \in Q,$$

where h_i is a form in the variables T_j of degree i. If a Gröbner Basis computation is possible over R then for the appropriate choice of term order such question is decidable.

Recognition

There is a hidden element in the construction of the integral closure of an ideal that highlights the difficulties we face: How does one tell whether $I = \bar{I}$? In other words, what are the general properties of integrally closed ideals? The following notion is very amenable for computer checking, particularly in the case of homogeneous ideals.

Definition 6.67. *Let $(R, \mathfrak{m})$ be a Noetherian local ring, and let I be an R–ideal.*

(a) *If $R/\mathfrak{m}$ is infinite, then I is called $\mathfrak{m}$–full if there exists an element $y \in R$ with $\mathfrak{m}I : y = I$.*

(b) *If $R/\mathfrak{m}$ is finite, then I is called $\mathfrak{m}$–full if there exists a Noetherian local ring $(R', \mathfrak{m}')$ with R' faithfully flat over R, $\mathfrak{m}' = \mathfrak{m}R'$, and $R'/\mathfrak{m}'$ infinite, such that IR' is $\mathfrak{m}'$–full.*

The element y here is a generic element of $\mathfrak{m}$. Every integrally closed ideal $I \neq \sqrt{0}$ is automatically $\mathfrak{m}$–full ([Got87, 2.4]), and every $\mathfrak{m}$–full ideal I has the *Rees property*, which means that $\nu(I) \geq \nu(J)$ for every R–ideal J with $I \subset J$ and $\ell(J/I) < \infty$ (see [Got87, 2.2], [Wat87, Theorem 3]).

This notion permits an advance towards the integral closure of an ideal I. Let J be any $\mathfrak{m}$–full ideal containing I, contained in $\bar{I}$. For a generic element $y \in \mathfrak{m}$,

$$I_1 = \mathfrak{m}I : y \subset \mathfrak{m}J : y = J.$$

Thus I_1 consists of elements which are integral over I, and we can apply to it the same process. This leads to the smallest $\mathfrak{m}$–full ideal containing I, which might be called the $\mathfrak{m}$–*full closure* of I.

A word of caution: After picking the element y, to ascertain that it is indeed generic for the purpose here, the ideal $\mathfrak{m}I{:}y$ must be certified to be contained in $\bar{I}$. Of course, with the same care, in the global case, we can more generally consider this notion with respect to the radical of I. In this case, it is helps when I is an unmixed ideal.

Generic Complete Intersection

Most of the criteria for an ideal to be integrally closed require detailed knowledge of its structure. The following has very general features and is effective (see [CHV98]).

Theorem 6.68. *Let I be a height unmixed ideal in a Cohen–Macaulay ring R that is generically a complete intersection. Set $L = I : \sqrt{I}$. Then I is integrally closed if and only if*

$$\sqrt{I} = I \cdot L{:}L^2. \qquad (6.14)$$

Observe that if R is a ring of polynomials over a large field, then all these expressions, beginning with $\sqrt{I} = I : J$, where J is the Jacobian ideal of I, are ready made for direct computation.

If I is not integrally closed, elements that go into this test can be used to produce some elements in its integral closure. This occurs as follow.

Corollary 6.69. *Let $\mathfrak{p}_1,\dots,\mathfrak{p}_n$ be the minimal prime ideals of I listed in such a way that $I_{\mathfrak{p}}$ is integrally closed for $\mathfrak{p} = \mathfrak{p}_i$ for $i \le s$, but not at the other primes. Set*

$$A = \sqrt{I} = \mathfrak{p}_1 \cap \cdots \cap \mathfrak{p}_n, \qquad B = \mathfrak{p}_1 \cap \cdots \cap \mathfrak{p}_s, \qquad C = \mathfrak{p}_{s+1} \cap \cdots \cap \mathfrak{p}_n.$$

Let $L = I{:}A$. Then

$$B = IL{:}L^2, \text{ and } C = A{:}B.$$

If I is not integrally closed, that is if $B \ne A$, then

$$H = I{:}C \ne I, \text{ and } H^2 = IH.$$

The assertion is that if I is not integrally closed, then $I \subsetneq H \subset \bar{I}$. Actually, if $I \subset A^{(s)}$ but $I \not\subset A^{(s+1)}$, taking the ideal $L_s = I{:}B^s$ will provide an ideal larger than L with the property that $L_s^2 = IL_s$ (see [CHV98] for more details).

A more general version of the preceding is (see [CHV98]):

Theorem 6.70. *Let R be a regular local and let I be a Gorenstein ideal. Then I is an integrally closed ideal if and only if*

$$\sqrt{I} = I \cdot L : L^2, \quad \text{where } L = I : \sqrt{I}.$$

The following is similar to Theorem 5.24. For simplicity we only state it in characteristic zero (see [EHV92]).

Theorem 6.71. *Let R be a ring of polynomials over a field of characteristic zero and let I be a height unmixed ideal. Denote by J the Jacobian ideal of I. If I is integrally closed then*

$$J \cdot I : J = I.$$

The converse holds if I is generically a complete intersection.

Proof. We only consider the converse. To show that I is integrally closed it suffices to show that its primary components are integrally closed. Localizing at minimal primes of I reduces the question to the previous proposition. But for any such prime $\wp$, $J_\wp$ is the generic socle of $I_\wp$.   □

Remark 6.72. If the test fails, that is if

$$I \cdot L : L = I' \neq I,$$

we could replace I by I' if the latter is height unmixed. Indeed it has the same radical as I so that its generic socle would be

$$L' = I' : \sqrt{I},$$

and we would test for

$$I' \cdot L' : L' = I',$$

and so on.

There may be difficulties to this process. For instance, it was an old question of Krull whether the integral closure of primary ideals are still primary. C. Huneke ([Hu87]) gave a counterexample in characteristic two but the characteristic zero case is still open.

In some very special cases however this is a reasonable procedure. Consider the monomial ideal

$$I = (x^3, y^4, z^3, xy^2z).$$

After 3 iterations we get the ideal

$$I' = (x^3, z^3, x^2z^2, xyz^2, y^2z^2, xy^3, x^2yz, xy^2z, x^2y^2, y^3z, y^4).$$

We leave to the reader to decide whether I' is the integral closure of I.

We do not know of any better choice of a test module than generic socle formulas.

The Rush–Ratliff Closure of an Ideal

One difficulty in getting $\bar{I}$ from I is to find a natural setting for using Proposition 6.64. The following is often helpful ([RR78]).

Definition 6.73. *Given an ideal I of a Noetherian ring R, the* Ratliff-Rush *closure of I is the ideal*

$$\tilde{I} = \bigcup_n I^{n+1} : I^n.$$

It follows that $\tilde{I} \subset \bar{I}$. There is a direct interpretation of this ideal. Let $A = R[It]$ be the Rees algebra of an ideal and let $X = \mathrm{Proj}(A)$. The module

$$C = \bigcup \mathrm{Hom}_A(I^n A, I^{n+1} A),$$

is graded and its component of degree 1 is $\tilde{I}$.

We also note that $\tilde{I} = C_1 = H^0(X, \mathcal{O}_X(-1))$ (see [Har77, Theorem 5.2]). For all large n, one has $\widetilde{I^n} = I^n$.

For a naive example, suppose $R = k[x,y]$ and $I = (x^2, y^2)$; then $I = \tilde{I} \subset \bar{I} = (x^2, xy, y^2)$.

Computing the Integral Closure the Hard Way

Let I be an ideal of a ring of polynomials R, and let $A = R[It]$ be the Rees algebra of I. There are cases in which the integral closure of A,

$$B = R + \bar{I}t + \cdots + \overline{I^n}t^n + \cdots,$$

is generated in degree 1, that is $\overline{I^n} = (\bar{I})^n$ for all n. When this happens it may still be worthwhile to obtain $\bar{I}$ by computing B.

Example 6.74. Let k be a field of characteristic 0 and let $J \subseteq R = k[x_1, x_2]$ be the codimension 2 complete intersection

$$I = (x_1^3 + x_2^6, x_1 x_2^3 - x_2^5).$$

Using Algorithm 6.38 we obtain:

$$I_1 = (x_1^3 + x_2^6, x_1 x_2^3 - x_2^5, x_2^8);$$

after a second iteration one gets

$$I_2 = (x_1^3 + x_2^6, x_1 x_2^3 - x_2^5, x_1^2 x_2^2 - x_2^6, x_2^7),$$

and finally, at the end of the last pass, one has that the integral closure of I is given by the ideal

$$I_3 = \bar{J} = (x_1 x_2^3 - x_2^5, x_2^6, x_1^3, x_1^2 x_2^2).$$

Exercise 6.75. Let $I \subset k[x_1, \ldots, x_m]$, and $J \subset k[y_1, \ldots, y_n]$ be ideals in distinct rings of polynomials. Their *join* is the ideal

$$I \star J = (I, J, x_i y_j, \ i = 1, \ldots, m, \ j = 1, \ldots, n) \subset k[x_1, \ldots, x_m, y_1, \ldots, y_n].$$

Prove that if I and J are integrally monomial ideals generated by squarefree monomials of the same degree then their join is also integrally closed.

Problem 6.76. Let P be a prime ideal of a ring of polynomials. Develop a method to test whether P^2 (and the higher powers of P as well) is integrally closed. Consider the case of codimension 2 Cohen–Macaulay ideals.

6.7 Integral Closure of a Morphism

Let A and B be affine integral domains defined over the field k. Suppose that

$$\psi : A \mapsto B$$

is a homomorphism of k–algebras. We seek to outline how to describe the set of elements of B which are integral over $\psi(A)$. It is the affine version of the question of how to carry out the Stein factorization of a morphism ([BV93]).

For clear reasons, we may assume that ψ identifies A with a subring of B. There are several variants among which we single out the following:

- Birational morphisms
- General morphisms
- Algebraic extensions

Birational Morphisms

We are first going to isolate the elements which will play the key roles in the constructions. Let $A \subset B$ be affine integral domains. Consider the diagram of inclusions:

$$
\begin{array}{ccc}
A \longrightarrow A' \longrightarrow B \\
\downarrow \qquad\qquad \downarrow \\
\overline{A} \longrightarrow \overline{B}
\end{array}
$$

where $\overline{A}$ and $\overline{B}$ are the integral closures of A and B respectively in their (common) quotient field and $A' = \overline{A} \cap B$. A' is the desired integral closure of A in B.

Since A and B have a common quotient field, and B is a k–algebra of finite type (and hence an A–algebra of finite type) we have an expression of B in the form:

$$k\left[f_1, \ldots, f_n, \frac{a_{n+1}}{b}, \ldots, \frac{a_m}{b}\right], \tag{6.15}$$

where A is the subalgebra generated by the set $\{f_1, \ldots, f_n\}$ and b is an element of A.

We actually take B as being given by

$$k[x_1,\ldots,x_m]/\mathfrak{P},$$

with A the k–subalgebra generated by the set $\{x_1,\ldots,x_n\}$. This new representation can be obtained in the following manner. For each x_i, $i = n+1,\ldots,m$, consider the k–subalgebra generated by the set $\{x_1,\ldots,x_n,x_i\}$. Choose an elimination term ordering on the indeterminates $\{x_1,\ldots,x_m\}$ such that $x_1,\ldots,x_n$ are the lowest indeterminates and x_i is the smallest of the remaining indeterminates (see [ShS86] and Proposition 7.26 for details). There will exist an element of the form $\beta_i x_i - \alpha_i$ in the Gröbner basis found by this process. Thus $x_i = \dfrac{\alpha_i}{\beta_i}$, and if one sets

$$a_i = \alpha_i \prod_{j \neq i} \beta_j,$$
$$b = \prod_{j=n+1}^{m} \beta_j,$$

one will then have the representation (6.15).

We shall now organize the two approaches to the determination of A'. They shall use some of the same elements. The first method, less general, takes place entirely in A, in a manner of speaking, while the other involves B.

From $A' = \overline{A} \cap B$, every element z of A' has an representation of the form:

$$z = \frac{w}{b^r}, \text{ where } w \text{ is in } \overline{(b^r)} \cap (a_{n+1},\ldots,a_m,b)^r.$$

This representation need not however be a minimal one in terms of the degree of the denominator.

For $i \in \mathbb{N}_0$, let us define S_i to be the A-submodule of B given by:

$$S_i = b^{-i}[\overline{(b^i)} \cap (a_{n+1},\ldots,a_m,b)^i]. \tag{6.16}$$

These remarks yield the following expression of A' as an ascending union of submodules.

Proposition 6.77. *The integral closure A' of A in B is given by:*

$$A' = \bigcup_{i \in \mathbb{N}_0} S_i. \tag{6.17}$$

This gives us the necessary vehicle for the computation of A'. Recall that an element b of an ideal I is *superficial* (of order 1) with *defect* c (see [Nag62]) if

$$(I^i:b) \cap I^c = I^{i-1} \; \forall \, i > c. \tag{6.18}$$

It is a consequence of the Artin–Rees lemma that if b is a regular element by enlarging c we may remove I^c from this equation (see [Mat80, Proposition 11.E]). We shall refer to such c as the *strong defect* of b.

Theorem 6.78. *If b is a superficial element of the ideal $(a_{n+1}, \ldots, a_m, b)$ with defect c then:*

(a) *There exists an $r \in \mathbb{N}_0$ such that $S_r = S_{r+k}$ for all $k \in \mathbb{N}_0$.*

(b) *If the strong defect is c and $r = \max\{c, \min\{i \mid \overline{(b^{i+1})} = b\overline{(b^i)}\}\}$ then $S_r = S_{r+k}$ for all $k \in \mathbb{N}_0$.*

Proof. (a) follows since A' is a Noetherian A–module, while (b) is a direct verification. $\qquad\qquad\square$

This means that if b is a superficial element of the ideal $(a_{n+1}, \ldots, a_m, b)$, the integral closure A' of A in B can be found by computing the submodule S_r, where r is as described in the theorem. The value of r can be found by the successive computation of the ideals

$$b^i \overline{A} \cap A = \overline{(b^i)},$$

and testing at each stage the condition

$$\overline{(b^i)} = b\overline{(b^{i-1})},$$

or equivalently

$$\overline{(b^i)} \subset (b).$$

We note that there are limits for this value which are independent of b. This is a consequence of the uniform Briançon–Skoda theorem (see [Hu92, Theorem 4.13]).

The defect of the superficial element b can be determined by computing the bound on the upper degree of an element in the annihilator of the image of b in the associated graded ring of A with respect to the $(a_{n+1}, \ldots, a_m, b)$-adic filtration. Hence if an oracle indicates to us that the element b is a superficial element of the ideal $(a_{n+1}, \ldots, a_m, b)$, the computation of the A–submodule S_r of B yields the integral closure of A in B. In the event that b is not superficial we must find an alternative path to the solution.

Theorem 6.79. *Let $A \subset B$ be finitely generated affine domains over a computable field k. Then the integral closure of A in B can be determined as follows. Let $\overline{A}$ be the integral closure of A in its field of fractions, and suppose B is generated over A by fractions having powers of b as denominators. Let r be an integer such that $\overline{A}b^{r+1} \cap A \subset (b)$. Then the integral closure of A in B is*

$$\boxed{A' = b^{-r}[b^r B \cap b^r \overline{A} \cap A].}$$

Now consider the elements of B which have representations of the form

$$z = \frac{w}{b^r}$$

then w is an element of the ideal $b^r \overline{A} \cap A$ of A. By representing the k–algebra B as:

$$k[x_1, \ldots, x_m]/P$$

with A the subalgebra generated by the set $\{x_1, \ldots, x_n\}$ of indeterminates, a generating set for the ideal $b^r B \cap A$ is obtained through a Gröbner basis computation on the earlier elimination order of the ideal generated by the elements of the ideal P and b^r.

To represent the algebra A', in addition to $x_1, \ldots, x_n$, we choose new variables $h_1, \ldots, h_s$ corresponding to the new elements in the intersection formula of the previous theorem, thus yielding:

$$B = k[x_1, \ldots, x_m, h_1, \ldots, h_s]/Q, \ Q = (P, h_i - g_i(x_1, \ldots, x_m), \ i = 1, \ldots, s)$$
$$A' = k[x_1, \ldots, x_n, h_1, \ldots, h_s]/Q \cap k[x_1, \ldots, x_n, h_1, \ldots, h_s].$$

General Morphisms

The computation of the integral closure in general relies on the reduction of the problem to the case where the morphism is birational. Again there are several approaches.

Suppose $A \hookrightarrow B$ is an injective morphism of affine domains and write $B = A[f_1, \ldots, f_r]$. Let

$$D = A[t, f_1 t, \ldots, f_r t] \subset B[t]$$

If $B[t]$ is graded so that $\deg t = 1$ and the elements of B have degree zero, then D is a graded subalgebra of $B[t]$. Note that $B[t]$ is a birational extension of D.

Consider the k–algebras A and B and the morphism $\varphi \colon A \to B$ to be represented as before. Then the k–algebra $B[t]$ has a representation

$$k[x_1, \ldots, x_n, t, u_{n+1}, \ldots, u_m, x_{n+1}, \ldots, x_m]/\mathfrak{P}.$$

The ideal $\mathfrak{P}$ in this representation is the ideal generated by the ideal P together with the polynomials $u_j - t x_j$ for $j = n+1, \ldots, m$. As for the subalgebra D, it is represented by the set $\{x_1, \ldots, x_n, t, u_{n+1}, \ldots, u_m\}$; D is birationally equivalent to $B[t]$.

Since the morphism $\psi \colon D \to B[t]$ is birational, the integral closure D' of D in $B[t]$ can be computed as in the previous section, with representation for $B[t]$ in the form

$$k[x_1, \ldots, x_m, t, u_{n+1}, \ldots, u_m, h_1, \ldots, h_r]/\mathfrak{Q},$$

where D' is the k–subalgebra generated by the set

$$\{x_1, \ldots, x_n, t, u_{n+1}, \ldots, u_m, h_1, \ldots, h_r\}.$$

Moreover as D is a graded k–subalgebra of $B[t]$ so is D'.

The k–algebra B therefore has a representation

$$k[x_1, \ldots, x_m, h_1, \ldots, h_r, t, u_{n+1}, \ldots, u_m]/\mathfrak{P},$$

where $\mathfrak{P}$ is generated by $\mathfrak{Q}$ and the element t. The k–subalgebra of B generated by the set

$$\{x_1, \ldots, x_n, h_1, \ldots, h_r, t, u_{n+1}, \ldots, u_m\}$$

is $D' \cap B$.

Theorem 6.80. *The k–algebra $D' \cap B$ is the integral closure A' of A in B.*

Proof. If x lies in A', then x lies in B and hence x is in $B[t]$ and is integral over A, and therefore x is integral over D so x lies in $D' \cap B$. Conversely, if x is in $D' \cap B$, then x is integral over D, and of degree 0. But x satisfies a monic homogeneous equation, which must be of degree 0. Hence x is integral over A, and so x lies in A'.   □

Algebraic Extensions

Let B be an algebraic extension of A. This is a very special case of the preceding The device that found the linear forms $\alpha_i x_i - \beta_i$ would now return a polynomial

$$c_s x_i^s + c_{s-1} x_i^{s-1} + \cdots + c_0 = 0,$$

of least degree s. The element $c_s x_i$ would be added to A to obtain A^*, an integral extension of A, with B a rational extension of A^*.

Problem 6.81. Here are some instances where it would be desirable to have more focused methods to find the integral closure:

- $A = k[f_1, \ldots, f_m] \subset k[x, y]$.
- $A = k[f, g, h] \subset k[x, y, z]$.
- A is a subring generated by monomials of $k[x_1, \ldots, x_n]$.

In the next chapter, when we discuss monomial algebras, their integral closures are also examined.

7

Ideal Transforms and Rings of Invariants

A difficult problem in the processing of rings is that of the specification of a morphism

$$\psi : A \mapsto B,$$

where A is a given affine ring but B is only given by generators or as the outcome of a process, or as a combination of both. In the first case, the study of B may require the knowledge of its relations, while in the second case the construction of the generators is the primary issue but with the availability of some of its generators and relations acting as the means to verify termination of the process. In the previous chapter we considered these issues for the case of the integral closure of certain rings. In this chapter we deal with other morphisms, such as those that occur in the theory of invariants in the guise of ideal transforms and in the theory of blowup algebras.

To give a glimpse of which kinds of constructions are involved, we consider one which played a key role, in the work of Rees and Nagata, to find counterexamples to Hilbert's 14th Problem and its generalization by Zariski. Let R be a Noetherian integral domain with field of quotients Q, and let I be an ideal. The ideal transform of I is the ring $T(I)$ of global sections of the structure sheaf of R on the open set defined by I. Some of its significance lies in the fact that rings of invariants of infinite groups can be expressed in this fashion. The ring $T(I)$ can be expressed by

$$\boxed{B = T(I) = \bigcap R_{\mathfrak{p}}, \ \mathfrak{p} \notin V(I) \subset \operatorname{Spec}(R).} \tag{7.1}$$

This computation is inherently non–terminating ([Nag64], [Ree58]), which leads to various *ad hoc* methods to deal with special cases. The basic approach is to consider the subalgebra $B_n = R[I^{-n}]$, and see whether it already equals B. It is time-consuming for n large. On the other hand, it cannot be disregarded, as it offers one of the few known paths to $T(I)$.

In this chapter, we discuss some general properties of this construction and of allied algebras: symmetric algebras, Rees algebras and symbolic power algebras.

The other main topic is the study of special subrings of rings of polynomials. Their understanding is key to the efficient processing of rings of invariants.

7.1 Divisorial Properties of Ideal Transforms

For simplicity we denote by $T(I)$ the ideal transform of I. Let us first deal with elementary properties of this construction, the proofs of which will be left to the reader.

Proposition 7.1. *Let R be a Noetherian integral domain and let I be a nonzero ideal. Then*

(a) $T(I) = T(\sqrt{I})$.
(b) *If I contains a regular sequence of two elements then $T(I) = R$.*
(c) *If $I = I_1 \cap I_2$, and I_2 contains a regular sequence of two elements, then $T(I) = T(I_1)$.*

Finiteness

Let us remark on one elementary finiteness criterion. A famous theorem of Zariski guarantees that, if R is an integrally closed affine domain of Krull dimension two then the condition is always realized.

Proposition 7.2. *Let R be an integral domain and let I be a finitely generated ideal. If $T(I) = I \cdot T(I)$, then $T(I)$ is a finitely generated R–algebra.*

Proof. From the equality $T(I) = I \cdot T(I)$, we get an equation

$$1 = q_1 a_1 + \cdots + q_r a_r, \ a_i \in I, \ q_i \in T(I).$$

We claim $T(I) = R[q_1, \ldots, q_r]$. For $q \in T(I)$, say $q \in I^{-n}$, raise the equation to the nth power and multiply both sides by q, to establish the claim. $\square$

Thus, if B_n is represented by $B_n = R[y_1, \ldots, y_m]/\mathfrak{p}$, we have $T(I) = B_n$ if and only if $1 \in (I, \mathfrak{p})$.

Condition (S_2) of Serre

We now consider a setting for computing ideal transforms more efficiently. For that we recall the conditions (S_r) of Serre.

Definition 7.3. *Let A be a Noetherian ring and let E be a finitely generated A–module, and let r be a non–negative integer. E satisfies the condition (S_r) if for every prime ideal $\mathfrak{p}$ of A*

$$\mathfrak{p}\text{–depth } E \geq \inf\{r, \text{height } \mathfrak{p}\}.$$

Thus the ring A satisfies (S_1) if it has no embedded primes. Further, A has (S_2) if it has no embedded prime and height$(\mathfrak{p}) = 1$ for all $\mathfrak{p} \in \text{Ass}(A/aA)$ for any regular element $a \in A$.

Detecting (S_r)

Let A be an affine algebra over a field k. We discuss two methods to ascertain whether A satisfies the condition (S_r) of Serre. When $A = k[x_1, \dots, x_n]/J$, it can be expressed in terms of projective dimensions.

Proposition 7.4. *Let*

$$R = k[x_1, \dots, x_d] \subset A$$

be a Noether normalization of A, and denote by

$$0 \to F_s \xrightarrow{\varphi_s} F_{s-1} \longrightarrow \cdots \longrightarrow F_1 \xrightarrow{\varphi_1} F_0 \longrightarrow A \to 0$$

a projective resolution of A as an R–module. Denote by $I_j(A)$ the ideal $I_{r_j}(\varphi_j)$, $r_j =$ rank φ_j. Then A satisfies (S_r) if and only if

$$\text{height } I_j(A) \geq j + r, \; j \geq 0.$$

Proof. It is an immediate consequence of the Auslander–Buchsbaum formula. $\qquad\square$

The other method does not require a Noether normalization and uses the ability of programs such as *Macaulay* to compute Ext's and annihilators of modules. (We use a standard convention: height$(R) = \infty$.)

Proposition 7.5. *Let B be the polynomial ring $k[x_1, \dots, x_n]$ and suppose $A = B/J$ is an equi–dimensional algebra of codimension g. Then A satisfies S_r if and only if the following two conditions hold:*

$$\text{height}(\text{ann}(\text{Ext}_B^i(A, B))) \geq r + i, \; i > g.$$

Remark 7.6. For an example on how this is used, suppose $I = (a_1, \dots, a_m)$ is an ideal of $R = k[x, y]$ and let $A = R[It] = R[T_1, \dots, T_m]/J = B/J$ be its Rees algebra. Suppose that I is (x, y)–primary. Set $n = m + 2$. Then we have:

(a) $R[It]$ has (S_2) if and only if

$$\text{Ext}_B^{n-1}(A, B) = 0$$
$$\text{height ann}(\text{Ext}_B^{n-2}(A, B)) \geq n.$$

(b) By cutting down the number of variables, the following trick is useful: Let $J_0 = (J, x)$ and put $A_0 = B/J_0$. Then $R[It]$ has (S_2) if and only if

$$\text{Ext}_B^n(A_0, B) = 0$$
$$\text{height ann}(\text{Ext}_B^{n-1}(A_0, B)) \geq n.$$

Let us comment on the proof of (b). Since the localization $R[It]_x$ is a polynomial ring, the Ext modules above are annihilated by a power of x. Consider the canonical exact sequence induced by multiplication by x:

$$0 \to B/J \xrightarrow{x} B/J \to B/J_0 \to 0,$$

and the long exact sequence of cohomology

$$\operatorname{Ext}_B^{n-2}(A,B) \xrightarrow{x} \operatorname{Ext}_B^{n-2}(A,B) \to \operatorname{Ext}_B^{n-1}(A_0,B) \to$$
$$\operatorname{Ext}_B^{n-1}(A,B) \xrightarrow{x} \operatorname{Ext}_B^{n-1}(A,B) \to \operatorname{Ext}_B^n(A_0,B) \to 0.$$

It follows that if $\operatorname{Ext}_B^n(A_0,B) = 0$, then $\operatorname{Ext}_B^{n-1}(A,B) = 0$ since multiplication by x is a nilpotent endomorphism. More interesting is the next step, that $\operatorname{Ext}_B^{n-1}(A_0,B)$ and $\operatorname{Ext}_B^{n-2}(A,B)$ have the same codimension given the vanishing of their higher Ext's. We leave this for the reader.

Reduction to Two–Generated Ideals

In the computation of ideal transforms, an useful step is the reduction to the case of ideals generated by two elements, a possibility that is present in many instances. A similar reduction is exploited in graded algebras arising from blowups.

Proposition 7.7. *Let R be a Noetherian integral domain with property (S_2) and let I be a nonzero ideal. There exist $f,g \in I$ such that (f,g) and I have the same ideal transforms.*

Proof. If I has codimension at least 2, the ideal transform of I is R by Proposition 7.1. Let then $I = I_1 \cap I_2$ be a decomposition of I in which I_1 has codimension 1 and I_2 has codimension at least 2. I_1 can be obtained from I as $I_1 = (I^{-1})^{-1}$. It follows by Proposition 7.1 that the ideal transforms of I and I_1 are the same.

Let us assume that I is unmixed, of codimension 1, and consider its primary decomposition

$$I = Q_1 \cap \cdots \cap Q_r,$$

where the Q_i is primary, of codimension 1. Let $0 \neq f \in I$ and consider the primary decomposition

$$(f) = Q_1' \cap \cdots \cap Q_r' \cap Q_{r+1} \cap \cdots \cap Q_s,$$

where

$$\sqrt{Q_i} = \sqrt{Q_i'}, \ 1 \le i \le r.$$

We can isolate

$$L = Q_{r+1} \cap \cdots \cap Q_s$$

by saturating (f) with respect to I:

$$L = (f):I^\infty.$$

Pass now to the ring $S = R/\sqrt{L}$, and use Proposition 2.3.2 to obtain an element g of I that is regular on S. Set $J = (f,g)$. We claim that I and J have the same ideal transforms. It will be enough, by Proposition 7.1, to show that these two ideals have the same associated prime ideals of codimension 1, a matter that is clear from the choices made. □

7.2 Equations of Blowup Algebras

We introduce now some algebras that play ubiquitous roles in commutative algebra. Their manipulation shows the face of the theory of syzygies that concerns itself with polynomial relations among the elements of a finite set of a ring.

Symmetric Algebras

The symmetric algebra of an R–module E will occur as an ancestor of all blowup algebras discussed here. For fuller details, see [Vas94, Chapter 1].

For simplicity we assume that R is an affine domain. If the R–module E is given by a presentation

$$R^n \xrightarrow{\varphi} R^m \longrightarrow E \to 0,$$

then the *symmetric algebra* of E is given by

$$S_R(E) = S(E) \simeq R[T_1,\ldots,T_m]/(f_1,\ldots,f_n),$$

where f_i is the linear form in the T_j's defined by the ith column of φ.

Geometrically, $S(E)$ is a family of hyperplanes parametrized by $\mathrm{Spec}(R)$. Despite its straightforward definition several of its properties are difficult to predict from the knowledge of its presentation. A case in point is to ascertain when $S(E)$ is an integral domain. Here we will just describe how the Krull dimension of $S(E)$ may be found.

Suppose that $\dim R = d$ and that r is the rank of E. Then $d + r$ is a lower bound for $\dim S(E)$ and the two numbers coincide when $S(E)$ is an integral domain. In general we make use of the Fitting ideals of E. Denote by $I_t(\varphi)$ the ideal generated by the $t \times t$ minors of φ. Define the following integer valued function on the interval $[1, \mathrm{rank}(\varphi)]$:

$$d(t) = \sup\{\,\mathrm{rank}(\varphi) - t + 1 - \mathrm{height}\, I_t(\varphi), 0\,\}.$$

Theorem 7.8. *Let R be an affine domain and let E be a module with a presentation given by the matrix φ. Then*

$$\dim S(E) = \dim R + \mathrm{rank}(E) + d(E),$$

where $d(E)$ is the maximum value of $d(t)$.

Presentation of the Rees Algebra

Let R be a ring, $\{f_1,\ldots,f_m\} \subset R$ and S is subring of R. A typical problem is to determine the kernel of the morphism

$$\varphi: S[T_1,\ldots,T_m] \mapsto R, \ \varphi(T_i) = f_i.$$

It is often convenient to convert this problem into another that involve finding the ordinary syzygies on sets of 'monomials' in the elements f_i's. When a graded structure exists the task becomes more amenable.

We discuss the cases of Rees algebras and tangent cones. Later in this chapter we treat subrings of rings of polynomials.

Let $I = (f_1,\ldots,f_m)$ be an ideal of the ring R. The *Rees algebra* of I is the Rees algebra defined the I–adic filtration,

$$R[It] = R + It + I^2 t^2 + \cdots \subset R[t].$$

It is generated by the "forms" $f_i t$ which gives rise to the R–algebra homomorphism

$$R[T_1,\ldots,T_m] \xrightarrow{\varphi} R[It], \ \varphi(T_i) = f_i t.$$

The *associated graded ring* of the ideal I is the algebra

$$\mathrm{gr}_I(R) = R/I \oplus I/I^2 \oplus \cdots = R[It] \otimes (R/I).$$

The problem is to find the presentation of these algebras as quotients of rings of polynomials. It suffices to deal with $R[It]$. Let $J = \ker(\varphi)$ be the kernel of φ, which by abuse of terminology one call the presentation of $R[It]$. It can be obtained as follows.

Proposition 7.9. *In the ring* $C = R[T_1,\ldots,T_m,t]$ *consider the ideal* L *generated by the polynomials* $T_j - t f_j, \ j = 1,\ldots,m$. *Then* $R[It] = B/J$, *where* J *is the contraction of* L *in the subring* $B = R[T_1,\ldots,T_m]$.

Proof. It is clear that $J \supset L \cap B$. Conversely, if $f(T_1,\ldots,T_m)$ is an element of J, we write

$$f(T_1,\ldots,T_m) = f(t f_1 + (T_1 - t f_1),\ldots,t f_m + (T_m - t f_m))$$

and use the Taylor expansion to show $f \in L$. $\qquad\square$

The ideal $J = \ker(\varphi)$ has a natural grading (in the T_i's variables)

$$J = J_1 + J_2 + \cdots,$$

where J_1 consists of the linear forms

$$\sum a_i T_i, \quad \sum a_i f_i = 0.$$

In other words, J_1 is determined by the syzygies of the f_i's. We use this as the starting point to the whole ideal J. Let

$$R^n \xrightarrow{\psi} R^m \longrightarrow I \to 0 \tag{7.2}$$

define the first order syzygies of I. Setting $\mathbf{T} = [T_1, \ldots, T_m]$, J_1 is generated by the entries of $\mathbf{T} \cdot \psi$. In the ring of polynomials $B = R[\mathbf{T}]$, J_1 is the defining ideal of another algebra, the so-called symmetric algebra of I, $S(I) = B/(J_1)$. Since there is a natural homomorphism

$$0 \to A = J/(J_1) \longrightarrow S(I) \longrightarrow R[It] \to 0, \tag{7.3}$$

comparisons between these algebra occur frequently. In the critical case, when $J = (J_1)$ we say that I is of *linear type*. In general however there may be generators of J that do not arise from J_1. Nevertheless J_1 can be used to derive J through the following device.

Proposition 7.10. *Let I be an ideal as above, and suppose $f \in I$ is a regular element. Then*

$$J = (J_1) : f^\infty.$$

In other words,

$$J = (J_1, 1 - f \cdot t) \bigcap R[T_1, \ldots, T_m].$$

Proof. Localizing the sequence (7.3) at R_f, we get the exact sequence

$$0 \to A_f \longrightarrow S(I_f) \longrightarrow R_f[I_f t] \to 0,$$

where $R_f[I_f t] = R_t[t]$ by the choice of f. In turn, $S(I_f)$ is the symmetric algebra of the ideal $I_f = R_f$ and therefore it is another ring of polynomials $S(I_f) = R_f[u]$, with u mapping to t. This means that $A_f = 0$ and thus $J \subset (J_1) : f^\infty$.

Finally, since f is a regular element of R it is clear the colon operation does not go beyond the ideal J. $\square$

Remark 7.11. Actually, we may replace J_1 by the ideal generated by the polynomials $f_i T_j - f_j T_1$, $j = 2, \ldots, m$, and use $f = f_1$ in the intersection. An appropriate name for this could be taking the *rational closure* of J_1.

Jacobian Dual

A bit of explicit elimination theory permits finding extra generators for the ideal of presentation of the Rees algebra $R[It]$. Using the presentation (7.2), we have that J_1 is given by the n 1–forms in the variables $T_1, \ldots, T_m$,

$$J_1 = [T_1, \ldots, T_m] \cdot \psi,$$

for a given matrix representation of ψ.

If we denote by $(x_1, \ldots, x_s)$ the ideal generated by the entries of ψ, we can also write

$$J_1 = [x_1, \ldots, x_s] \cdot B(\psi), \tag{7.4}$$

where $B(\psi)$ is a $s \times m$ matrix of linear forms in the variables T_i's. This matrix is called a *Jacobian dual* of ψ. It may not be uniquely defined. In case the entries of ψ are 1–forms in a set of indeterminates over some field, then $B(\psi)$ and ψ are the constituent blocks of the Jacobian matrix attached to the polynomials generating J_1.

This construction extends to modules as well and it is particularly useful when the module E is defined over the ring of polynomials $R = k[x_1, \ldots, x_s]$,

$$R^n \xrightarrow{\varphi} R^m \longrightarrow E \to 0,$$

where φ is a matrix whose entries are linear forms in the x_i's. In this case we can write

$$J_1 = [T_1, \ldots, T_m] \cdot \varphi = [x_1, \ldots, x_s] \cdot B(\varphi),$$

where $B(\varphi)$ is a matrix of linear forms in the T_j's. We can use $B(\varphi)$ to define a module over $A = k[T_1, \ldots, T_m]$,

$$A^n \xrightarrow{B(\varphi)} A^s \longrightarrow F \to 0.$$

This gives the representations

$$S_R(E) \simeq S_A(F).$$

Another way to put it is by saying that $S_R(E)$ is defined over the ring $k[x_i's, T_j's]$ by quadratic forms whose terms only involve monomials of the kind $x_i T_j$ (see [Vas94] for more details).

Part of the justification for the introduction of these modules lies in the fact that they carry part of the burden of doing elimination in the sense of the following:

Proposition 7.12. *Suppose that R is an integral domain. Then the ideal $I_s(B(\psi))$ generated by the $s \times s$ minors of the matrix $B(\psi)$ is contained in J.*

Proof. From (7.4), by Cramer's rule we have that

$$(x_1, \ldots, x_s) \cdot I_s(B(\psi)) \subset J.$$

Since J is a prime ideal which does not contain nonzero elements of R, we must have $I_s(B(\psi)) \subset J$. $\square$

Dirty Syzygies

The previous observations lead to the following method to find the syzygies of an ideal by any program able to compute Gröbner bases (but is not set up to pack the syzygies intelligently).

Remark 7.13. Suppose $I = (a_1, \ldots, a_m) \subset k[x_1, \ldots, x_n]$. Set

$$L = (T_1 - a_1 t, \ldots, T_m - a_m t, T_i T_j, \ 1 \le i, j \le m) \subset k[x_1, \ldots, x_n, T_1, \ldots, T_m, t].$$

The linear forms in the T_i in $L \cap k[x_1, \ldots, x_n, T_1, \ldots, T_m]$ generate the syzygies of the a_i.

Tangent Cones and Analytic Spread

An important barometer for the behavior of an ideal I are the equations of algebraic dependence among the elements in a set of generators. The key matrix of such relations is the ideal of presentation of its Rees algebra. Often we need a possibly simpler set of relations such as those defined by the fiber cone of I at a prime ideal,

$$F(I_{\mathfrak{p}}) = R[It] \otimes k(\mathfrak{p}), \text{ where } k(\mathfrak{p}) \text{ is the residue field of } R_{\mathfrak{p}}.$$

The dimension of this ring is the analytic spread $\ell(I_{\mathfrak{p}})$ of I at $\mathfrak{p}$. It is obviously difficult to read in general but is more amenable to get at if $\mathfrak{p}$ is a maximal ideal (when no actual localization is needed) and even simpler when I is an ideal generated by forms in a graded ring and $\mathfrak{p}$ is the irrelevant maximal ideal, when the analytic spread is denoted simply by $\ell(I)$.

Proposition 7.14. *Let I be an ideal of the ring of polynomials R generated by the forms $f_1, \ldots, f_m$. Denote by $\mathfrak{m}$ the irrelevant maximal ideal of R and let*

$$R[It] = R[T_1, \ldots, T_m]/J$$

be the presentation of the Rees algebra of I. Then

$$\ell(I) = \begin{cases} \dim R + m - \text{height}(J, \mathfrak{m}) \\ \dim k[f_1, \ldots, f_m], & \text{if the } f_i\text{'s are forms of the same degree.} \end{cases}$$

Proof. The first formula is the definition of $\ell(I)$. If I is an ideal of $k[x_1, \ldots, x_n]$ generated by forms $f_1, \ldots, f_m$ of the same degree, with $\mathfrak{m} = (x_1, \ldots, x_n)$, we have

$$R[It] = k[f_1t, \ldots, f_mt] \oplus \mathfrak{m}R[It],$$

so that the fiber cone $F(I) \simeq k[f_1, \ldots, f_m]$. $\qquad\square$

Example 7.15. Suppose that each f_i is a monomial $\mathbf{x}^{v_i}$ (where $v_i = (a_{i1}, \ldots, a_{in})$ is a vector of exponents). In this case, it is easy to see that the dimension of the algebra $k[\mathbf{x}^{v_1}, \ldots, \mathbf{x}^{v_m}]$ is the rank of the matrix $[v_1, \ldots, v_m]$.

It is a lot more challenging to find the analytic spread of ideal generated by binomials. There may be some glimmer of hope if the ideal is toric (e.g. prime).

Factorial Closure and Symbolic Blowups

There are two classes of ideal transforms that have very natural approaches, factorial closures and symbolic blowups.

The first of these, is the following algebra. Let R be a factorial domain and let E be a finitely generated R–module. Let $S(E)$ be the symmetric algebra of E. The algebra $B(E) =$ graded bi–dual of $S(E)$, defined by: if

$$S(E) = \bigoplus_{t \geq 0} S_t(E)$$

then

$$B(E) = \bigoplus_{t \geq 0} S_t(E)^{**},$$

where (**) denotes the bi-dual of an R-module. $B(E)$ is in fact an algebra, not necessarily Noetherian, but it is always factorial. We refer to it as the factorial closure of $S(E)$ ([Vas94]). Note that $B(E)$ is an ideal transform of $S(E)$. The other algebra is the symbolic blowup

$$R_s(I) = \sum_{n \geq 0} I^{(n)} t^n.$$

It also can be interpreted as an ideal transform.

In these examples, if the ground ring R is a normal domain (or at least (S_2)), that is the degree zero component of a graded domain,

$$A = R \oplus A_1 \oplus A_2 \oplus \cdots,$$

whose components are finitely generated R-modules. The transform

$$B = \bigcup_{t \geq 1} (A:I^t)$$

was taken relative to an ideal I with generators of degree zero. Actually, one could take $I = (f,g)$, with f and g forming a regular sequence on R.

The ring B is a graded algebra

$$R \oplus B_1 \oplus B_2 \oplus \cdots,$$

with each B_i a finitely generated R-module. B_i is contained in the bi-dual of A_i as an R-module according to Proposition 6.24.

We denote by $B(r)$ the subalgebra of B generated by its homogeneous components up to degree r. B is also the ideal transform, with respect to the same ideal, of any of the $B(r)$'s ($r \geq 1$).

We need the following elementary observation:

Proposition 7.16. *Let R be a Noetherian domain and let I be an ideal. Let T be the I–ideal transform of R, and let C be a Noetherian ring $R \subset C \subset T$. If grade $IC \geq 2$ then $C = T$.*

Proof. This is clear since T is also the IC–ideal transform of C.   □

It can be turned into a termination criterion:

Proposition 7.17. $B = B(r)$ *if and only if* grade $I \cdot B(r) \geq 2$.

Algorithms

There are some general approaches on using the theory of Gröbner bases to compute ideal transforms of the blowup algebras. Those discussed here could also be used to deal with other ideal transforms. It is a rather special in that each of the rings is a graded algebra, the ideal is generated by elements of degree zero, and the ideal transform lives naturally in a polynomial ring. We make use of each of these features.

The starting point is the symmetric algebra of an R–module with a presentation

$$R^m \xrightarrow{\varphi} R^n \longrightarrow E \to 0.$$

If we denote by

$$\mathbf{T} = [T_1, \ldots, T_n]$$

a vector of fresh variables, we obtain the representation of the symmetric algebra of the module E

$$S(E) = R[\mathbf{T}]/J(E),$$

where $J(E)$ is the ideal of the polynomial ring $R[\mathbf{T}]$ generated by the m linear forms given by the entries of $\mathbf{T} \cdot \varphi$.

The ground algebra is actually $S(E)$ modulo the ideal of torsion elements. We denote it by $D(E)$. To get started, let us indicate how to obtain the algebra $D(E)$. We assume the module E is given as the cokernel of the matrix $\varphi = (a_{ij})$, an $n \times m$ matrix. Let f be a nonzero element of R chosen so that the localization of E at f is a free module. If E has rank e, f can be taken in the $(n-e)$th Fitting ideal of E; if E is a prime ideal, f may be an element of E itself. Set

$$L(E) = \bigcup_{t \geq 1} (J(E) : f^t).$$

Then $D(E) = R[T_1, \ldots, T_n]/L(E)$. Note that if E is an ideal, $D(E)$ is the corresponding Rees algebra. In particular, it provides a presentation of the module E_0 obtained by moding out the R-torsion of E.

If we are looking for the factorial closure, the double dual of E must be determined first (In the case of the symbolic blowup, $D(P) = B(1)$.). Instead of using the method to be discussed later, at this point the following alternative procedure may be applied. From the preceding one may assume that E is a torsion-free R-module. In fact, we suppose that E is given as a submodule of a free module $RT_1 \oplus \cdots \oplus RT_e$, generated by the forms

$$g_k = b_{k1}T_1 + \cdots + b_{ke}T_e, \quad k = 1, \ldots, s.$$

$D(E)$ is then the subring of the polynomial ring $R[T_1, \ldots, T_n]$ generated by the forms g_k's over R. To obtain the bi–dual of E it is useful to compute the ideal transform of (f, g), chosen earlier, with respect to the subring generated by the g_k–forms and all the products T_iT_j. This is somewhat cumbersome but effective.

We represent each $B(r)$ as a quotient $P(r)/J(r)$, where $P(r)$ is a graded polynomial ring. Thus $J(E) = J(1)$ is the ideal of 1-forms given by the presentation of the module E. We seek to construct $J(r+1)$ from $J(r)$ and to test whether the equality $B = B(r)$ already holds.

There are three parts to obtaining $J(r+1)$ from $J(r)$. Assume that R is a polynomial ring in the variables $x_1, \dots, x_d$, and that E is a module with a presentation as before.

- Let f, g be a regular sequence as in [Vas89, Corollary 1.1.2]. To determine whether new generators must be added to $B(r)$ in order to obtain $B(r+1)$, we must compute the component of degree $r+1$ of the ideal transform $T_{B(r)}(f,g)$, that is

$$\bigcup_{t \geq 1} B(r)_{r+1} : (f,g)^t.$$

This module is nothing but B_{r+1}. It is convenient to keep track of degrees and work in the full ring $A = B(r)$. Observe that for any ideal $L = (f_1, \dots, f_p)$ of A,

$$A:L = \bigcap_{1 \leq i \leq p} (Af_1 :_A f_i) f_1^{-1}.$$

When applied to $L = (f,g)^t$, we obtain:

$$A(t) = A : (f,g)^t = \left(\bigcap_{1 \leq i \leq t} (Af^t :_A g^i f^{t-i}) \right) f^{-t} = \left(\bigcap_{1 \leq i \leq t} (Af^i :_A g^i) \right) f^{-t}.$$

Phrased in terms of the ring $P(r)$ this would be:

$$\left(\bigcap_{1 \leq i \leq t} (L_{r+1}, f^i) :_{P(r)} g^i \right) f^{-t},$$

where L_{r+1} denotes the subideal of $J(r)$ generated by the elements of degree at most $r+1$. (Note: One can also use the full $J(r)$ instead of L_{r+1}, a move that is useful when $B(r)$ turns out to be $B(r+1)$.) Because each $B(r)_j$, $j \leq r$, is a reflexive module, $A(t)$ and $A(t+1)$ can only begin to differ in degree $r+1$. Stability is described by the equality:

$$f\left(\bigcap_{1 \leq i \leq t} Af^i :_A g^i \right)_{r+1} = \left(\bigcap_{1 \leq i \leq t+1} Af^i :_A g^i \right)_{r+1}.$$

Suppose $u_i = e_i f^{-t}, i = 1, \dots, s$, generate $A(t)_{r+1}/A_{r+1}$; we now get to the determination of the equations for the u_i's.

- The brute force application of the Gröbner basis algorithm permits mapping the polynomial ring $P(r+1)$ onto the new generators to get the ideal $J(r+1)$. It does not work well at all. It is preferable making use of the fact that B is a rational extension of $B(r)$, by seeking the $B(r)$-conductors of each of the u_i. It provides for a number of equations of the form

$$A_{ji}U_i - B_{ji}, \ j = 1, \ldots, r_i.$$

This is useful when $B(r)$ is normal and $B(r+1)$ is singly generated over it; the set above would be all that is needed. In general one needs one extra move.
As a matter of fact, it suffices to consider the R-conductors of the elements u_i whenever the next step is going to be used.

- Denote $P(r+1) = P(r)[U_1, \ldots, U_s]$ and let $L(r+1) = J(r)$ together with all the linear equations above. If $L(r+1)$ is a prime ideal, then it obviously equals $J(r+1)$. Let h be an element such that the localization $S(E_h) = B_h$, e.g. pick $h = f$ above. It is easy to see that

$$J(r+1) = \bigcup_{t \geq 1} (L(r+1) :_{P(r+1)} h^t).$$

Remark 7.18. In view of the complexity of Gröbner basis algorithms, considerable experimentation must be exercised. For instance, (i) the choice of f, g may be changed from one approximation to the next, and (ii) the monomial order has to be played with to permit the computation to go through. It always worked better when f and g were actual variables.

Conjecture 7.19. Let E be a finitely generated module over a regular local ring R. If the symmetric algebra $S_R(E)$ is factorial then the projective dimension of E is at most 1.

Problem 7.20. Let k be a field and let $\psi : A \mapsto C$ and $\phi : B \mapsto C$ be homomorphisms of affine k–algebras. The pullback of this pair of homomorphisms is the ring

$$D = \{(a,b) \in A \times B \mid \psi(a) = \phi(b)\}.$$

Give reasonably general conditions for D to be an affine algebra over k and outline a method to determine it.

Exercise 7.21. Let E be the module over $R = k[x,y,z]$ defined by the matrix

$$\varphi = \begin{bmatrix} x & y & z & 0 \\ x^2 & zy & 0 & z^2 \\ y^2 - z^2 & 0 & xy & x^2 \end{bmatrix}.$$

Find the Krull dimension of the symmetric algebra $S_R(E)$.

Exercise 7.22. Find the factorial closure of the symmetric algebra of the module over $R = k[x,y,z]$ given by $E = R^3/R(yz, xz, xy)$.

Exercise 7.23. Let R be an integral domain and let E be a finitely generated R–module. Prove that $S_R(E)$ is an integral domain if and only if it has a unique associated prime ideal.

Exercise 7.24. Let $(R, \mathfrak{m})$ be a localization at a prime ideal of an affine domain over a field, and let I be a nonzero ideal of R. Set $G = \mathrm{gr}_I(R)$, the associated graded ring of I, and denote by $H = H^0_\mathfrak{m}(G)$ the ideal of G of all elements which are annihilated by some power of $\mathfrak{m}$. Show that

$$\ell(I) < \dim R \Leftrightarrow H \text{ is nilpotent}.$$

The hypothesis on R is intended to assure that each minimal prime of G has dimension equal to $\dim R$: $G = R[It, t^{-1}]/(t^{-1})$.

Exercise 7.25 (Huneke). Let R be a regular local ring and let $\mathfrak{p}$ be a prime ideal of R. Denote by G the associated graded ring of $\mathfrak{p}$, $G = \mathrm{gr}_\mathfrak{p}(R)$. Prove that the symbolic powers and the ordinary powers of $\mathfrak{p}$ coincide if and only if the nil radical of G is a prime ideal.

7.3 Subrings

The processing of subrings is an important aspect of computational algebra. Very often, for a given subring R_0 of a ring of polynomials $R = k[x_1,\ldots,x_n]$ is specified by generators $f_1,\ldots,f_m \in R$, one changes the setting to that of a presentation

$$k[T_1,\ldots,T_m]/I \simeq k[f_1,\ldots,f_m] \hookrightarrow R.$$

In general this reduction is only advisable when m is small and the ideal I has a simple structure. Many problems require that the treatment of R_0 be addressed directly.

We begin by discussing how the reduction to a presentation can be used, and later consider the issue of direct processing.

Using a Presentation

Typical of the techniques were those introduced in [ShS86] (see also [Stu93]), of which we single out those of algebraic dependence and of subring membership.

Proposition 7.26. *Let* $f_1,\ldots,f_m,g$ *be elements of* $k[x_1,\ldots,x_n]$.

(a) *Let* $y_1,\ldots,y_m$ *be a new set of variables and let* G *be the Gröbner basis of* $\{y_1 - f_1,\ldots,y_m - f_m\}$ *with respect to the elimination order given by* $x_1 > \cdots > x_n > y_1 > \cdots > y_m$. *Then the* f_i's *are algebraically independent if and only if* $G \cap k[y_1,\ldots,y_m] = \emptyset$.

(b) *In this setting,* $g \in k[f_1,\ldots,f_m]$ *if and only if the normal form* H *of* g *with respect to* G *is contained in* $k[y_1,\ldots,y_m]$, *in which case* $g = H(f_1,\ldots,f_m)$.

Proof. (a) First we note that for any monomial $y_i \mathbf{y}^\alpha$, one has

$$y_i \mathbf{y}^\alpha - g_i \mathbf{g}^\alpha = y_i(\mathbf{y}^\alpha - \mathbf{g}^\alpha) + (y_i - g_i)\mathbf{g}^\alpha,$$

so if $F(y_1,\ldots,y_m)$ is a polynomial such that $F(f_1,\ldots,f_m) = 0$, then

$$F(y_1,\ldots,y_m) = F(y_1,\ldots,y_m) - F(f_1,\ldots,f_m) \in (y_1 - f_1,\ldots,y_m - f_m).$$

The converse and Part (b) are clear. $\square$

Semigroup Rings

A class of rings that has specific features in their processing is that of *semigroup rings* or *monomial subrings*. They are subrings of the ring of Laurent polynomials

$$R = k[x_1, x_1^{-1}, \ldots, x_n, x_n^{-1}]$$

generated by a finite set of monomials $A = k[f_1, \ldots, f_m]$, $f = \mathbf{x}^{\mathbf{a}} = x_1^{a_1} \cdots x_n^{a_n}$, $\mathbf{a} = (a_1, \ldots, a_n) \in \mathbb{Z}^n$. Each such ring is coded by the integral matrix whose rows are the exponent vectors $\mathbf{a}_i$ of the f_j's,

$$\varphi = \begin{bmatrix} \mathbf{a}_1 \\ \vdots \\ \mathbf{a}_q \end{bmatrix}.$$

For ease of terminology we refer to the rows of φ as "vectors" or "monomials", and the ring is denoted $k[\varphi]$ (see [Gil84], [He70]).

Krull Dimension

The set $\mathbf{x} = \{x_1, \ldots, x_n\}$ will be fixed throughout unless local conditions change it.

Proposition 7.27. *Let φ be a matrix with integral entries. Then for any field k,*

$$\dim k[\varphi] = \operatorname{rank}(\varphi).$$

Proof. We determine the Krull dimension of $A = k[\varphi] = k[f_1, \ldots, f_q]$ as the transcendence degree of its field of fractions. To this end, note that effecting any elementary row operation (over the integers) on φ produces an integral matrix whose associated semigroup ring has the same transcendence degree. Carrying out full row reduction (possibly involving exchange of columns) leaves a matrix of the form

$$\begin{bmatrix} d_1 & 0 & \cdots & 0 & * & * \\ 0 & d_2 & \cdots & 0 & * & * \\ \vdots & \vdots & \ddots & \vdots & \vdots & \vdots \\ 0 & 0 & \cdots & d_r & * & * \end{bmatrix}$$

with $d_i \neq 0$, whose attached semigroup ring is generated by r algebraically independent elements. $\qquad\square$

The Syzygy Graph of an Integral Matrix

Although Proposition 7.27 is excellent for explicit cases, there are other approaches that are useful for theoretical calculations. We consider one of these devices here ([SVV94]).

To each matrix φ with non-negative integral entries, one associates a graph in the following manner.

Definition 7.28. *The* linear syzygy graph $S(\varphi)$ *of* φ *is the graph with vertices* **x***, where* (x_i, x_j) *is an edge if and only if there are monomials* $x_i \cdot f$ *and* $x_j \cdot f$ *in* φ.

Proposition 7.29. *Let* φ *be a* $n \times m$ *matrix with positive integral entries and let* $S(\varphi)$ *be its linear syzygy graph. Then*

$$\operatorname{rank}(\varphi) \geq n - c + 1,$$

where c *is the number of connected components of* $S(\varphi)$. *In particular the matrix* φ *has full rank if the graph* $S(\varphi)$ *is connected.*

Proof. We construct an algebra from the monomials of φ in the following manner. For each monomial f of one degree d we form the element $ft^d \in R[t]$. Let R be the subring of $R[t]$ generated by R and all the ft^d's for all the monomials of φ. It is clear that the algebra R decomposes as

$$R = \mathfrak{m}R \oplus k[ft^d, \ f \in \varphi],$$

where $\mathfrak{m} = (x_1, \ldots, x_n)$. This means that $\mathfrak{m}R$ is a prime ideal of R. On the other hand, $k[ft^d, \ f \in \varphi]$ is also a monomial subring in the variables $x_1, \ldots, x_n, t$, whose associated matrix ψ is obtained from φ by adding a new column equal to the sum of all the columns of φ. By Proposition 7.27 the two algebras have the same dimension.

Since $\dim R = n + 1$ (we assume φ is not the null matrix), to find $\dim k[\varphi]$ it will suffice to determine the height of $\mathfrak{m}R$, or equivalently the dimension of the localization $R_{\mathfrak{m}R}$. At this point the graph structure of $S(\varphi)$ kicks in: If $x_i g t^d$ and $x_j g t^d$ are two monomials defining an edge of $S(\varphi)$, since they are not in $\mathfrak{m}R$, we may divide one by the other so that the element $\frac{x_i}{x_j} \in R_{\mathfrak{m}R}$ which shows that all the x_i's in the same connected component are associated to each other. This means that the ideal $\mathfrak{m}R_{\mathfrak{m}R}$ can be generated by c elements, one from each connected component of $S(\varphi)$. By Krull's principal ideal theorem, the height of $\mathfrak{m}R$ is at most c, and therefore $\dim R / \mathfrak{m}R \geq n - c + 1$, since R is equidimensional. $\qquad\square$

Graphs and Triangulations

Many instances of semigroup rings arise as triangulations of surfaces. We discuss some algebras related to graphs.

Let G be a graph with vertex set **x** and edge set E. For a fixed integer ℓ, denote by G_ℓ the set of all paths in G of length ℓ; thus $G_2 = E$. These sets define monomial ideals $I(G_\ell)$ and semigroup rings $k[G_\ell]$

Corollary 7.30. *If* G *is a connected graph, then*

$$\dim k[G_2] = \begin{cases} n-1, & \text{if } G \text{ is bipartite,} \\ n, & \text{otherwise.} \end{cases}$$

Proof. If G is bipartite, the quadratic monomials $f_1, f_2, \ldots, f_{2s-1}, f_{2s}$ that define an even cycle satisfy the relation

$$f_1 f_3 \cdots f_{2s-1} = f_2 f_4 \cdots f_{2s},$$

which means one of them can be dropped for the calculation of the Krull dimension of $k[G_2]$. Iterating we break all the cycles until a connected tree G' is reached. Since G' has $n-1$ edges and adding one variable x_i to $k[G'_2]$ gives a ring of Krull dimension n, this establishes the first assertion.

Suppose now G has an odd cycle which we fix. Note that all vertices of an odd cycle lie in the same connected component of $S(G_2)$. Since any other vertex of G can be reached by a path with an even number of edges from one vertex in the cycle, $S(G_2)$ is connected. $\square$

Integral Closure of Semigroup Rings

Let $R = k[x_1, \ldots, x_n]$ be a polynomial ring over a field k, and let $F = \{f_1, \ldots, f_q\}$ be a finite set of monomials in R. We outline the steps of a process that computes the integral closure of $k[F]$. (A comprehensive treatment will appear in [Vil01].) The presentation approach, in which we find first a representation $k[F] = R[T_1, \ldots, T_q]/I$, is not very appropriate since q is usually large.

We set $R = k[\mathbf{x}]$ and for each monomial we write $f = \mathbf{x}^\alpha$. For the given generators of F we denote $L = \{\alpha_1, \ldots, \alpha_q\}$. The integral closure of $k[F]$ is easy to describe in principle. Indeed, $k[F]$ will still have the fine grading, which will imply that the conductor of $\overline{k[F]}$ over $k[F]$ is a homogeneous ideal of $k[F]$ for the fine grading and therefore it will be generated by monomials in the f_i's. It follows that $\overline{k[F]}$ is generated by monomials $f = \mathbf{x}^\beta$ with the following two properties:

- $f = \prod f_i^{a_i}, \quad a_i \in \mathbb{Z},$
- $f^m = \prod f_i^{b_i}, \quad m, b_i \in \mathbb{Z}_+.$

The first condition asserts that f lies in the field of fractions of $k[F]$, the other being the special form the integrality condition takes for such elements. In other words, we have:

Proposition 7.31. *The integral closure of $k[F]$ is given by:*

$$\overline{k[F]} = k[\mathbf{x}^a \mid a \in \mathbb{Z}L \cap \mathbb{R}_+L],$$

where $\mathbb{R}_+L$ is the cone generated by L.

The following sequence of steps, suggested by Winfried Bruns, make this description more concrete:

- By means of an affine transformation we may assume that $\mathbb{Z}L = \mathbb{Z}^n$. As a polyhedron the cone $\mathbb{R}_+L$ can be described as the set of feasible solutions of a finite set of linear inequalities, that is, there is an $m \times n$ matrix A so that

$$\mathbb{R}_+L = \{x \in \mathbb{R}^n \mid Ax^T \le 0\}.$$

The computation of the matrix A is closely related to the computation of the vertices of a polyhedron (see [Ch83, Chapters 16, 17, 18]). The matrix A is obtained using the vectors in L and the matrix A completely determines $\mathbb{R}_+L$.

- Set N equal to the maximum of the absolute value of the entries of the vectors in L. Since

$$\mathbb{R}_+L \cap \mathbb{Z}^n = N v_1 + \cdots + N v_m,$$

where the entries of $v_1,\ldots,v_m$ are bounded by $M = qN$ (see [BH93, Proposition 6.1.2(b)]), one can use the matrix A to find the v_i's, by simply testing the elements in $[-M,M]^n \cap \mathbb{N}^n$ which belong to $\mathbb{R}_+L$.

Noether Normalizations

Let F be a finite set of monomials of $k[x_1,\ldots,x_n]$ and let $A = \overline{k[F]}$ be the integral closure of $k[F]$.

The graded Noether normalizations of $k[F]$ determine very sharp bounds for the degrees of the generators of A ([Ho72]).

Theorem 7.32. *Let $R = k[z_1,\ldots,z_d]$ be a Noether normalization of $k[F]$. Then A is a free R–module whose generators have degree at most $\sum_i \deg(z_i)$.*

Graphs

Let G be a graph with vertex set $V = \{x_1,\ldots,x_n\}$. Let F be the set of all monomials $x_i x_j$ in R, such that $\{x_i x_j\}$ is an edge of G. The integral closure of $k[G]$ has a very explicit description according to [HO98], [SVV98].

One begins by exhibiting some monomials in the integral closure of $k[G]$. Given an induced subgraph w of G consisting of two edge disjoint odd cycles

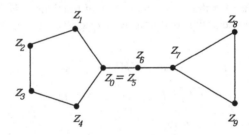

$$Z_1 = \{z_0, z_1, \ldots, z_r = z_0\} \text{ and } Z_2 = \{z_s, z_{s+1}, \ldots, z_t = z_s\},$$

joined by a path (those subgraphs will be called *bow ties*), associate the monomial $M_w = z_1 \cdots z_r z_{s+1} \cdots z_t$. The cycles Z_1 and Z_2 are allowed to intersect and that only the variables in the cycles occur in M_w, not those in the path itself. It is easy to see that M_w lies in the field of fractions of $k[G]$ and obviously $M_w^2 \in k[G]$.

Theorem 7.33. *Let G be a graph and let k be a field. Then the integral closure $\overline{k[G]}$ of $k[G]$ is generated as a k-algebra by the set*

$$B = \{f_1, \ldots, f_q\} \cup \{M_w \mid w \text{ is a bow tie}\},$$

where $f_1, \ldots, f_q$ denote the monomials defining the edges of G.

Remark 7.34. Observe how the monomials M_w occur in the multiplication subrings of ideals generated by products of some f_i. For instance, in a diagram as above, let I be the ideal defined by the product of the odd and even segments of the path between the two cycles, respectively,

$$f = (z_s z_{s+1})(z_{s+2} z_{s+3}) \cdots,$$
$$g = (z_{s+1} z_{s+2})(z_{s+3} z_{s+4}) \cdots$$

Clearly $M_w \in \mathrm{End}(I)$.

Sagbi Bases

Sagbi bases were invented by Robbiano and Sweedler ([RoS90]), who named them, and independently by Kapur and Madlener ([KM89]), to permit the processing in the subrings of a ring of polynomials $R = k[x_1, \ldots, x_n]$ in a manner that mirrors the way Gröbner bases facilitates the processing of the ideals of R.

Let A be a subring of R. For a given term ordering $>$ of R, denote by $in_>(A)$ the k–subalgebra of R generated by all the initial monomials $in(f)$, $f \in A$.

Definition 7.35. *A Sagbi basis of A is a set of polynomials $\{f_\alpha, \alpha \in \Gamma\} \subset A$ such that*

$$in_>(A) = k[in(f_\alpha), \alpha \in \Gamma].$$

Example 7.36. Let $I \subset R$ be an ideal generated by homogeneous polynomials and denote by $S = R[It]$ the Rees algebra of I. Given a term ordering of R extend it to $R[t]$ in the natural manner (see [CHV96]),

$$\mathbf{x}^\alpha t^i < \mathbf{x}^\beta t^j \iff i < j \quad \text{or } i = j \quad \text{and} \quad \alpha < \beta.$$

Then
$$in(S) = \bigoplus_{i \geq 0} in(I^i) t^i.$$

The following result gives the basic test for a Sagbi basis (see [CHV96], [RoS90]).

Let $f_1, \ldots, f_m$ be a set of polynomials of $k[x_1, \ldots, x_n]$ generating the subalgebra A. Let

$$\varphi : k[T_1, \ldots, T_m] \longrightarrow k[in(f_1), \ldots, in(f_m)]$$

be a presentation with kernel A.

Proposition 7.37 (Robbiano-Sweedler). *Let* $h_1, \ldots, h_r$ *be a set of binomial generators of* A. *Then* $f_1, \ldots, f_m$ *is a Sagbi basis of* A *if and only if for* $j = 1, \ldots, r$ *there exist* $\lambda_\nu^{(j)} \in k$ *such that*

$$h_j(f_1, \ldots, f_m) = \sum_\nu \lambda_\nu^{(j)} f^\nu, \quad \text{with} \quad in(f^\nu) \leq in(h_j(f_1, \ldots, f_m)),$$

for all $\lambda_\nu^{(j)} \neq 0$, *where* $f^\nu = f_1^{\nu_1} \cdots f_m^{\nu_m}$ *for the multi-index* $\nu = (\nu_1, \ldots, \nu_m)$.

A Sagbi basis will generate the algebra A. It may occur however that the Sagbi basis of a finitely generated algebra is not finitely generated. A stark example is that of the subalgebra

$$A = k[x + y, xy, xy^2] \subset k[x, y],$$

which has the property that for any term ordering $>$, the subalgebra $in_>(A)$ is not finitely generated. For example, if we choose an ordering so that $x > y$ then

$$in_>(A) = k[x + y, xy, xy^2, xy^3, \ldots].$$

Sagbi Bases versus Gröbner Bases

Let $f_1, \ldots, f_m$ be a set of polynomials in $R = k[x_1, \ldots, x_n]$, and form the subalgebra and ideal they generate, $A = k[f_1, \ldots, f_m]$, $I = (f_1, \ldots, f_m)$. Suppose a term order $>$ for R is chosen. Under some conditions whether the f_i form a Sagbi basis for A is related to their being a Gröbner basis for I and the power products $\{f_{i_1} \cdots f_{i_k}\}$ giving a Gröbner basis of I^k, for all k. The following formulation was indicated to us by Aldo Conca (see [CHV96]).

Theorem 7.38. *Assume that* $\{f_1, \ldots, f_m\}$ *is a Gröbner basis of* I. *Then the following conditions are equivalent:*

(a) *The power products of degree* k *in the* f_i's *form a Gröbner basis of* I^k *for all* k.
(b) $in(I)^k = in(I^k)$ *for all* k.
(c) *For any extension of* $>$ *to a term order of* $R[T]$, *the initial subalgebra of the Rees algebra* $R[IT] \subset k[x_1, \ldots, x_n, T]$ *satisfies* $in(R[IT]) = R[in(I)T]$.
(d) *The polynomials* $x_1, \ldots, x_n, f_1 T, \ldots, f_m T$ *form a Sagbi basis of* $R[IT]$ *with respect to the extended* $>$.
Furthermore if the f_i *are forms of the same degree and* (a),(b),(c),(d) *hold then*

$$in(k[f_1, \ldots, f_m]) = k[in(f_1), \ldots, in(f_m)].$$

When available a Sagbi basis opens the possibility of using methods of integer programming to study the properties of A: dimension, integral closure etc. Even a partial Sagbi basis may provide useful information about the algebra. The subject has been hampered by the lack of implementations of the algorithms described in [RoS90]. On the other hand, there are already several theoretical applications of the method, of which [CHV96] is an excellent example.

Exercise 7.39. Let $A = k[f_1,\ldots,f_m]$ be a subring of a ring of polynomials over the field k. If $\dim A \leq 2$, develop a method to find a Noether normalization of A that uses Sagbi bases.

Exercise 7.40. Let $f_1,\ldots,f_m$, $m \geq 2$, be a set of distinct monomials of $k[x_1,\ldots,x_n]$ of the same degree. Let f be the least common multiple of these monomials and set $g_i = f/f_i$. Show that the assignment $f_i \mapsto g_i$ defines an isomorphism between $k[f_1,\ldots,f_m]$ and $k[g_1,\ldots,g_m]$.

Exercise 7.41. Let G be a graph with vertex set $\mathbf{x} = \{x_1,\ldots,x_n\}$ and edge set E. Denote by $k[G]$ the subring of $R = k[\mathbf{x}]$ generated by the monomials $x_i x_j$ with $(x_i, x_j) \in E$. Let $C(G)$ be the cone over G, that is the graph with vertices given by the x_i and t whose edges are those in E together with all (x_i, t). Show that $k[C(G)]$ is isomorphic to the Rees algebra of the ideal of R generated by the monomials defined by E.

7.4 Rings of Invariants

The aim is this section is to give an overview of some algorithmic pathways to the determination of rings of invariants. The exposition will be brief and filtered by the discussion in [DK97], [Ke91], [Spr77], and particularly [Stu93].

Let V be a vector space of dimension n over the field k, and let $R = k[x,\ldots,x_n]$ be the ring of polynomial functions on V. The general linear group $Gl(V)$ acts on R by

$$\sigma(f)(\mathbf{x}) = f(\sigma(\mathbf{x})), \quad \sigma \in Gl(V).$$

For a given subgroup $G \subset Gl(V)$, the polynomials

$$R^G = \{\, f \mid \sigma(f) = f, \forall \sigma \in G \,\} \tag{7.5}$$

define a graded k–subalgebra of R, the *ring of invariants* of G.

A fundamental set of problems about these subrings is:

- When is R^G Noetherian, and if so what are its properties?
- How can R^G be determined *theoretically*?
- Which approaches are available to *effectively* find invariants?

At issue is to find in principle, when not actually, a *short* set of invariants $f_1, \ldots, f_m$ generating R^G,

$$R^G = k[f_1,\ldots,f_m].$$

To simplify our discussion, we assume throughout that k is a field of characteristic zero.

The premier example is that of the symmetric group S_n acting on the set $\{x_1, \ldots, x_n\}$ of indeterminates in the usual manner. The ring of invariants is another 'ring of polynomials', generated by the standard symmetric polynomials

$$s_1 = \sum_i x_i$$

$$s_2 = \sum_{i<j} x_i x_j$$

$$\vdots$$

$$s_n = x_1 x_2 \cdots x_n$$

$$R^{S_n} = k[s_1, \ldots, s_n].$$

Alternatively, when the characteristic of k is larger than n, R^{S_n} is generated by Newton symmetric powers,

$$p_1 = \sum_i x_i$$

$$p_2 = \sum_i x_i^2$$

$$\vdots$$

$$p_n = \sum_i x_i^n.$$

In both cases, the ring of invariants is obtained from the averaging of key monomials of the ring R. More generally, for any finite group G, whose order is not divisible by the characteristic of k, the averaging

$$f \in R \mapsto \mathcal{R}(f) = \frac{1}{|G|} \sum_{\sigma \in G} \sigma(f) \in R^G, \tag{7.6}$$

gives a mechanism to find the invariants of G.

An aspect of this process is exhibited in the examples: the *average* of a simple monomial may turn out to be a very *long* polynomial, as the group action spreads the monomial over the representation space.

Theoretical Framework

The overall strategy for describing the ring of invariants R^G were laid out in Hilbert's original papers ([Hil90], [Hil93]) and its modern sequels. The topic itself is now subsumed in the theory of representations of algebraic groups.

Reductive Groups

Definition 7.42. A *Reynolds operator* on R is a homomorphism of graded R^G–modules

$$\mathcal{R} : R \mapsto R^G, \tag{7.7}$$

that is the identity on R^G.

This operator is guaranteed to exist for a large class of groups and their representations. Noteworthy are the *reductive groups*: those where each representation V decomposes into a direct sum of irreducible ones. When applied to the induced representation on the space of forms of a given degree d of R, the sum of the trivial subrepresentations—those corresponding to R_d^G—split off R_d, thus providing for a Reynolds operator. In Hilbert's original use, when G was the special linear group, the role of this operator was played by the Ω–process. The existence of such operator leads to the truly beautiful:

Theorem 7.43 (Hilbert–Nagata). *Let I be the ideal of R generated by all the forms in R^G of degree > 0. If $I = (f_1, \ldots, f_m)$, where the f_i's are forms of R^G, then $R^G = k[f_1, \ldots, f_m]$.*

Its proof *only* requires the first historical application of the Hilbert basis theorem:

Theorem 7.44. *Let B be a graded k–subring of $R = k[x_1, \ldots, x_n]$. Denote by I the ideal of R generated by all the forms of B of positive degree. If*

$$I = (f_1, \ldots, f_m), \quad f_i \in B, \tag{7.8}$$

then $B = k[f_1, \ldots, f_m]$.

The ideal I_N of R generated by the invariant forms of degree > 0 defines the so called *nullcone* N.

The basic structure of R^G is of much more recent vintage ([HR74]):

Theorem 7.45 (Hochster–Roberts). *The ring R^G is Cohen–Macaulay. In particular, there is a set of homogeneous, algebraically independent invariants $h_1, \ldots, h_d$ such that R^G is a finitely generated free $k[h_1, \ldots, h_d]$–module.*

The statement means that there are finitely many invariants, $\theta_1, \ldots, \theta_r$ such that each θ_i is integral over $A = k[h_1, \ldots, h_d]$ and

$$R^G = \bigoplus_{1 \leq j \leq r} A\theta_j. \tag{7.9}$$

As a consequence, the Hilbert–Poincaré series of R^G is given by

$$H_{R^G}(\mathbf{t}) = \frac{\sum_{1 \leq j \leq r} \mathbf{t}^{e_j}}{\prod_{1 \leq i \leq d}(1 - \mathbf{t}^{d_i})}, \tag{7.10}$$

where $d_i = \deg h_i$ and $e_j = \deg \theta_j$.

One aspect of this function that is significant for the computation of invariants is the following. We recall (see Appendix B) that the integer

$$a(R^G) = \max_j \{ e_j \} - \sum_{1 \leq i \leq d} d_i$$

is the a–invariant of R^G. The fact that R^G is a direct summand of the ring of polynomials R makes it a rational singularity ([Ke79]); such rings have non-positive a–invariants. In fact, according to [Kno89], $a(R^G) \leq -\dim R^G = -d$. In particular,

$$e_j \leq \sum_{1 \leq i \leq d} d_i - d, \quad \forall j, \tag{7.11}$$

that show how a ceiling for the degrees of sets of generators of R^G is defined by the degrees of the h_i's.

Definition 7.46. *The polynomials $h_1, \ldots, h_d$ form a set of* fundamental invariants *of R^G, or simply of G.*

Finding such a set is path to the determination of the ring R^G according to:

Proposition 7.47. *Let B be a finitely generated graded k–algebra, let A be a k–subalgebra generated by the forms of positive degree $h_1, \ldots, h_s$, and denote by A_+ the ideal of A generated by all elements of positive degree. Then B is a finitely generated A–module if and only if*

$$B_r \subset A_+ B, \quad \text{for } r > r_0. \tag{7.12}$$

Proof. Suppose $B = k[f_1, \ldots, f_m]$, with the f_i's forms of positive degree. Denote by $b_1, \ldots, b_p$ the set of all the products

$$f_1^{e_1} \cdots f_m^{e_m}, \quad \sum e_i = r_0.$$

Set

$$C = B_0 + B_1 + \cdots + B_{r_0 - 1} + \sum_{j=1}^{p} A b_j.$$

C is a finitely generated A–submodule of B. From (7.12) it follows that $C = B$. The converse is clear. $\qquad \square$

Hilbert's approach to the determination of subrings of $B = R^G$, with this property, is his other pearl:

Theorem 7.48. *Let I be the ideal of R generated by all forms of R^G of positive degree. Suppose $h_1, \ldots, h_r \in R^G$ have the property that*

$$\sqrt{I} = \sqrt{(h_1, \ldots, h_r)}.$$

Then R^G is a finitely generated $k[h_1, \ldots, h_r]$–module.

Proof. To enable the condition (7.12), for a given set $f_1, \ldots, f_m$ of generators of R^G (which, by (7.8) is equivalent to being a set of elements of R^G generating the ideal I), let p be an integer such that

$$(\sqrt{I})^p \subset (h_1, \ldots, h_r)R.$$

In the case of R^G, once the h_i's have been found, there are two pathways to the full R^G. Although these forms are not a fundamental set of invariants in the sense above, there exists a Noether normalization (k is infinite)

$$k[h'_1, \ldots, h'_d] \hookrightarrow k[h_1, \ldots, h_r]$$

such that

$$\deg(h'_i) \leq \max\{ \deg(h_j) \}.$$

As a consequence we have (after notation change):

Corollary 7.49. *Given a set $h_1, \ldots, h_d$ of fundamental invariants, R^G can be described as:*

(a) R^G *is the integral closure of $k[h_1, \ldots, h_d]$ in $k[x_1, \ldots, x_n]$.*
(b) R^G *is generated over $k[h_1, \ldots, h_d]$ by the invariants of degree less than*

$$\sum_{i=1}^{d} \deg(h_i).$$

To sum up, to find R^G requires sharp bounds on the degrees of a full set of generators, or a bound for the degrees of a certifiable fundamental set of generators. In the latter, to be followed either by the computation of integral closure or for the search for a full set of generators using the guide provided by the degrees of the invariants in the fundamental set.

What is Given

The calculus of invariants is very dependent on the specification of the group action. We begin with a general reformulation of R^G.

Proposition 7.50. *Suppose G is a connected affine algebraic group and V is a finite dimensional G–module. The action corresponds to a k–homomorphism*

$$\psi : k[V] = k[x_1, \ldots, x_n] \mapsto k[G] \otimes k[V] = k[y_1, \ldots, y_m]/I(G) \otimes k[x_1, \ldots, x_n],$$

where $I(G)$ is the defining ideal of G in the affine space $\mathbb{A}^m$. One then has

$$\boxed{k[V]^G = k[\psi(x_1), \ldots, \psi(x_n)] \cap k[x_1, \ldots, x_n].} \tag{7.13}$$

Any Gröbner basis for the ideal $I(G)$ gives the means to test whether a polynomial is invariant and by using undetermined coefficients to find in principle all invariants of a given degree. Note that this Gröbner basis is computed in $k[y_1,\ldots,y_m]$ and extended to $k[x_1,\ldots,x_n,y_1,\ldots,y_m]$ in the obvious manner.

Proposition 7.51. *Let G be a Gröbner basis of $I(G)$ and NF its corresponding Normal Form algorithm. Then $f(\mathbf{x}) \in k[x_1,\ldots,x_n]$ is an invariant if and only if*

$$\text{NormalForm}(f(\psi(\mathbf{x})) - f(\mathbf{x})) = 0.$$

Derksen's Algorithm

The following ([Der99], [DK97]) provides an algorithmic path to the ring of invariants of a reductive group. The case of tori is treated already in [Stu93, Algorithm 1.4.5].

Theorem 7.52. *Let G be a connected, reductive algebraic group over a field of characteristic zero and let V be a representation of dimension n. Denote by B the closure of the image of the mapping*

$$\phi : G \times V \longrightarrow V \times V, \quad \phi(g,v) = (gv,v),$$

and let I be the homogeneous ideal of

$$k[V \times V] = k[\mathbf{z},\mathbf{w}]$$

defining B. If $h_1(\mathbf{z},\mathbf{w}),\ldots,h_s(\mathbf{z},\mathbf{w})$ are homogeneous generators of I then

$$k[V]^G = k[z_1,\ldots,z_n]^G = k[\mathcal{R}(h_1(\mathbf{z},0)),\ldots,\mathcal{R}(h_s(\mathbf{z},0))],$$

where $\mathcal{R}$ is the Reynolds operator.

Its proof (cf. [Der99]) is made up of two steps:

- If I is the defining ideal of B, then

$$(I,(w_1,\ldots,w_n)) \cap k[z_1,\ldots,z_n] = I_N.$$

- For any homogeneous set of generators $\{p_1,\ldots,p_s\}$ of I_N,

$$k[V]^G = k[\mathcal{R}(p_1),\ldots,\mathcal{R}(p_s)].$$

The handling of the ideal I using elimination proceeds as follows. Let

$$\psi : k[V] \longrightarrow k[G \times V]$$

be the homomorphism defining the action of the group G as above. To find the image of

$$\phi : G \times V \longrightarrow V \times V,$$

we follow Proposition 2.6. Let $\mathbf{z} = z_1, \ldots, z_n$ and $\mathbf{w} = w_1, \ldots, w_n$ be two new sets of indeterminates. In the ring

$$k[G \times V][\mathbf{z}, \mathbf{w}] = k[G \times V][z_1, \ldots, z_n, w_1, \ldots, w_n],$$

consider the ideal

$$J = (I(G), (z_1 - \psi(x_1), \ldots, z_n - \psi(x_n), w_1 - x_1, \ldots, w_n - x_n)).$$

Then

$$I = J \cap k[z_1, \ldots, z_n, w_1, \ldots, w_n] = (h_1(\mathbf{z}, \mathbf{w}), \ldots, h_s(\mathbf{z}, \mathbf{w})).$$

Obviously, if n is large, this is not entirely practical.

Reynolds Operator and Lie Algebras

Let $\psi : G \mapsto Gl(V)$ be a linear representation of a reductive group. Except for very special cases–the noteworthy exception being finite groups–the Reynolds operator associated to such representation is not given explicitly. This is partly remedied by finding the nullspace of the action of some operators on the graded components of $R = k[V] = k[x_1, \ldots, x_n]$.

Denote by $\mathfrak{g}$ the image of the Lie algebra of G in $M_n(k)$, the Lie algebra of $Gl(V)$. We assume that the details of the representation affords the access to $\mathfrak{g}$. For a given $\sigma \in \mathfrak{g}$, with a matrix representation $[a_{ij}] \in M_n(k)$, consider the differential operator on R,

$$d_\sigma = \sum a_{ij} x_i \frac{\partial}{\partial x_j}.$$

d_σ acts on each homogeneous component R_d of R.

Proposition 7.53. *Let $\sigma_1, \ldots, \sigma_r$ be a set of generators of $\mathfrak{g}$. Then*

$$R^G = \bigcap_{1 \leq i \leq r} d_{\sigma_i}^{-1}(0).$$

Of course we only use this Proposition on the components where the expected degrees occur.

Other Groups and Special Representations

We look at some groups and their special representations.

Additive Group

When G is the group G_a, $k[G] = k[t]$, this intersection is given as in (7.13) by a reduced Sagbi basis of the subring

$$k[\psi(x_1), \ldots, \psi(x_n)] \subset k[x_1, \ldots, x_n, t],$$

for any elimination term ordering with $t > x_i$, $\forall i$. In this case, probably the theorem of Weitzenböck guarantees the finiteness of such Sagbi bases.

When G is a direct product of copies of G_a we may not have the finiteness of the Sagbi basis, as in Nagata's famous counterexample to Hilbert's 14th Problem.

Adjoint Representation

A special action is the *adjoint* representation of a group G. Denoting by $\mathfrak{g}$ its Lie algebra, which we also view as a vector space, let

$$\mathrm{Ad} : G \longrightarrow Gl(\mathfrak{g}),$$

be given by conjugation.

Denote by R the ring of polynomials

$$R = S(\mathfrak{g}) = k[x_1, \ldots, x_n].$$

We describe the method of [Vas94, Chapter 9], to which we refer for more details, dealing with the case when $\mathfrak{g}$ is a simple Lie algebra, to find R^G.

Let $B(\cdot, \cdot)$ be the Killing of $\mathfrak{g}$. We can use this non–degenerate quadratic form on $\mathfrak{g}$ to define directional derivatives in the standard manner. In particular, for any polynomial $f(x) \in R$, define its *gradient*, $\triangledown f(x)$, by

$$B(\triangledown f(x), y) = df_x(y).$$

Since B is unimodular, $\triangledown f(x)$ is actually an element of R^n.

The issue is what are the syzygies of $\triangledown f(x)$? One source of syzygies is given by [Vas94, Proposition 9.3.3]. Let $e_1, \ldots, e_n$ be a basis of $\mathfrak{g}$ and define

$$\mathbf{x} = x_1 e_1 + \cdots + x_n e_n,$$

a generic element of the Lie ring $R \otimes \mathfrak{g}$. We can also view $\triangledown f(x)$ as an element of this ring.

Proposition 7.54. *Let $f(x)$ be an invariant polynomial under the action of G. Then*

$$[\triangledown f(x), \mathbf{x}] = 0. \tag{7.14}$$

In the case of a simple group the syzygies obtained in this manner all correspond to invariant polynomials ([Vas94, Theorem 9.3.4]):

Theorem 7.55. *Let* $\mathfrak{g}$ *be a simple Lie algebra of rank* ℓ *over an algebraically closed field of characteristic zero. Let* $p_1(x),\ldots,p_\ell(x)$ *be the homogeneous polynomials generating* R^G. *Then*

$$\nabla p_1(x),\ldots,\nabla p_\ell(x) \tag{7.15}$$

is a basis of the module of relations of the mapping

$$\psi : R^n \longrightarrow R^n, \quad \mathbf{z} \mapsto [\mathbf{z},\mathbf{x}]. \tag{7.16}$$

Once a syzygy is found, the corresponding polynomial $f(x)$ *can be reconstituted by Euler formula*

$$\deg f(x) \cdot f(x) = B(\nabla f(x),\mathbf{x}). \tag{7.17}$$

Except for $p_1(x) = B(\mathbf{x},\mathbf{x})$, the others tend to be rather dense. For instance, in the case of a group of type G_2, when the Killing form was diagonalized the second invariant polynomial $p_2(x)$ extended over 200 printed pages.

Conjecture 7.56. Let G be a connected, reductive affine algebraic group and let $I(G) \subset k[y_1,\ldots,y_m]$ be its ideal of definition. For some term order $>$ on the ring of polynomials the ideal $in_>(I(G))$ is Cohen–Macaulay. [1]

---

[1] B. Sturmfels [Stu01] has settled this question affirmately if G is an abelian group.

8

Computation of Cohomology

By David Eisenbud

The problem considered here is: Given a finitely generated graded $S = k[x_0, \ldots, x_r]$–module M, compute the cohomology of the sheaf $\widetilde{M}$ on $\mathbb{P}^r$. A desirable refinement: starting from a bigraded module, compute the cohomology of the corresponding sheaf on $\mathbb{P}^r \times \mathbb{P}^s$ (this should be much easier to do directly than to first take the Segre embedding, since then one has $(r+1)(s+1)$ variables instead of $r+s+2$ variables). Of course we could also ask for the multigraded case, and more generally the case of local cohomology with arbitrary supports.

What does it mean to compute the cohomology? Depending on circumstances, it might mean anything from finding the dimension $h^i(\widetilde{M}(n))$ of one cohomology vector space $H^i\widetilde{M}(n)$ to finding the cohomology modules $\sum_n H^i\widetilde{M}(n)$, as modules over the ring $S = \sum_n H^0(\mathcal{O}_{\mathbb{P}^r}(n))$.

We will deal with three related methods, appropriate in different circumstances. Here is a summary of them, and their relative virtues and defects:

- **'Eyeballing' a free resolution of M:**
  - Often the cheapest in really simple cases.
  - Only computes H^0 up to an extension.
  - Really a naive version of the following method.
  - Does not work well in the bigraded case (unless a certain spectral sequence degenerates).
1. **Local Duality:** For $i \geq 1$ we have

$$\sum_n H^i\widetilde{M}(n) \simeq \operatorname{Ext}_S^{r-i}(M, \omega)^\vee,$$

where ω denotes $S(-r-1)$ and $L^\vee$, for a graded module $L = \sum_n L_n$, denotes $\operatorname{Hom}_{\text{graded } k}(L, k) = \sum_n \operatorname{Hom}(L_n, k)$, the graded vector space dual of the graded module L, as a module over S. For $i = 0$ we have the exact sequence

$$0 \to \mathrm{Ext}_S^{r+1}(M,\omega)^\vee \longrightarrow M \longrightarrow \sum_n H^0\widetilde{M}(n) \longrightarrow \mathrm{Ext}_S^r(M,\omega)^\vee \to 0.$$

- This form of (1) is better suited for automation, not so suited for "eye-balling".
- It's hard to use this to get the module structures, since $N^\vee$ is hard to compute as a module (it is not even finitely generated in general).
- The method of choice for getting information about

$$\sum_n H^d(O_X(n)) \quad \text{or} \quad \omega_X = \left(\sum_n H^d(O_X(n))\right)^\vee$$

  when X is arithmetically Cohen–Macaulay subvariety of $\mathbb{P}^r$.
- Does not work in the bigraded case.

2. **Approximation:** For all $i \geq 0$ and any $b \in \mathbb{Z}$ we have

$$\sum_{n \geq b} H^i\widetilde{M}(n) \simeq \mathrm{Ext}_S^i(J,M)_{n \geq b},$$

J is any homogeneous ideal primary to $(x_0,\ldots,x_r)$ and that is contained in

$$(x_0,\ldots,x_r)^a,$$

and where $a = a(M)$ is the maximum of the degrees of syzygies of M, diminished by $r + b$; that is, if

$$0 \to F_m \longrightarrow \cdots \longrightarrow F_0 \longrightarrow M \to 0$$

is a minimal free resolution of M, and $F_i = \sum S[-a_{ij}]$, then we may take

$$a = \max_{ij}\{\, a_{ij} \,\} - r - b.$$

We think of this as approximation because the modules $\mathrm{Ext}_S^i(J,M)$ "converge" to $\sum_n H^i\widetilde{M}(n)$ as J moves into a higher and higher power of the maximal ideal, for example for the sequence of J's given by $J_a = (x_0^a,\ldots,x_r^a)$.

- Well suited for automatic computation of the structure of the cohomology module in interesting cases.
- Can be extended to a multigraded setting, and in fact to the computation of local cohomology with given supports.
- Can be relatively expensive computationally, but seems best if one wants to understand $\sum_n H^i\widetilde{M}(n)$ as a module over S.

8.1 Eyeballing

The idea is to compare M through exact sequences with modules whose cohomology are easy to understand—for example the graded free modules, which are sums of modules of the form $S(m)$. Of course $\widetilde{S}(m) = O_{\mathbb{P}^r}(m)$, and we have

$$\sum_n H^i(\mathcal{O}_{\mathbb{P}^r}(n)) = \begin{cases} S & \text{if } i = 0 \\ 0 & \text{if } 0 < i < r \text{ or } i > r \\ S(-r-1)^\vee = \omega^\vee & \text{if } i = r, \end{cases}$$

(This last formula is somewhat misleading, since it might lead one incorrectly to think that H^r was a contravariant functor! It really should be exhibited functorially, saying that for any graded free module G we have

$$H^r(\widetilde{G}) = \operatorname{Hom}_S(G, \omega)^\vee.$$

If $G = S$ we the get the formula above.)

An example will probably be more helpful here than a theorem:

Cohomology of the Smooth Rational Quartic in $\mathbb{P}^3$

Let $X \subset \mathbb{P}^3$ be the image of $\mathbb{P}^1$ under the map $(s,t) \mapsto (s^4, s^3t, st^3, t^4)$. Using *Macaulay*, we may easily compute the ideal of X and its free resolution:

$$\begin{array}{cccc} 1 & - & - & - \\ - & 1 & - & - \\ - & 3 & 4 & 1 \end{array}$$

in the notation used by *Macaulay* "betti" command. The last map in the resolution is given by the matrix (x_0, x_1, x_2, x_3). This information in the diagram suffices to compute some cohomology. For example, if I is the ideal sheaf of X and K is the first syzygy sheaf of I, then

$$\begin{aligned}
H^1(\mathcal{O}_X) &= H^2(I) \\
&\quad \text{since } H^i(\mathcal{O}_{\mathbb{P}^3}(m)) = 0 \text{ for } i = 1, 2, \text{all } m \\
&= H^3(K) \\
&\quad \text{since } H^i(\mathcal{O}_{\mathbb{P}^3}(-2)) \oplus H^i(\mathcal{O}_{\mathbb{P}^3}(-3)^3) = 0 \text{ for } i = 2, 3 \\
&= \operatorname{coker} H^3(\mathcal{O}_{\mathbb{P}^3}(-5)) \longrightarrow H^3(\mathcal{O}_{\mathbb{P}^3}(-4)^4) \\
&\quad \text{since } H^4(\mathcal{O}_{\mathbb{P}^3}(-5)) = 0 \\
&= \operatorname{coker} S_1^\vee \longrightarrow (S_0^4)^\vee
\end{aligned}$$

under the map induced by $(x_0, \ldots, x_3)$,

by the computation of the cohomology $H^i \mathcal{O}_{\mathbb{P}^3}(n)$ above, where we have written S_m for the vector space of forms of degree m in S.

Now the map induced by $(x_0, \ldots, x_3)$ is none other than the dual, in the appropriate degree, of the map $S^4 \longrightarrow S(1)$ induced by $(x_0, \ldots, x_3)$. Since the map is a monomorphism (in fact an isomorphism) in degree 0, the dual is an epimorphism in degree 0; that is,

$$H^1(\mathcal{O}_X) = \operatorname{coker} S_1^\vee \longrightarrow (S_0^4)^\vee = 0.$$

This is indeed correct, since X is a rational curve!

A more subtle example is furnished by the computation of $H^0(\mathcal{O}_X(1))$. From the resolution we see that X lies on no hyperplanes, that is $H^0(I(1)) = 0$, and we deduce the exact sequence

$$0 = H^0(I(1)) \to H^0(\mathcal{O}_{\mathbb{P}^3}(1)) \to H^0(\mathcal{O}_X(1)) \to H^1(I(1)) \to H^1(\mathcal{O}_{\mathbb{P}^3}(1)) = 0.$$

Since we know that $H^0(\mathcal{O}_{\mathbb{P}^3}(1)) = S_1$, we can at least see that

$$h^0(\mathcal{O}_X(1)) = 4 + h^1(I(1)).$$

But as above

$$
\begin{aligned}
H^1(I(1)) &= H^2(K(1)) \\
&= \ker\left(H^3(\mathcal{O}_X(-4)) \longrightarrow H^3(\mathcal{O}_X(-3)^4) = 0\right) \\
&= S_0^\vee
\end{aligned}
$$

so $h^1(I(1)) = 1$, (corresponding to the fact that X is the projection of a curve in $\mathbb{P}^4$). What makes this example subtle is that we were not able to compute $H^0(\mathcal{O}_{\mathbb{P}^3}(1))$ except as an extension of computable objects. This suffices for dimension computations, but would not suffice for finding the module structure (let alone the ring structure) on $\sum_n H^0(\mathcal{O}_X(n))$.

8.2 Local Duality

Theorem 8.1. *For $i \geq 1$ we have*

$$\sum_n H^i \widetilde{M}(n) \simeq \mathrm{Ext}_S^{r-i}(M, \omega)^\vee,$$

where ω denotes $S(-r-1)$ and $N^\vee$ denotes $\mathrm{Hom}_{\mathrm{graded}\,k}(N,k)$ the graded vector space dual of the graded module N as a module over S. For $i = 0$ we have only the exact sequence

$$0 \to \mathrm{Ext}_S^{r+1}(M, \omega)^\vee \longrightarrow M \longrightarrow \sum_n H^0 \widetilde{M}(n) \longrightarrow \mathrm{Ext}_S^r(M, \omega)^\vee \to 0.$$

This theorem is really just a systematization of the method above. Here is a quick proof which makes this clear and which also shows how the method breaks down in the case of multigradings. We use the language of spectral sequences, but the reader with a distaste for this (suites spectrales ≡ spectral sweets ≡ ghost candy, and who wants to be a ghost?) may easily provide a proof by chasing long exact sequences, following the procedure in the example of "eyeballing" above.

Proof. The basis of the proof is the fact that cohomology may be defined from the Čech complex. Recall that the Čech complex of S is:

$$\mathfrak{F}: \quad 0 \to \prod_i S[x_i^{-1}] \to \prod_{i<j} S[x_i^{-1} x_j^{-1}] \to \cdots \to S[x_0^{-1} \cdots x_r^{-1}] \to 0.$$

$\mathfrak{F}$ is easily seen to be the direct limit of the (truncated) Koszul complexes

$$\mathfrak{F}_m: \quad 0 \to S^n(m) \to \wedge^2 S^n(2m) \to \cdots \to \wedge^n S^n(nm) \to 0,$$

of the ideals $(x_0^m, \ldots, x_r^m)$, (where for compactness we have written n for $r+1$), under the maps

$$f: \mathfrak{F}_m \longrightarrow \mathfrak{F}_{m+1}$$

sending

$$\wedge^k S^n(km) \longrightarrow \wedge^k S^n(k(m+1))$$

by the kth exterior power of the diagonal map with entries $(x_0, \ldots, x_r)$. From this description, and a knowledge of the homology of the Koszul complexes, the cohomology of the Čech complex is obvious: Its value is given in the table for the cohomology of $\mathcal{O}_{\mathbb{P}^r}$ given above. By definition the cohomology module $\sum_n H^i \widetilde{M}(n)$, for any graded module M, is just the cohomology (actually homology, in the usual way of looking at things) of the complex $\mathfrak{F} \otimes M$ at the term

$$\prod_{j_0 < \cdots < j_i} S[x_{j_0}^{-1} \cdots x_{j_i}^{-1}] \otimes M.$$

Now given a free resolution F of M, we get a double complex $\mathfrak{F} \otimes F$. As usual, two spectral sequences converge to the total homology of this complex. Since the complexes

$$\prod_{j_0 < \cdots < j_i} S[x_{j_0}^{-1} \cdots x_{j_i}^{-1}] \otimes F$$

have as their only homology

$$\prod_{j_0 < \cdots < j_i} S[x_{j_0}^{-1} \cdots x_{j_i}^{-1}] \otimes M,$$

one of these spectral sequences degenerates and shows that the homology modules of the total complex $\mathfrak{F} \otimes F$ are precisely the cohomology modules of $\widetilde{M}$. Thus the other spectral sequence gives

$$H_q(H^p(\widetilde{F})) \Rightarrow H^{p-q}(\widetilde{M}),$$

where $H_q(H^p(\widetilde{M}))$ denotes the homology of the complex made from F by sheafifying and replacing each term by its pth cohomology module.

Now F is a complex of free modules, so $H^p(\widetilde{F}) = 0$ unless $p = 0$ or $p = r$. Further, the complex $H^0(\widetilde{F})$ is just the original complex F, and thus has homology only at the 0th place. Thus the E_2 term of the spectral sequence has nonzero entries in only two rows: in one of these we get only the module M, in the 0th position, and in the other, r rows above, we get the homology of the complex $H^r(\widetilde{F})$. Now for any graded module G, $H^r(\widetilde{G})$ is $\mathrm{Hom}(G, \omega)^\vee$, and $^\vee$ is an exact functor, so the cohomology of $H^r(\widetilde{F})$ is $\mathrm{Ext}_S^{r-i}(M, \omega)^\vee$. The spectral sequence can thus be represented as follows:

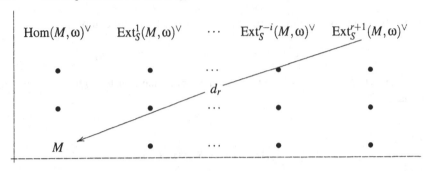

where we have drawn the only non vanishing higher differential. Now the ith cohomology of M is computed from the terms on the $(r-i)$th diagonal of this diagram, so we see that for $i > 0$ we have $\sum_n H^i \widetilde{M}(n) \simeq \mathrm{Ext}_S^{r-i}(M,\omega)^\vee$ as required by the theorem, whereas for $i = 0$ we get a filtration of $\sum_n H^i \widetilde{M}(n)$ consisting of a submodule which is the cokernel of d_r, and a quotient module which is $\mathrm{Ext}_S^r(M,\omega)^\vee$; that is, we get an exact sequence of the form required for the theorem (d_r is a monomorphism because its kernel is $\sum_n H^{r+1} \widetilde{M}(n)$ which is zero since we are on $\mathbb{P}^r$.) $\square$

The spectral sequence argument works just as well in the case of a product of projective spaces. The difference is that there are 4 nonzero cohomology modules of a sheaf on $\mathbb{P}^r \times \mathbb{P}^s$ associated to a bigraded module (computed from the Künneth formula) instead of 2. Thus the spectral sequence has 4 nontrivial rows and lots of nontrivial differentials, so methods 1 and 2 cannot be applied in general in the bigraded (and even less in the multigraded) case.

8.3 Approximation

Theorem 8.2. *For all $i \geq 0$ and any $b \in \mathbb{Z}$ we have*

$$\sum_{n \geq b} H^i \widetilde{M}(n) \simeq \mathrm{Ext}_S^i(J,M)_{n \geq b},$$

where J is any homogeneous ideal primary to $(x_0, \ldots, x_r)$ contained in $(x_0, \ldots, x_r)^a$, where $a = a(M)$ is the maximum of the degrees of the syzygies of M, diminished by $r + b$; that is, if

$$0 \to F_m \longrightarrow \cdots \longrightarrow F_0 \longrightarrow M \to 0$$

is a free resolution of M, and $F_i = \sum S(-a_{ij})$, then we may take

$$a = (\max_{ij} a_{ij}) - r - b.$$

Proof. Because both cohomology and Ext fit into long exact sequences, it is enough to check that the formula holds for the modules $M = S(m)$, where it is trivial, and use the short exact sequences that make up the resolution for M. $\square$

Since the homology of the truncated Koszul complex $\mathfrak{F}_m$ is $\mathrm{Ext}_S^i(J_m, M)$, and since homology commutes with direct limits, we get

$$\sum_n H^i\widetilde{M}(n) = \lim_m \mathrm{Ext}_S^i(J_m, M)$$

from the definition above of Čech cohomology. Since the J_m are cofinal with any sequence of ideals I_m such that I_m is primary to $(x_0, \ldots, x_r)$ and $I_m \subset (x_0, \ldots, x_r)^m$, we can replace the J_m in this formula with I_m, and the formula will remain true. The result above is simply a quantitative version giving "uniform convergence on compact subsets": in the degrees above any given degree, equality already holds at some finite stage, which is explicitly estimated in terms of the twists in the resolution of M. This is really the best one can do: the cohomology modules $\sum_n H^i\widetilde{M}(n)$ themselves are often not even finitely generated modules, whereas each $\mathrm{Ext}_S^i(J_m, M)$ is finitely generated! On the other hand the given estimate for a, though sometimes sharp, is not always so; it would be quite interesting to have other, more subtle results along these lines. For example, is it better computationally to use the ideals J_m above or to use the powers $I_m = (x_0, \ldots, x_r)^m$?

In one case there is a sharper convergence result, which seems worth mentioning:

Theorem 8.3. *If i is the smallest integer > 0 for which $\sum_n H^i\widetilde{M}(n) \neq 0$, then the natural map*

$$\mathrm{Ext}_S^i(J, M) \longrightarrow \sum_n H^i\widetilde{M}(n)$$

is injective, with image the submodule of all elements annihilated by J.

Proof. We may rephrase (a stronger version of) this in terms of local cohomology: if i is the smallest integer such that the local cohomology module $H_\mathfrak{m}^{i+1}(M) \neq 0$, then the natural map

$$\mathrm{Ext}_S^{i+1}(S/J, M) \longrightarrow H_\mathfrak{m}^{i+1}(M)$$

is injective with image the submodule annihilated by J. This is obvious if $i + 1 = 0$. If $i + 1 > 0$ then M has depth > 0, and we may finish by an induction using a nonzero divisor x on M and the associated exact sequence

$$0 \to M \xrightarrow{\cdot x} M \longrightarrow M/xM \to 0.$$

In general however, the maps $\mathrm{Ext}_S^i(J, M) \to \sum_n H^i\widetilde{M}(n)$ are not well–behaved away from the range given in Theorem 8.2.

The adaptation of this method to the bigraded case is the following:

Theorem 8.4. *If M is a bigraded module on $S = k[x_0, \ldots, x_r, y_0, \ldots, y_s]$ and $\widetilde{M}$ is the corresponding sheaf on $\mathbb{P}^r \times \mathbb{P}^s$, then for all $i \geq 0$ and any $(b_1, b_2) \in \mathbb{Z}^2$ we have*

$$\sum_{(m,n) \geq (b_1, b_2)} H^i\widetilde{M}(m, n) \simeq \mathrm{Ext}_S^i(J, M)_{\geq (b_1, b_2)},$$

where $J = J_{b_1} J_{b_2}$ *is any product of homogeneous ideals*

$$J_{b_1} \subset k[x_0, \ldots, x_r], \ J_{b_2} \subset k[y_0, \ldots, y_s]$$

primary to $(x_0, \ldots, x_r)$ *and* $(y_0, \ldots, y_s)$, *and contained in* $(x_0, \ldots, x_r)^{a_1}$, $(y_0, \ldots, y_s)^{a_2}$ *respectively, where* (a_1, a_2) *is the maximum of the degrees of the syzygies of M, diminished by* $(r + b_1, s + b_2)$.

Proof. Because of the existence of bigraded free resolutions, it suffices as before to prove that the theorem holds in case $M = S(m, n)$. But both $\mathrm{Ext}_S^i(J, M)$ and

$$\sum_{(m,n)} H^i \widetilde{M}(m, n)$$

satisfy a Künneth formula, so we reduce everything to Theorem 8.2. $\square$

Problem 8.5. It might be more convenient if we could use arbitrary ideals contained in higher and higher powers of the ideal of bilinear forms

$$(x_0, \ldots, x_r)(y_0, \ldots, y_s),$$

but this is not the case—if we take ideals with embedded components corresponding to the prime ideal $(x_0, \ldots, x_r, y_0, \ldots, y_s)$ then counterexamples can be manufactured. However, one suspects that it is enough to take ideals primary to $(x_0, \ldots, x_r, y_0, \ldots, y_s)$ and contained in large powers of it; is this true? Are there any particularly simple such ideals (for example, with few, nice, generators), to make it really worthwhile computationally?

9

Degrees of Complexity of a Graded Module

There are several measures of size of a graded module M over a standard graded ring A (or of a module M over a local ring R). To name a few: multiplicity, arithmetic and geometric degrees, Castelnuovo–Mumford regularity. There are also related measures of 'good behavior' (or of regularity): depth and projective dimension. Each of these has been used as a *complexity*, that is as a measure of the cost of extracting information about the module. When used for this purpose, some of these numbers tend to be very large and to regularly overshoot the real cost.

Here we will discuss these and other notions of the degree of a module and make comparisons amongst these numbers and those arising from associated modules of initial terms.

To illustrate our discussion of attaching *degrees* to a module, consider the following simple situation. Let M be a finitely generated module over a discrete valuation domain $(R, \mathfrak{m})$ and let

$$M = R^r \oplus R/\mathfrak{m}^{e_1} \oplus \cdots \oplus R/\mathfrak{m}^{e_s}, \ e_i \geq 1,$$

be its decomposition as a direct sum of cyclic modules.

A degree, as understood here, is an answer to the question: which *single* integer is best representative of M? Of course, the use of this integer determines the character of the answer. The rank r of M is probably the best measure of gross size. In turn, the integer $\sup\{e_i\}$ is a kind of irregularity of M. The so-called *arithmetic degree* will be given by

$$r + e_1 + \cdots + e_s.$$

It seeks to represent contributions from all prime ideals. In some ways, it would be better to code the numerical data differently, for example, in this case by the polynomial

$$r + (e_1 + \cdots + e_s)\mathbf{t},$$

that keeps track of the provenance of the integers but we have not found any application for such a representation.

The general philosophy is to attach to a graded module M (or to a module over a local ring) a *cycle*

$$\text{div}(M) = \sum_{i=1}^{s} \ell_i[\mathfrak{p}_i],$$

where ℓ_i are positive integers and $[\mathfrak{p}_i]$ are prime ideals, both derived from M, and associating *degrees* to $\text{div}(M)$. To illustrate, in the case of the arithmetic degree, ℓ_i will be the multiplicity of the prime $\mathfrak{p}_i$ in the primary decomposition of M, weighted by the degree of $\mathfrak{p}_i$,

$$\text{adeg}(M) = \sum_{i=1}^{s} \text{mult}_M(\mathfrak{p}_i) \deg(A/\mathfrak{p}_i).$$

In contrast the *multiplicity* $\deg(M)$ and the *geometric degree* $\text{gdeg}(M)$ are given by a similar expression in which only the primes $\mathfrak{p}_i$ of dimension $\dim(M)$ are taken in $\deg(M)$, and only the minimal primes in the support of M are considered for $\text{gdeg}(M)$.

We point out two features of these constructions of cycles attached to a given module M. First, it will be convenient to define more than one degree attached to an ideal. The degree itself is tailored to the need. Second, it is usual to focus on the prime ideals that are associated to M, as in some loose sense they represent obstructions to carrying out on M operations that are typical of vector spaces (read free modules). Viewed in this manner, it seems also natural to consider other prime ideals that occur as obstructions to such constructions but *on the level of the cohomology of M*. We have in mind those that appear as associated prime ideals of $\text{Ext}_A^i(M,A)$ (now we assume that A is a Gorenstein ring). The prime ideals

$$\bigcup_{i \geq 0} \text{Ass}(\text{Ext}_A^i(M,A)) \setminus \text{Ass}(M)$$

will play a significant role in our introduction of a family of degrees with a homological character. These primes themselves will be called *the hidden associated primes of M*. (Of course, this will lead to the notion of *well hidden associated prime!*)

Our guide to define new degrees is the following rule: That it generalizes ordinary length and behaves well under generic hyperplane section. In the case of $\text{adeg}(M)$, one has for a regular hyperplane section $\text{adeg}(M) \leq \text{adeg}(M/hM)$, with often a strict inequality. What is needed is another notion of degree $\text{Deg}(\cdot)$, which reduces to the usual multiplicity in the Cohen–Macaulay case, with the following properties:

(i)   If $L = \Gamma_{\mathfrak{m}}(M)$ is the submodule of elements of finite support of M and $\overline{M} = M/L$, then

$$\text{Deg}(M) = \text{Deg}(\overline{M}) + \ell(L).$$

(ii)   If $h \in S$ is a regular, generic hyperplane section on M, then

$$\text{Deg}(M) \geq \text{Deg}(M/hM).$$

(iii) If M is Cohen–Macaulay, then

$$\text{Deg}(M) = \deg(M).$$

Condition (ii) appears to be counter–intuitive until one looks at $\mathrm{Deg}(M)$ as carrying a composite of cohomological information: surely M is richer than M/hM and their degrees should reflect it.

There are several possible definitions for such degrees. One advantage is that they would offer *a priori* estimates for bounds one finds in Cohen–Macaulay modules as expressed by the ordinary multiplicity. Unfortunately, if M is not Cohen–Macaulay, $\deg(M)$ leaks all over as a predictor of properties of M, and to a lesser degree so do $\mathrm{adeg}(M)$ and the Castelnuovo–Mumford regularity $\mathrm{reg}(M)$ as well.

Let us now outline the contents of the chapter. In section 9.1 we recall the definitions of all major degrees we discuss, except for the homological degrees of section 9.4 Three features of the degrees are emphasized:

- comparisons between the degrees of an ideal I and $in(I)$ obtained from some term ordering,
- behavior under hyperplane sections and
- computation of degrees under *CoCoA* or *Macaulay*.

Section 9.2 discusses *degrees of nilpotency*, that is the estimation of the exponent in the Nullstellensatz. More precisely, given an ideal $I = (f_1, \ldots, f_m)$ of a polynomial ring $k[x_1, \ldots, x_n]$, determine an integer s such that

$$(\sqrt{I})^s \subset I.$$

The *index of nilpotency* or *degree of nilpotency* $\mathrm{nil}(I)$ of an ideal I is the smallest such integer s. A related index of nilpotency is $\mathrm{nil}_0(I)$, the smallest integer t such that $x^t \in I \ \forall x \in \sqrt{I}$. We give an internal analysis of the factors that contribute to $\mathrm{nil}(I)$. It will be seen that the degrees of the associated primes of I do not play a role and that the notion of multiplicity has to be replaced by another one, not larger, that carries structure better.

The purpose of section 9.3 is to find ways to make predictions about the outcome of carrying out Noether normalization on a graded algebra. Let A be a finitely generated, positively graded algebra over a field k,

$$A = k + A_1 + A_2 + \cdots = k + A_+,$$

where A_i denotes the space of homogeneous elements of degree i. We further assume that A is generated by its 1–forms, $A = k[A_1]$, in which case $A_i = A_1{}^i$. If k is sufficiently large and $\dim A = d$, there are forms $z_1, \ldots, z_d \in A_1$, such that

$$R = k[z_1, \ldots, z_d] = k[\mathbf{z}] \hookrightarrow A = S/I, \ S = k[z_1, \ldots, z_d, x_{d+1}, \ldots, x_n]$$

is a Noether normalization, that is, $x_1, \ldots, x_d$ are algebraically independent over k and A is a finite R–module. Let $b_1, b_2, \ldots, b_s$ be a minimal set of homogeneous generators of A as an R–module

$$A = \sum_{1 \leq i \leq s} R b_i, \quad \deg(b_i) = r_i.$$

We will be looking at the distribution of the r_i, particularly at the largest $r_R(A)$ of these degrees. For the 'best' of all Noether normalizations, it will be called the *reduction number* of A: $r(A)$. It turns out that $\operatorname{reg}(A)$ (in any characteristic) and $\operatorname{adeg}(A)$ (in characteristic zero) give estimates for $r(A)$.

Section 9.4 describes several possible definitions of a *homological degree* of a module M. If A is a ring of polynomials in d indeterminates over a field and M is a graded module, they all take the recursive form

$$\operatorname{Deg}(M) = \deg(M) + \sum_{i \geq 1}^{d} c_i \cdot \operatorname{Deg}(\operatorname{Ext}_A^i(M,A)),$$

where the c_i's are binomial coefficients chosen to accommodate the good behavior of $\operatorname{Deg}(\cdot)$ under hyperplane sections. They are reminiscent of formulas that occur in the theory of Buchsbaum modules. They suffer from some deficiencies, the most glaring being that $\operatorname{Deg}(M)$ and $\operatorname{Deg}(M/hM)$ may differ when h is regular.

Finally, section 9.5 shows how a degree function $\operatorname{Deg}(\cdot)$ leads to a technique to compare degrees and number of generators of ideals and modules.

9.1 Degrees of Modules

In this section S will be a standard graded algebra, which may just be a ring of polynomials with the usual grading, or a Noetherian local ring.

To a given finitely generated S–module M, graded when required, we will attach several degrees whose purpose is to size M. We discuss the degrees derived from the numerical data coded in the primary decomposition of the trivial submodule of M. In this section we consider those degrees that depend only on the local contributions of the components—multiplicities, arithmetic and geometric degrees. In section 9.4 we discuss another degree that seeks to capture among other things the interaction between the components as well.

We shall draw repeatedly from general facts on Hilbert functions and local cohomology; these can be found in Appendix B.

Classical Multiplicity of a Module

We begin by recalling the definition of multiplicity in local algebra, which opens a path for all refinements that follow.

Let $(R, \mathfrak{m})$ be a local ring and let M be a finitely generated R–module. The Hilbert function of M is

$$H_M : n \mapsto \ell(M/\mathfrak{m}^n M),$$

which for $n \gg 0$ is given by a polynomial (the Hilbert polynomial of M):

$$H_M(n) = \frac{\deg(M)}{d!}n^d + \text{lower order terms.} \qquad (9.1)$$

Definition 9.1. *The integer* $\deg(M)$ *is the* multiplicity *of the module M, and the integer $d = \dim M$ is its* dimension. *When no confusion arises, $\deg(M)$ is called the* degree *of M. The degree of a prime ideal $\mathfrak{p}$ is the degree of the module $A/\mathfrak{p}$. The degree of R will be also denoted by $e(R)$.*

If I is an $\mathfrak{m}$–primary ideal, using the filtration defined by the powers of I these notions lead to the functions of Hilbert–Samuel and associated degrees. Among these, the *multiplicity of the ideal I*, $e(I) = e(I;R)$, is the multiplicity of the associated graded ring $\mathrm{gr}_I(R)$.

The behavior of $\deg(M)$ and $\dim M$ with regard to submodules and primary decomposition arises from the following property.

Proposition 9.2. *Let R be a local ring and let*

$$0 \to L \longrightarrow M \longrightarrow N \to 0$$

be an exact sequence of R–modules. Then

$$\deg(M) = \begin{cases} \deg(L) + \deg(N) & \text{if } \dim L = \dim N = \dim M, \\ \deg(L) & \text{if } \dim N < \dim M, \\ \deg(N) & \text{if } \dim L < \dim M. \end{cases}$$

Proof. Let $\mathfrak{m}$ be the maximal ideal of R. For each integer $n \geq 0$ we have the exact sequence

$$0 \to L \cap \mathfrak{m}^n M / \mathfrak{m}^n L \longrightarrow L/\mathfrak{m}^n L \longrightarrow M/\mathfrak{m}^n M \longrightarrow N/\mathfrak{m}^n N \to 0,$$

which we split into two short exact sequences,

$$0 \to L \cap \mathfrak{m}^n M / \mathfrak{m}^n L \longrightarrow L/\mathfrak{m}^n L \longrightarrow L'_n \to 0,$$

$$0 \to L'_n \longrightarrow M/\mathfrak{m}^n M \longrightarrow N/\mathfrak{m}^n N \to 0.$$

They give rise to the equality of Hilbert functions

$$H_n(M) = H_N(n) + \ell(L'_n) \qquad (9.2)$$

$$\ell(L'_n) = H_L(n) - \ell(L \cap \mathfrak{m}^n M / \mathfrak{m}^n L). \qquad (9.3)$$

On the other hand, by the Artin-Rees lemma, there exists an integer r such that for all $n > r$

$$L \cap \mathfrak{m}^n M \subset \mathfrak{m}^{n-r} L.$$

This implies that

$$\ell(L \cap \mathfrak{m}^n M/\mathfrak{m}^n L) \leq \ell(\mathfrak{m}^{n-r}L/\mathfrak{m}^n L) = H_L(n) - H_L(n-r).$$

Note that for large n the right hand side of this equation is a polynomial of degree equal to $\dim L - 1$. When taken into (9.3), it follows that $\ell(L'_n)$ grows as a polynomial of degree equal to $\dim L$. Finally we use this information in (9.2) to get the desired assertion.   □

Remark 9.3. This proposition shows that $\deg(\cdot)$ is not very good at dealing with apples and oranges at the same time. The rules it obeys are however very good for computation according to the following observations.

(a) Let

$$M = M_0 \supset M_1 \supset \cdots \supset M_s = 0,$$

be a filtration of M whose factors $M_i/M_{i+1} \simeq R/\mathfrak{p}_i$, with $\mathfrak{p}_i$ a prime ideal. Applying Proposition 9.2 we obtain

$$\deg(M) = \sum_{\mathfrak{p}} \ell_{\mathfrak{p}} \cdot \deg(R/\mathfrak{p}),$$

where the sum is over the set of prime ideals $\mathfrak{p}$ with $\dim R/\mathfrak{p} = \dim M$ that occur in the filtration. All such primes are associated to M and do not contain any other such. As for the integers $\ell_{\mathfrak{p}}$, they measure the length of the Artinian $R_{\mathfrak{p}}$–module $M_{\mathfrak{p}}$,

$$\ell_{\mathfrak{p}} = \ell(\Gamma_{\mathfrak{p}}(M_{\mathfrak{p}})),$$

where $\Gamma_{\mathfrak{p}}(\cdot)$ is the local cohomology functor $H^0_{\mathfrak{p}}(\cdot)$ (see Appendix A).
(b) If M is a graded module over a standard graded algebra A, $\deg(M)$ is taken relative to the maximal homogeneous ideal of A.
(c) If R is a local ring, then for any ideal I one has $\deg(I) \leq \deg(R)$ and equality holds if I is not contained in any minimal prime.

There are other notions of degree, of which we recall the following. If R is an integral domain of field of fractions K, then for any finitely generated module M a natural candidate for a degree is

$$\dim_K(M \otimes_R K).$$

Regrettably this will be zero if M has a nontrivial annihilator. This difficulty is dealt with in the following manner.

Definition 9.4. *Let R be a Noetherian ring and M a finitely generated R–module. For a prime ideal $\mathfrak{p} \subset R$, the integer*

$$\mathrm{mult}_M(\mathfrak{p}) = \ell(\Gamma_{\mathfrak{p}}(M_{\mathfrak{p}}))$$

is the length multiplicity *of $\mathfrak{p}$ with respect to M.*

This number $\mathrm{mult}_M(\mathfrak{p})$, which vanishes if $\mathfrak{p}$ is not an associated prime of M, is a measure of the contribution of $\mathfrak{p}$ to the primary decomposition of the null submodule of M. It is not usually accessible through computation. On the other hand, the multiplicity of the module, $\deg(M)$, is connected to some of these numbers and the degrees of the corresponding primes by the expression

$$\deg(M) = \sum_{\mathrm{height}(\mathfrak{p})=\dim(M)} \mathrm{mult}_M(\mathfrak{p}) \cdot \deg R/\mathfrak{p}. \tag{9.4}$$

This expression works very well in case $R/\mathfrak{p}$ is a local ring or a graded ring. Another case is that of an affine domain $R = k[x_1, \ldots, x_n]/\mathfrak{p}$. Among all possible Noether normalizations,

$$S = k[z_1, \ldots, z_d] \hookrightarrow R,$$

define

$$\deg(R) = \min_S \{ \deg(R \otimes_S k(z_1, \ldots, z_d)) \}.$$

Arithmetic Degree of a Module

For most of the remainder of this section, we consider graded modules over graded algebras in order to emphasize computational issues. Occasionally we make assertions about local rings as well.

Definition 9.5. *Let R be a local ring (resp. a standard graded algebra) and let M be a finitely generated R–module (resp. finitely generated graded R–module). The arithmetic degree of M is the integer*

$$\mathrm{adeg}(M) = \sum_{\mathfrak{p} \in \mathrm{Ass}(M)} \mathrm{mult}_M(\mathfrak{p}) \cdot \deg R/\mathfrak{p}. \tag{9.5}$$

The definition applies to more general modules, not just graded modules, although its main use is for graded modules over a standard algebra A. If all the associated primes of M have the same dimension, then $\mathrm{adeg}(M)$ is just $\deg(M)$.

In general, $\mathrm{adeg}(M)$ can be put together as follows. Collect the associated primes of M by their dimensions:

$$\dim M = d_1 > d_2 > \cdots > d_n,$$

$$\mathrm{adeg}(M) = a_{d_1}(M) + a_{d_2}(M) + \cdots + a_{d_n}(M),$$

where $a_{d_i}(M)$ is contribution of all primes in $\mathrm{Ass}(M)$ of dimension d_i. More precisely, define $\mathrm{adeg}_r(M)$ by

$$\boxed{\mathrm{adeg}_r(M) = \sum_{\dim A/\mathfrak{p}=r} \mathrm{mult}_M(\mathfrak{p}) \cdot \deg A/\mathfrak{p}.}$$

$$(9.6)$$

The integer $a_{d_1}(M)$, in the case of a graded module M, is its multiplicity $\deg(M)$.

Geometric Degree of a Module

Bayer and Mumford ([BM93]) have refined $\deg(M)$ into the following integer:

Definition 9.6. *Let A be a local ring (resp. a standard graded algebra) and let M be a finitely generated A–module (resp. finitely generated graded A–module). The geometric degree of M is the integer*

$$\boxed{\mathrm{gdeg}(M) = \sum_{\mathfrak{p} \text{ minimal } \in \mathrm{Ass}(M)} \mathrm{mult}_M(\mathfrak{p}) \cdot \deg A/\mathfrak{p}.}$$

$$(9.7)$$

If $A = S/I$, we want to view $\mathrm{reg}(A)$ and $\mathrm{adeg}(A)$ as the two basic measures of complexity of A. (In addition, [BM93] is a far-flung survey of $\mathrm{reg}(A)$ and $\mathrm{adeg}(A)$ in terms of the degree data of a presentation $A = k[x_1,\dots,x_n]/I$.)

Computation of the Arithmetic Degree of a Module

We show how a program with the capabilities of *Macaulay* or *CoCoA* can be used to compute the arithmetic degree of a graded module M without availing itself of any primary decomposition. Let $S = k[x_1,\dots,x_n]$ and suppose $\dim M = d \le n$. It suffices to construct graded modules M_i, $i = 1,\dots,n$, such that

$$a_i(M) = \deg(M_i).$$

For each integer $i \ge 0$, denote $L_i = \mathrm{Ext}_S^i(M,S)$. By local duality, a prime ideal $\mathfrak{p} \subset S$ of height i is associated to M if and only if $(L_i)_\mathfrak{p} \ne 0$; furthermore $\ell((L_i)_\mathfrak{p}) = \mathrm{mult}_M(\mathfrak{p})$.

This gives what is required: Compute for each L_i its degree $e_0(L_i)$ and codimension c_i. Then choose M_i according to

$$M_i = \begin{cases} 0 & \text{if } c_i > i \\ L_i & \text{otherwise.} \end{cases}$$

In other words:

Proposition 9.7. *For a graded S–module M and for each integer i denote by c_i the codimension of* $\text{Ext}_S^i(M,S)$. *Then*

$$\text{adeg}(M) = \sum_{i=0}^{n} \left\lfloor \frac{i}{c_i} \right\rfloor \, \deg(\text{Ext}_S^i(M,S)). \tag{9.8}$$

Equivalently,

$$\text{adeg}(M) = \sum_{i=0}^{n} \deg(\text{Ext}_S^i(\text{Ext}_S^i(M,S),S)). \tag{9.9}$$

Note that the second formula involves more computation. The amusing thing about this formula (one sets $\lfloor \frac{0}{0} \rfloor = 1$) is that it gives a sum of sums of terms some of which are not always available.

The module

$$\text{Hull}_i(M) = \text{Ext}_S^i(\text{Ext}_S^i(M,S),S)$$

appeared in Section 3.2 in the definition of *i*th hull of an ideal:

$$\text{hull}_i(I) = \text{annihilator of } \text{Hull}_i(S/I).$$

The formula for adeg(M) can be rewritten as

$$\text{adeg}(M) = \sum_{i \geq 0} \deg(\text{Hull}_i(M)).$$

Theorem 9.8. *Let S be a Gorenstein local ring and let I be an ideal of codimension m with a decomposition*

$$I = I_m \bigcap_j Q_j,$$

where I_m is an unmixed ideal of codimension m and the Q_j's are irreducible ideals with distinct radicals and codimension greater than m. Then

$$\text{adeg}(S/I) = \sum_i \deg(S/\text{hull}_i(I)).$$

Proof. The hull$_i(I)$ have the same associated primes as the modules Hull$_i(S/I)$, so the point is whether the multiplicities of the associated primes are equal.

For a prime of codimension m this is clear, since

$$\deg(\text{Hull}_m(S/I)) = \deg(S/I).$$

So we may assume, by localization, that we have a local ring $(S, \mathfrak{m})$ of dimension d, and an ideal

$$I = J \cap Q,$$

where Q is an irreducible $\mathfrak{m}$–primary ideal and J is an ideal of lower codimension. From the embedding

$$0 \to S/I \longrightarrow S/J \oplus S/Q,$$

we get the surjection

$$\mathrm{Ext}_S^d(S/Q, S) \longrightarrow \mathrm{Ext}_S^d(S/I, S) \to 0,$$

and finally by duality the embedding

$$\mathrm{Ext}_S^d(\mathrm{Ext}_S^d(S/I, S), S) = \mathrm{Hull}_d(S/I) \hookrightarrow S/Q = \mathrm{Ext}_S^d(\mathrm{Ext}_S^d(S/Q, S), S).$$

Now we make use of the assumption that $A = S/Q$ is a Gorenstein Artin algebra. For the ideal $L = \mathrm{Hull}_d(S/I)$ of A, we have that

$$\ell(A) = \ell(L) + \ell(A/L) = \ell(L) + \ell(0 : L),$$

and therefore $\ell(L) = \ell(A/0 : L)$. To complete the proof, it suffices to note that $A/0 : L$ is just the original $S/\mathrm{hull}_d(S/I)$ localized at one of its minimal primes. $\square$

Computation of the Geometric Degree of a Module

Given a graded module M, the computation of $\mathrm{gdeg}(M)$ requires more screening of the modules $M_i = \mathrm{Ext}_S^i(\mathrm{Ext}_S^i(M, S), S))$ than what was used in (9.9). We now outline the steps.

- Let C_i be the annihilator of M_i. Since M_i is an equidimensional module, C_i is unmixed. Its radical, $I_i = \sqrt{C_i}$, is the intersection (empty perhaps) of two sets of primes, those which are minimal over the support of M and those which are embedded primes of M. We set $I_i = J_i \cap L_i$ for the corresponding decomposition.

- If r is the codimension of M as an S–module, $M_i = 0$ for $i < r$ and $I_r = J_r$. If there are minimal primes of M in the next codimension, J_{r+1} is an intersection of prime ideals none of which contains J_r, while if L_{r+1} is a proper ideal each of its minimal primes contains a minimal prime of J_r and therefore $J_r \subset L_{r+1}$. We can get hold of J_{r+1} from

$$I_{r+1} : J_r = (J_{r+1} \cap L_{r+1}) : J_r = (J_{r+1} : J_r) \cap (L_{r+1} : J_r) = J_{r+1} : J_r = J_{r+1}.$$

To get hold of higher J_i's the method is similar,

$$J_i = I_i : (J_1 \cap \cdots \cap J_{i-1}).$$

- For each M_i, denote by $(M_i)_0$ the elements of M_i annihilated by some high power of J_i,

$$(M_i)_0 = \Gamma_{J_i}(M_i).$$

Proposition 9.9. *For a graded S–module M, its geometric degree is given by*

$$\text{gdeg}(M) = \sum_{i=0}^{n} \deg((\text{Ext}_S^i(\text{Ext}_S^i(M,S),S))_0). \tag{9.10}$$

Degrees and Initial Ideals

We are going to follow an instructive technique of [STV95] to describe some relationships in the form of inequalities between the arithmetic and geometric degrees of an ideal and the corresponding degree of any of its initial ideals. They are useful estimates for $\text{adeg}(A)$. The technique itself is just as significant.

Arithmetic Degree

It is a fact that the arithmetic degree of a homogeneous ideal is bounded above by that of any initial ideal, and, moreover this inequality holds for the contributions in each dimension. This result is a special case of a more general result due to R. Hartshorne. Indeed, in [Har66, Theorem 2.10] it is shown that $\text{adeg}_r(\cdot)$ is upper-semicontinuous with respect to flat families of projective schemes, and a well-known result of Gröbner basis theory states that, for any term order, the initial ideal $in(I)$ is a flat deformation of I (see e.g. Appendix B, [BM93], [Ei95]).

Theorem 9.10. *Fix any term order on S, and let $I \subset S$ be any homogeneous ideal. Then*

$$\text{adeg}_r(I) \leq \text{adeg}_r(in(I)) \qquad \text{for all } r = 0, 1, \ldots, n.$$

The following proof is taken in its entirety from [STV95]. Let I be a homogeneous ideal in S. For each integer $r = 0, 1, \ldots, n$ we consider

$$I_{\geq r} := \{ f \in S : \dim(I : f) < r \}. \tag{9.11}$$

It is clear that the set $I_{\geq r}$ is an ideal in S, containing I, and that it equals the intersection of all primary components of dimension $\geq r$ in any primary decomposition of I. We also find that the Hilbert polynomial of the graded S-module $I_{\geq r}/I$ has the form

$$H_{I_{\geq r}/I}(d) = \frac{\text{adeg}_{r-1}(I)}{(r-1)!} \cdot d^{r-1} + O(d^{r-2}). \tag{9.12}$$

Proof. We claim the following inclusion of monomial ideals:

$$in(I_{\geq r}) \subseteq (in(I))_{\geq r}. \tag{9.13}$$

Indeed, suppose $m \in in(I_{\geq r})$. Then $m = in(f)$ for some $f \in I_{\geq r}$, and therefore $r > \dim(I : f) = \dim in(I : f)$. Since $in(I : f)$ is contained in $(in(I) : in(f)) = (in(I) : m)$, we conclude $\dim(in(I) : m) < r$, and therefore $m \in (in(I))_{\geq r}$.

Whenever we have an inclusion of homogeneous ideals $I \subseteq I'$ in S, then their quotient I'/I is isomorphic as a graded vector space to $in(I')/in(I)$. It suffices to note that the canonical monomial basis for $in(I')/in(I)$ lifts to a basis for I'/I. Applying this observation to $I' = I_{\geq r}$, and using (9.13), we obtain the inclusion of graded vector spaces:

$$I_{\geq r}/I \simeq in(I_{\geq r})/in(I) \hookrightarrow \left(in(I)\right)_{\geq r}/in(I). \tag{9.14}$$

The leading term of the Hilbert polynomial of the right hand side in (9.14) exceeds that of the left hand side. By (9.12), this proves Theorem 9.10.   □

Geometric Degree

The geometric degree behaves quite differently from the arithmetic degree.

Theorem 9.11. *Let I be a homogeneous ideal in $k[x_1,\ldots,x_n]$ and $in(I)$ its initial ideal with respect to any term order. Then*

$$\mathrm{gdeg}(I) \geq \mathrm{gdeg}(in(I)). \tag{9.15}$$

Proof. First suppose that I is pure d-dimensional, that is, each isolated prime of I has dimension d. Then $in(I)$ is pure d-dimensional as well (see e.g. [KS95]). Since the degree is preserved under taking initial ideals, and since degree and geometric degree coincide for pure ideals, we have $\mathrm{gdeg}(I) = \deg(I) = \deg(in(I)) = \mathrm{gdeg}(in(I))$.

Suppose now that I has dimension d but is *not* pure. Write $I = J \cap K$, where J is pure d-dimensional and each isolated prime of K has dimension $\leq d-1$ and is isolated in I as well. We have $\deg(I) = \deg(J)$ and $\mathrm{gdeg}(I) = \deg(J) + \mathrm{gdeg}(K)$. If $in(I)$ is pure, then we are done since $\mathrm{gdeg}(in(I)) = \deg(in(I)) = \deg(I) \leq \mathrm{gdeg}(I)$. If $in(I)$ is *not* pure, then write $in(I) = J' \cap K'$ where J' is a pure d-dimensional monomial ideal and each isolated prime of K' has dimension $\leq d-1$ and is isolated in $in(I)$. By induction on the dimension we have $\mathrm{gdeg}(in(K)) \leq \mathrm{gdeg}(K)$, and therefore it suffices to show

$$\mathrm{gdeg}(K') \leq \mathrm{gdeg}(in(K)).$$

We use the formula

$$\sqrt{in(J)} = \sqrt{J'} \cap \sqrt{K'}.$$

Let P be an isolated prime of K'. By the above formula, P is an isolated prime of either $in(J)$ or $in(K)$. But J, and hence $in(J)$, is pure of dimension d. Thus P is an isolated prime of $in(K)$. To complete the proof, it suffices to show that

$$\mathrm{mult}_{K'}P \leq \mathrm{mult}_{in(K)}P.$$

To see this, note that

$$in(J) \cdot in(K) \subseteq in(J \cdot K) \subseteq in(I) = J' \cap K'.$$

Localizing at P, this gives

$$in(K)S_P \subseteq K'S_P.$$

This yields the desired inequality.   □

Remark 9.12. There is another explanation for the inequality

$$\operatorname{adeg}(S/in(I)) \geq \operatorname{adeg}(S/I).$$

It arises from Proposition 9.7 and the fact that for any degree filtration F, the module $\operatorname{Ext}_S^r(M,S)$ is a subfactor of $\operatorname{Ext}_S^r(\operatorname{gr}_F(M),S)$ (see [Gr89], [Sjo73]).

Degrees and Hyperplane Sections

A fruitful method to probe a ring A is to compare the properties of A with those of $A/(h)$ for some hyperplane section h. The great Italian algebraists used this repeatedly.

Definition 9.13. *Let k be a field and let $A = k + A_1 + \cdots + A_i + \cdots$ be a standard algebra. A* hyperplane section h of A *is a form in A_1 that is not contained in any minimal prime of A (sometimes stricter conditions are imposed). If M is a graded A–module, a* hyperplane section *for it is an element $h \in A_1$ such that $\dim(M/hM) = \dim M - 1$.*

If k is an infinite field, most elements of A_1 are hyperplane sections.

We contrast this with the notion of a *superficial element*.

Definition 9.14. *Let A and M be as above. An element $h \in A_1$ is a* superficial element *for M if the submodule $(0 :_M h)$ is trivial in almost all degrees, that is, $(0 :_M h)_n = 0$ for $n \gg 0$.*

From the exact sequence

$$0 \to 0 :_M h \longrightarrow M \xrightarrow{\ .h\ } M \longrightarrow M/hM \to 0,$$

one then obtains the equality of Hilbert functions

$$H_{M/hM}(n) = H_M(n) - H_M(n-1), \quad n \gg 0.$$

This allows to recover many properties of M from those of M/hM.

If $(A, \mathfrak{m})$ is a local ring and M is a finitely generated A–module, the notion of superficial element adapts as follows. Let $G = \operatorname{gr}_{\mathfrak{m}}(A)$ be the associated graded ring of A and let $M = \operatorname{gr}_{\mathfrak{m}}(M)$ be the corresponding associated graded module. A superficial element for M is an element $x \in \mathfrak{m} \setminus \mathfrak{m}^2$ whose image h in $G_1 = \mathfrak{m}/\mathfrak{m}^2$ is superficial for M. There is a similar behavior for the Hilbert functions given by $\ell(M/(x,\mathfrak{m}^n)M)$ and $\ell(M/\mathfrak{m}^{n+1}M)$.

Theorem 9.15. *Let $(R, \mathfrak{m})$ be a Noetherian local ring of dimension $d > 0$ and let M be a finitely generated R–module. Let $x \in \mathfrak{m}$ and consider the short exact sequence induced by multiplication by x,*

$$0 \to L \longrightarrow M \xrightarrow{x} M \longrightarrow G \to 0.$$

If L is a module of finite length, then $\ell(H^0_{\mathfrak{m}}(G)) \geq \ell(L)$. Moreover, if $d = 1$ then

$$\ell(H^0_{\mathfrak{m}}(G)) = \ell(L) + \ell(\overline{M}/x\overline{M}),$$

where $\overline{M} = M/\text{torsion}$.

Note that the last formula follows immediately if M decomposes into a direct sum of a torsionfree plus torsion summands. In case R is a standard graded algebra over a field of characteristic zero, the first assertion is contained in the refined statement of [HU93, Proposition 3.5].

Proof. By going over to the completion of R, we may assume that R is a Gorenstein local ring. Furthermore, by modding out a regular sequence in R we may assume that $\dim R = \dim M$.

We break up the sequence above into two short exact sequences

$$0 \to L \longrightarrow M \longrightarrow N \to 0$$
$$0 \to N \longrightarrow M \longrightarrow G \to 0$$

and take their cohomology. We get the long exact sequences

$$0 \to N_0 \xrightarrow{\alpha} M_0 \to L_0 \to N_1 \to \cdots \to L_{d-1} \to N_d \to M_d \to L_d \to 0$$
$$0 \to G_0 \to M_0 \xrightarrow{\beta} N_0 \to G_1 \to \cdots \to N_{d-1} \to G_d \to M_d \to N_d \to 0$$

where for a module C we denote $C_i = \text{Ext}^i_R(C,R)$.

We observe that $G_0 = 0$ and that $L_i = 0$ for $i < d$. If $d > 1$, the first sequence says that $\ell(M_d) = \ell(N_d) + \ell(L_d)$, while the second yields $\ell(M_d) \leq \ell(G_d) + \ell(N_d)$. The first assertion follows from $\ell(L_d) = \ell(H^0_{\mathfrak{m}}(L))$ and $\ell(G_d) = \ell(H^0_{\mathfrak{m}}(G))$ by local duality.

If $d = 1$, we get the equalities

$$\ell(G_1) + \ell(N_1) = \ell(M_1) + \ell(\text{coker}(\beta))$$
$$\ell(M_1) = \ell(N_1) + \ell(L_1).$$

Thus $\ell(G_1) = \ell(L_1) + \ell(\text{coker}(\beta))$. Finally, note that β composed with the isomorphism α is multiplication by x on the module M_0, which has the same multiplicity as $\overline{M}$.

Taking local cohomology of the two basic sequences we have the exact sequences of modules of finite length,

$$0 \to H^0_{\mathfrak{m}}(L) = L \longrightarrow H^0_{\mathfrak{m}}(M) \longrightarrow H^0_{\mathfrak{m}}(N) \to H^1_{\mathfrak{m}}(L) = 0$$
$$0 \to H^0_{\mathfrak{m}}(N) \longrightarrow H^0_{\mathfrak{m}}(M) \longrightarrow H^0_{\mathfrak{m}}(G)$$

from which the inequality $\ell(L) \leq \ell(H^0_{\mathfrak{m}}(G))$ follows. $\square$

The preceding yields the following criterion for a module to be almost Cohen–Macaulay.

Corollary 9.16. *Let $(R, \mathfrak{m})$ be a Noetherian local ring and let M be a finitely generated R–module of dimension $d \geq 2$. Let $\mathbf{x} = x_1, \ldots, x_{d-2} \in \mathfrak{m}$ be a superficial sequence of elements for M. If depth $M/(\mathbf{x})M > 0$ then depth $M \geq d - 1$.*

Proof. We may assume $d \geq 3$ because for $d = 2$ this is trivial. Theorem 9.15 (with $M = M/(x_1, \ldots, x_{d-3})M$ and $x = x_{d-2}$) implies that x_{d-2} is a regular element on $M/(x_1, \ldots, x_{d-3})M$, so that depth $M/(x_1, \ldots, x_{d-3})M > 0$. By induction, $x_1, \ldots, x_{d-3}$ is a regular sequence on M, whence so is $x_1, \ldots, x_{d-2}$; and since depth $M/\mathbf{x}M > 0$, we conclude that depth $M \geq d - 1$. ∎

Arithmetic Degree

To examine the behavior of $\mathrm{adeg}(M)$ under hyperplane sections we focus on the following:

Theorem 9.17. *Let A be a standard graded algebra, let M be a graded module and let $h \in A_1$ be a regular element on M. Then*

$$\mathrm{adeg}(M/hM) \geq \mathrm{adeg}(M). \tag{9.16}$$

We now begin to assemble the elements for a proof of Theorem 9.17.

Proposition 9.18. *Let M be a finitely generated R–module and let $\mathfrak{p}$ be an associated prime of M. If $h \in R$ is regular on M then any minimal prime P of $(\mathfrak{p}, h)$ is an associated prime of M/hM.*

Proof. We may assume that P is the unique maximal ideal of R. Let $L = \Gamma_{\mathfrak{p}}(M)$; note that we cannot have $L \subset hM$ as otherwise since h is regular on M we would have $L = hL$. This means that there is a mapping

$$R/\mathfrak{p} \longrightarrow M,$$

which on reduction modulo h does not have a trivial image. Since $R/(\mathfrak{p}, h)$ is annihilated by a power of P, its image will also be of finite length and non-trivial, and therefore $P \in \mathrm{Ass}(M/hM)$. ∎

Let us make a brief comparison between the associated primes of a module M and the associated primes of M/hM, where h is a regular element on M. According to the previous Proposition, for each associated prime $\mathfrak{p}$ of M, for which $(\mathfrak{p}, h)$ is not the unity ideal (which will be the case when M is a graded module and h is a homogeneous form of positive degree) there is at least one associated prime of M/hM that contains $\mathfrak{p}$: it is easy to see that any minimal prime of $(\mathfrak{p}, h)$ will do. There may be however associated primes of M/hM, such as $\mathfrak{n}$, which do not arise in this manner. We indicate this in the diagram below:

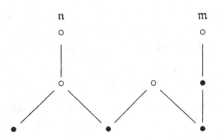

In the expression for adeg(M/hM) we are going to keep apart primes such as $\mathfrak{m}$ (which we call associated primes of the first kind) and primes as $\mathfrak{n}$, which are not minimal over $(\mathfrak{p}, h)$, for any $\mathfrak{p} \in \text{Ass}(M)$.

Proof of Theorem 9.17. For $\mathfrak{m} \in \text{Ass}(M/hM)$ of the first kind, let

$$A(\mathfrak{m}) = \{\mathfrak{p}_1, \ldots, \mathfrak{p}_r\}$$

be the set of associated primes of M such that $\mathfrak{m}$ is minimal over each ideal $(\mathfrak{p}_i, h)$. Note that these prime ideals have the same dimension, $\dim(A/\mathfrak{m}) + 1$. Denote by I the product of the $\mathfrak{p}_i$ and let $L(\mathfrak{m}) = L = \Gamma_I(M)$. We have an embedding

$$L/hL \hookrightarrow M/hM$$

since $L \cap hM = hL$. As a consequence, $\text{mult}_{L/hL}(\mathfrak{m}) \leq \text{mult}_{M/hM}(\mathfrak{m})$. On the other hand,

$$\text{mult}_L(\mathfrak{p}_i) = \text{mult}_M(\mathfrak{p}_i) \tag{9.17}$$

for any prime in the set $A(\mathfrak{m})$.

Now let

$$L = L_0 \supset L_1 \supset L_2 \supset \cdots \tag{9.18}$$

be a filtration of L whose factors are of the form $A/\mathfrak{q}$ for some homogeneous prime ideal $\mathfrak{q}$. The number of times in which the prime $\mathfrak{p} = \mathfrak{p}_i$ occurs in the filtration is $\text{mult}_L(\mathfrak{p})$.

Consider the effect of multiplication by h on the terms of this filtration: By the snake lemma, we have exact sequences

$$0 \to {}_hL_{i+1} \longrightarrow {}_hL_i \longrightarrow {}_hA/\mathfrak{q} \longrightarrow L_{i+1}/hL_{i+1} \longrightarrow L_i/hL_i \longrightarrow A/(\mathfrak{q}, h) \to 0.$$

Localizing at $\mathfrak{m}$ we get sequences of modules of finite length over $A_\mathfrak{m}$. Adding these lengths, and taking into account the collapsing, we get

$$\ell(L_\mathfrak{m}/hL_\mathfrak{m}) = \sum \ell((A/(\mathfrak{q}, h))_\mathfrak{m}) - \ell(({}_hA/\mathfrak{q})_\mathfrak{m}), \tag{9.19}$$

where ${}_hA/\mathfrak{q}$ is $A/\mathfrak{q}$ if $h \in \mathfrak{q}$ and zero otherwise. Note that some of the $\mathfrak{q}$ may occur repeated according to the previous observation. After the localization at $\mathfrak{m}$ only the $\mathfrak{q}$ corresponding to the $\mathfrak{p}_i$'s survive:

$$\text{mult}_{L/hL}(\mathfrak{m}) = \ell(L_\mathfrak{m}/hL_\mathfrak{m}) = \sum_{\mathfrak{p}_i \subset \mathfrak{m}} \text{mult}_M(\mathfrak{p}_i)\ell((A/(\mathfrak{p}_i,h))_\mathfrak{m}). \qquad (9.20)$$

We thus have

$$
\begin{aligned}
\text{adeg}(M/hM) &= \sum_\mathfrak{m} \text{mult}_{M/hM}(\mathfrak{m})\deg(A/\mathfrak{m}) \\
&\quad + \sum_\mathfrak{n} \text{mult}_{M/hM}(\mathfrak{n})\deg(A/\mathfrak{n}) \\
&\geq \sum_\mathfrak{m} \text{mult}_{M/hM}(\mathfrak{m})\deg(A/\mathfrak{m}) \\
&\geq \sum_\mathfrak{m} \text{mult}_{L(\mathfrak{m})/hL(\mathfrak{m})}(\mathfrak{m})\deg(A/\mathfrak{m}) \qquad (9.21) \\
&= \sum_\mathfrak{m}\Big(\sum_{\mathfrak{p}_i \in A(\mathfrak{m})} \text{mult}_M(\mathfrak{p}_i)\ell((A/(\mathfrak{p}_i,h))_\mathfrak{m}) \Big)\deg(A/\mathfrak{m}) \\
&= \sum_{\mathfrak{p}_i} \text{mult}_M(\mathfrak{p}_i)\Big(\sum_{\mathfrak{p}_i \in A(\mathfrak{m})} \ell((A/(\mathfrak{p}_i,h))_\mathfrak{m})\deg(A/\mathfrak{m}) \Big) \\
&= \sum_{\mathfrak{p}_i} \text{mult}_M(\mathfrak{p}_i)\deg(A/\mathfrak{p}_i) \\
&= \text{adeg}(M),
\end{aligned}
$$

where in equation (9.21) we first used the equality (9.17), and then derived from each $L(\mathfrak{m})$ the equality provided by the computation of multiplicities in (9.19) and (9.20), while taking into account that

$$
\begin{aligned}
\deg(A/\mathfrak{p}_i) &= \deg(A/(\mathfrak{p}_i,h)) \\
&= \sum_{\mathfrak{p}_i \in A(\mathfrak{m})} \ell((A/(\mathfrak{p}_i,h))_\mathfrak{m})\deg(A/\mathfrak{m}),
\end{aligned}
$$

the last equality by [BH93, Corollary 4.6.8]. $\qquad\square$

The computation shows all that is required in order to obtain equality.

Example 9.19. A simple example where strict inequality holds is the ideal of two disjoint lines in $\mathbb{P}^3$: $A = k[x,y,z,w]/I$, with $I = (x,y) \cap (z,w)$ and $h = x+z$, then $\text{adeg}(A) = 2 < 3 = \text{adeg}(A/hA)$.

Problem 9.20 (Gunston). Let M be a module of finite length. The Dilworth number of M, $\text{dil}(M)$, is the supremum of the numbers of generators of the submodules of M. If M is a graded module, develop a method to find $\text{dil}(M)$.

Problem 9.21. Develop algorithms to compute the arithmetic and geometric degrees of rings defined by monomial ideals, and which do no appeal directly to Ext functors.

9.2 Index of Nilpotency

One of the purposes of primary decomposition is to provide a path to the estimation of the exponent in the Nullstellensatz. More precisely, given an ideal $I = (f_1, \ldots, f_m)$ of a polynomial ring $k[x_1, \ldots, x_n]$, determine an integer s such that

$$(\sqrt{I})^s \subset I.$$

The *index of nilpotency* or *degree of nilpotency* $\mathrm{nil}(I)$ of an ideal I is the smallest such integer s. A related index of nilpotency is $\mathrm{nil}_0(I)$, the smallest integer t such that $x^t \in I$, $\forall x \in \sqrt{I}$. Note that $\mathrm{nil}_0(I) = \mathrm{nil}(I)$ if the characteristic of the field k is larger than $\mathrm{nil}_0(I)$.

The examination of these integers (we focus on $\mathrm{nil}(I)$) is framed by two benchmark examples, the first an expression of an old theorem of Macaulay ([Mac16]), the other a modern general estimate of the degrees in the exponent of the Nullstellensatz by Brownawell ([Br87]) and Kollár ([Ko88]).

The first example is the case of the monomial ideal

$$I = (x_1^{a_1}, \ldots, x_n^{a_n}).$$

Then

$$\mathrm{nil}(I) = \sum_{i=1}^{n} a_i - n + 1,$$

the familiar estimate of everybody's first experience with sums of nilpotent elements! It is an instance of a formula of Macaulay ([Mac16, Theorem 58]), asserting that if the zero-dimensional ideal I is generated by a regular sequence $f_1, \ldots, f_n$ of forms of degrees $d_1, \ldots, d_n$,

$$\mathrm{nil}(f_1, \ldots, f_n) = d_1 + \cdots + d_n - n + 1. \tag{9.22}$$

The other example is the binomial ideal of Mayr and Meyer type, which is also a regular sequence (see [MMe82] and [Ko88]):

$$I = (x_1^{a_1}, x_1 x_n^{a_2 - 1} - x_2^{a_2}, \ldots, x_{n-2} x_n^{a_{n-1} - 1} - x_{n-1}^{a_{n-1}}, x_{n-1} x_n^{a_n - 1} - x_0^{a_n})$$

has a single associated prime $\mathfrak{p} = (x_0, \ldots, x_{n-1})$. Furthermore, specializing $x_n = 1$, gives that

$$k[x_0, \ldots, x_n]/(I, x_n - 1) \simeq k[x_0]/(x_0)^{\prod a_i},$$

and therefore

$$\mathrm{nil}(I) = \prod_{i=1}^{n} a_i,$$

a lamentable state of affairs!

Classically, degree bounds for $\mathrm{nil}_0(I)$ which are doubly-exponential in n are found in [Her26]. It came as a surprise that one can do a lot better when Brownawell [Br87] and Kollár [Ko88] established the beautiful (see also [BM93]):

Theorem 9.22 (Brownawell, Kollár). *Let k be any field, let*

$$I = (f_1, \ldots, f_m) \subset k[x_1, \ldots, x_n]$$

and let $d = \max\{\deg(f_i),\ i = 1, \ldots, m; 3\}$. If $n = 1$, replace d by $2d - 1$. If $g \in \sqrt{I}$, then there is an expression

$$g^s = \sum_{i=1}^{m} h_i f_i$$

where $s \leq d^n$ and $\deg(h_i) \leq (1 + \deg(g))d^n$. Moreover, $\mathrm{nil}(I) \leq d^n$.

There are cases in which this result is even sharp. Our purpose here is to understand the factors that lead to this still non-elementary bound. The estimates above have the character of the degree of a variety, of the kind expressed by Bézout theorem and its generalizations (see [STV95]). In the analysis to follow, we are going to see that the multiplicities of the primary components—isolated and embedded—have a role in determining the exponent in the Nullstellensatz, but that the degrees of the associated primes do not. Furthermore, the notion of multiplicity has to be modified to reflect structure rather than length.

It is important to obtain estimates for these integers, the better if they can be found from a generating set of the ideal I. A strict rule in the search for such formulas is that the index of nilpotency of a radical ideal should be 1.

Radical Embeddings

Internally, an estimate can be approached as follows. Let $I = J_1 \cap \cdots \cap J_n$ be some decomposition, not necessarily primary or irredundant. The homomorphism of rings

$$0 \to R/I \longrightarrow \prod_{i=1}^{n} R/J_i$$

shows that

$$\sqrt{I}/I \hookrightarrow \prod_{i=1}^{n} \sqrt{J_i}/J_i,$$

and therefore that

$$\mathrm{nil}(I) \leq \max\{\mathrm{nil}(J_i),\ i = 1, \ldots, n\}. \tag{9.23}$$

This approach to $\mathrm{nil}(I)$ is tied, to some extent, to the choice of the J_i's. These choices should have attached numerical data depending entirely on I. Thus taking for J_i the components of a primary decomposition of I will not always meet this criterion.

It is useful to view this question in another setting. Let R be a ring and let φ be a nilpotent endomorphism of a finitely generated R–module,

$$\varphi : E \longrightarrow E, \quad \varphi^r = 0.$$

The least such exponent r is unchanged if R and E are localized at the multiplicative set S defined by the associated primes $\mathfrak{p}_i$ of E, $S = R \setminus \bigcup \mathfrak{p}_i$. Further, if E were to decompose into a direct of submodules invariants under φ, then r would be the maximum of the index of nilpotency of φ on the submodules. This is further improved by considering the localizations $R_{\mathfrak{p}_i}$ and replacing direct sums by certain filtrations.

Loewy Length of a Module

Let us consider first the question of expressing the index of nilpotency for a $\mathfrak{p}$–primary ideal J. Since R/J has only $\mathfrak{p}$ for associated prime, it follows that $\mathrm{nil}(J) = \mathrm{nil}(J_\mathfrak{p})$. For later reference:

Remark 9.23. If I is a $\mathfrak{p}$–primary ideal then $\mathrm{nil}(I) \leq \ell(R_\mathfrak{p}/I_\mathfrak{p})$.

This is the classical notion of multiplicity of a primary ideal. For the purposes here, it overshoots almost always the true value of $\mathrm{nil}(I)$. We make a brief analysis of the factors that affect $\mathrm{nil}(I)$.

Definition 9.24. *Let $(R, \mathfrak{m})$ be a local ring and let M be an R-module of finite length. The* Loewy length *of M is the smallest integer s such that $\mathfrak{m}^s M = 0$. It is denoted $\ell\ell(M)$.*

For a non trivial vector space V, $\ell\ell(V) = 1$, while if $N \subset M$, $\ell\ell(N) \leq \ell\ell(M)$. On the other hand, the function $\ell\ell(\cdot)$ is not additive on short exact sequences, which limits its general usefulness.

Remark 9.25. If I is a $\mathfrak{p}$–primary ideal then $\mathrm{nil}(I) = \ell\ell(R_\mathfrak{p}/I_\mathfrak{p})$.

Another strategy consists in using irreducible ideals. If $I = J_1 \cap \cdots \cap J_n$, $n \geq 2$, is an irreducible decomposition of the $\mathfrak{p}$–primary ideal I, note that the embedding

$$0 \to R/I \longrightarrow \prod_{i=1}^{n} R/J_i$$

shows that

$$\mathrm{nil}(I) = \ell\ell(R_\mathfrak{p}/I_\mathfrak{p}) \leq \ell(R_\mathfrak{p}/I_\mathfrak{p}) - n,$$

where the inequality arises from the fact that the maximal ideal of $R_\mathfrak{p}/I_\mathfrak{p}$ is not principal if $n \geq 2$.

Example 9.26. Let us show how resorting to irreducible decompositions works very well in the case of monomial ideals. Let

$$I = (f_1, \ldots, f_s) \subset k[x_1, \ldots, x_n]$$

be a monomial ideal, $f_i = \mathbf{x}^{a_i}$. I admits a decomposition by irreducible monomial ideals,

$$I = \bigcap_{j=1}^{m} I_j,$$

each of which is easily seen to be generated by powers of variables,

$$I_j = (x_{j_1}^{b_{j_1}}, \ldots, x_{j_{r_j}}^{b_{j_{r_j}}}).$$

It follows from (9.22) and (9.23) that

$$\mathrm{nil}(I) \leq \max\Big\{ \sum_{k=1}^{r_j} b_{j_k} - r_j + 1, \ k = 1, \ldots, m \Big\}.$$

There remains to estimate the sums $d_j = \sum_{k=1}^{r_j} b_{j_k}$ in terms of the $\deg(f_i)$. It is clear that

$$d_j \leq \sum_{p=1}^{n} \sup\{\deg_{x_p}(f_i), \ i = 1, \ldots, s\}.$$

Classical and Loewy Multiplicities

Let M be a finitely generated R–module and let $\mathfrak{p}$ be a prime ideal. One can measure the contribution of $\mathfrak{p}$ as an associated prime of M in the following ways.

Definition 9.27. *Let* $L(M;\mathfrak{p}) = \Gamma_{\mathfrak{p}}(M_{\mathfrak{p}})$

(a) *The multiplicity of M at $\mathfrak{p}$ is* $\mathrm{mult}_M(\mathfrak{p}) = \ell(L(M;\mathfrak{p}))$.
(b) *The Loewy multiplicity of M at $\mathfrak{p}$ is* $\mathrm{Lmult}_M(\mathfrak{p}) = \ell\ell(L(M;\mathfrak{p}))$.

A natural name for this length is also the *local contribution of the prime* $\mathfrak{p}$. Note that these integers are zero unless $\mathfrak{p} \in \mathrm{Ass}(M)$. Let us formulate a first estimate for the index of nilpotency of a general ideal I. It is patterned after one in [STV95], whose argument we adapt, but gives much lower bounds (the formula in [STV95] had other aims though).

Proposition 9.28. *Suppose* $I = J \cap L$ *where L is a $\mathfrak{p}$–primary ideal and all the associated primes of J are strictly contained in $\mathfrak{p}$. Then*

$$\mathrm{nil}(I) \leq \mathrm{nil}(J) + \mathrm{Lmult}_{R/I}(\mathfrak{p}). \tag{9.24}$$

Proof. We may assume that $\mathfrak{p}$ is the unique maximal ideal. Write the right hand side of (9.24) as $N + h$, where N is the value of the formula for the ideal J, and $h = \mathrm{Lmult}_{R/I}(\mathfrak{p})$. (We may assume $\mathfrak{p}$ is not the only associated prime.) Noting that

$$\sqrt{I} = \sqrt{J}$$

and

$$J/I = H_{\mathfrak{p}}^{0}(R/I),$$

one has

$$(\sqrt{I})^{N+h} \subset \mathfrak{p}^h \cdot (\sqrt{J})^N$$
$$\subset \mathfrak{p}^h \cdot J$$
$$\subset I,$$

to prove the assertion.   □

The argument still misses the 'interaction' between J and L. Also, $(\sqrt{J})^N$ may land deep into J so that some lower power of p could be used.

If we now apply this bound repeatedly on the chains of associated primes of an ideal we obtain:

Proposition 9.29. *For an ideal I of a Noetherian ring R let C denote the set of chains of primes in* $\mathrm{Ass}(R/I)$

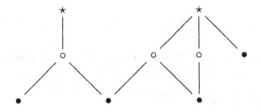

Then we have

$$\mathrm{nil}(I) \leq \max_C \big\{ \textstyle\sum_{\mathfrak{p}\in C} \mathrm{Lmult}_{R/I}(\mathfrak{p}) \big\}. \tag{9.25}$$

Proof. Let $I = \bigcap Q_i$ be a primary decomposition of I. For each maximal associated prime ideal $\mathfrak{p}_k$, set

$$J_k = \bigcap Q_i, \ Q_i \subset \mathfrak{p}_k,$$

which can be written

$$J_k = L_k \bigcap H_k,$$

where L_k is the $\mathfrak{p}_k$–primary component among the Q_i and H_k is the intersection of all other Q_i's contained in $\mathfrak{p}_k$. Observe that for any associated prime of J_k,

$$\mathrm{Lmult}_{R/J_k}(\mathfrak{p}) = \mathrm{Lmult}_{R/I}(\mathfrak{p}).$$

From Proposition 9.28 we have

$$\mathrm{nil}(J_k) \leq \mathrm{nil}(H_k) + \mathrm{Lmult}_{R/I}(\mathfrak{p}),$$

and apply induction on the chains associated to H_k.   □

Remark 9.30. Unfortunately we don't know how to convert this intrinsic efficiency into *a priori* estimates for nil(I). One has ([STV95])

$$\text{adeg}(S/I) \geq \text{nil}(I).$$

The great advantage of the left hand side expression lies in that it can be computed without the actual knowledge of the associated primes, according to Proposition 9.7. It differs significantly from (9.25) by the presence of the "degree" terms, by the substitution of the multiplicities by intrinsically smaller ones, and finally by taking partial sums. Each of these actions may significantly affect the estimate for nil(I).

Remark 9.31. It will be interesting to find estimates for the Loewy multiplicities in terms of the other "descriptors" of I, such as Betti numbers. Small gains can be further gotten by using irreducible decompositions and chains of primary components.

Question 9.32. An unsettled issue is the relationship between nil(I) and nil(I'), where $I' = in(I)$ for some term order. Does

$$\text{nil}(I) \leq \text{nil}(I')$$

hold always?

Exercise 9.33. Let I be a homogeneous ideal and h a generic hyperplane set of R/I. Make comparisons between nil(I,h) and nil(I).

9.3 Qualitative Aspects of Noether Normalization

For a graded ring $A = k[x_1,\ldots,x_n]/I$, a great many of its properties with regard to one of its graded, Noetherian normalizations R can be read from the Hilbert function of A (e.g. the rank of A as an R–module and in certain cases the degrees of the module generators. Our aim in this section is to find ways to make predictions about the outcome of carrying out Noether normalization on a graded algebra. (We follow [Vas96] and related references.)

The Reduction Number of an Algebra

Let A be a finitely generated, positively graded algebra over a field k,

$$A = k + A_1 + A_2 + \cdots = k + A_+,$$

where A_i denotes the space of homogeneous elements of degree i. We further assume that A is generated by its 1–forms, $A = k[A_1]$, in which case $A_i = A_1{}^i$. If k is sufficiently large and $\dim A = d$, there are forms $x_1,\ldots,x_d \in A_1$, such that

$$R = k[x_1,\ldots,x_d] = k[\mathbf{z}] \hookrightarrow A = S/I, \; S = k[x_1,\ldots,x_d,x_{d+1},\ldots,x_n]$$

is a Noether normalization, that is, the x_i are algebraically independent over k and A is a finite R–module. Let $b_1, b_2, \ldots, b_s$ be a minimal set of homogeneous generators of A as an R–module

$$A = \sum_{1 \leq i \leq s} Rb_i, \ \deg(b_i) = r_i.$$

We will be looking at the distribution of the r_i, particularly at the following integer.

Definition 9.34. *The* reduction number $r_R(A)$ *of A with respect to R is the supremum of all* $\deg(b_i)$. *The* global reduction number $r(A)$ *of A is the infimum of $r_R(A)$ over all possible Noether normalizations of A.*

One of our aims is to make predictions about these integers, but without availing ourselves of any Noether normalization. We emphasize this by saying that the Noether normalizations are invisible to us, and the information we may have about A comes from the presentation $A = S/I$.

Hilbert Function and Reduction Number

Let A be an standard graded algebra over a field k and let

$$HP_A(\mathbf{t}) = \frac{f(\mathbf{t})}{(1-\mathbf{t})^d}, \ d = \dim A, \ f(\mathbf{t}) = 1 + a_1 \mathbf{t} + \cdots + a_b \mathbf{t}^b, \ a_b \neq 0$$

be its Hilbert–Poincaré series.

If A is Cohen–Macaulay, its reduction number can be read from $HP_A(\mathbf{t})$,

$$r(A) = b.$$

We would like to derive estimates for $r(A)$ using the data that describes $HP_A(t)$ under more general conditions.

It is natural to view $r(A)$ as a measure of complexity of the algebra A. Taken this way, it has been compared to another index of complexity of A, the Castelnuovo regularity of A and the various degrees that were discussed in the previous section.

Example 9.35. (a) In dimension 1 we have $\operatorname{reg}(A) < \operatorname{adeg}(A)$. Indeed, let $R = k[x]$ be a Noether normalization of A and let

$$A \simeq \overset{e}{\underset{i=1}{\bigoplus}} R(-a_i) \overset{f}{\underset{j=1}{\bigoplus}} (R/(x^{c_j}))(-b_j)$$

be the decomposition of A as the direct sum of cyclic R–modules.

We have

$$r(A) = \sup\{a_i, b_j\},$$
$$\operatorname{reg}(A) = \sup\{a_i, b_j + c_j - 1\},$$
$$\operatorname{adeg}(A) = e + \sum_{j=1}^{f} c_j,$$

while the minimal number of generators of A as an R–module is

$$v(A) = e + f.$$

Because A is generated by elements of degree 1, there must be no gaps in the sequence of degrees of its module generators, which implies $b_j < v(A)$, and no gaps either in the degrees of its torsion-free part so that $a_i < e$. This proves that for each integer ℓ,

$$b_\ell \leq v(A) - 1 \leq e + \sum_{j \neq \ell} c_j,$$

and therefore

$$r(A) \leq \mathrm{reg}(A) < \mathrm{adeg}(A).$$

(b) In dimension greater than 1, the inequality $\mathrm{reg}(A) < \mathrm{adeg}(A)$ does not always hold, according to the following example of B. Ulrich. Let $A = k[x,y,u,v]/I$, with $I = ((x,y)^2, xu^t + yv^t)$. If $t \geq 3$,

$$r(A) = 1 < \mathrm{adeg}(A) = 2 < \mathrm{reg}(A) = t.$$

Regular Sequences

A simple but often effective approach to finding the reduction number of a graded algebra A is the following. Suppose $A = k[x_1, \dots, x_n]/I$, where I is a homogeneous ideal of height r. Suppose we can find a regular sequence $\mathbf{f} = f_1, \dots, f_r$ of forms of degrees $d_1, \dots, d_r$, contained in I. They define a graded algebra homomorphism

$$A_0 = k[x_1, \dots, x_n]/(\mathbf{f}) \longrightarrow A \to 0.$$

Since $\dim A_0 = \dim A$, the image of any Noether normalization of A_0 will provide a Noether normalization for A also. It will follow that $r(A_0) \geq r(A)$.

Since A_0 is a complete intersection, the distributions of the degrees that occur under Noether normalization is known: The Hilbert function of A_0 is

$$H_{A_0}(t) = \frac{\prod_{i=1}^{r}(1 + t + \cdots + t^{d_i - 1})}{(1 - t)^{n-r}},$$

and therefore

$$r(A_0) = \sum_{i=1}^{r} d_i - r.$$

Let us sum up this observation in the following general bound. (It reveals that we must be wary of multiplicative bounds.)

Proposition 9.36. *let A be a standard graded algebra over an infinite field. If the presentation ideal I is generated by forms of degree $\leq d$, then $r(A) \leq dr - r$. In particular if I is generated by forms of degree 2 then $r(A) \leq \mathrm{height}(I)$.*

Example 9.37. Test cases for finding reduction numbers are two families of algebras associated to graphs. If G is a graph on the set $\mathbf{x} = \{x_1, \ldots, x_n\}$ of vertices, one associates to its edge set E the following algebras. (Assume k is an infinite field.)

(a) The semigroup algebra of G,

$$k[G] = k[x_i x_j \mid \{x_i, x_j\} \in E].$$

(b) The edge ring of G,

$$k[E_G] = k[x_1, \ldots, x_n]/I(G), \quad I(G) = (x_i x_j \mid \{x_i, x_j\} \in E).$$

These algebras present different difficulties to estimating the reduction numbers. (To make both algebras standard, in the first ring set $\deg(x_i) = \frac{1}{2}$ for each variable.)

In the case of $k[E_G]$, the result above implies that

$$r(k[E_G]) \leq \text{height } I(G).$$

In turn height $I(G)$ is the length of the shortest vertex cover of the graph G.

For the algebra $k[G]$, the problem is much more delicate since its equations are going to be defined by the cycles of G, but it is not clear when long cycles are needed in obtaining the required complete intersection. (There is a study of this for certain planar graphs in [DG96].)

Castelnuovo–Mumford Regularity and Reduction Number

When the details of a projective resolution of A as an R–module are known, the following use of regularity provides a bound for the reduction number of A.

Theorem 9.38. *Let A be a standard graded algebra over an infinite field k and let $R \hookrightarrow A$ be a standard Noether normalization. Then $r_R(A) \leq \text{reg}(A)$.*

Proof. A more general statement is given in [Tru87] and we content ourselves here for algebras such as A. Let $R \hookrightarrow A$ be a (graded) Noether normalization of A. From Theorem B.27(b), $\text{reg}(A)$ can be determined by the degrees where the local cohomology modules $H_{\mathfrak{m}}^i(A)$ vanish: If we set

$$a_i(A) = \alpha_+(H_{\mathfrak{m}}^i(A)) = \sup\{n \mid H_{\mathfrak{m}}^i(A)_n \neq 0\},$$

then

$$\text{reg}(A) = \sup\{a_i(A) + i\}.$$

Note that these modules are also given by $H_{\mathfrak{p}}^i(A)$, where $\mathfrak{p}$ is the irrelevant maximal ideal of R. If we now apply the definition of $\text{reg}(A)$, to a minimal resolution of A as an R-module, we gather that $r_R(A) \leq \text{reg}(A)$. $\square$

This result indicates why $\text{reg}(A)$ tends to overshoot $r_R(A)$.

Estimation *Sauvage*

The number $a(A) = b - d$ defines the so-called *index of regularity* or *a–invariant* of A: For $n > a(A)$, the Hilbert function of A, $H_A(n)$, and its Hilbert polynomial, $P_A(n)$, agree (see [BH93, Chapter 4] for fuller details). Since $P_A(n)$ can be obtained from $f(t)$ and d,

$$P_A(n) = \sum_{i=0}^{d-1} (-1)^i e_i \binom{n+d-i-1}{d-i-1}, \quad e_i = \frac{f^{(i)}(1)}{i!},$$

we can derive a very crude bound for $r(A)$ from an abstract yet sharp result about reductions given by the following theorem of Eakin and Sathaye ([ES76]), which we state in our setting as follows:

Theorem 9.39 (Eakin–Sathaye). *Let A be a standard graded algebra over an infinite field k. For positive integers n and r, suppose $\dim_k A_n < \binom{n+r}{r}$. Then there exist $z_1, \ldots, z_r \in A_1$ such that*

$$A_n = (z_1, \ldots, z_r)A_{n-1}.$$

Moreover, if $x_1, \ldots, x_p \in A_1$ are such that $(x_1, \ldots, x_p)^n = A_n$, then r generic linear combinations of $x_1, \ldots, x_p$ will define such sets.

Definition 9.40. *Let $(R, \mathfrak{m}, k)$ be a local ring of residue field k. We say that an ideal $I \subseteq R$ satisfies condition $S(r, n)$ if it satisfies each of the following:*

- $\mathfrak{m}I^n = 0$.
- $\dim_k I^n < \binom{n+r}{r}$.
- *There exist $\{y_i\}_{i=1}^{\infty} \subseteq I$ and an integer p such that for all i, we have*

$$I = (y_i, y_{i+1}, \ldots, y_{i+p-1}).$$

- *S_ω (the symmetric group on the natural numbers) acts on R as a group of automorphisms such that $\sigma(y_i) = y_{\sigma(i)}$ for all $\sigma \in S_\omega$.*

Theorem 9.41. *If $I \subseteq R$ is an ideal satisfying $S(r, n)$, then*

$$(y_1, \ldots, y_r)I^{n-1} = I^n.$$

To prove the theorem, we reduce to the following lemma.

Lemma 9.42. *If $I \subseteq R$ is an ideal satisfying $S(r, n)$, then*

$$(y_1, \ldots, y_{r+1})I^{n-1} = I^n \implies (y_1, \ldots, y_r)I^{n-1} = I^n.$$

Proof. Assume the lemma is false. Let r be minimal such that the lemma does not hold. Let n be minimal for this r.

Note that for each $\sigma \in S_\omega$, $\sigma(I) = I$. Thus for each $\sigma \in S_\omega$, $y_i I^{n-1} \cong y_{\sigma(i)} I^{n-1}$. Let d be the common dimension, $d = \dim_k y_i I^{n-1}$.

Case 1. $d \geq \binom{n+r-1}{r}$. If $r = 1$, then $\dim_k y_1 I^{n-1} \geq \binom{n}{1} = n$, and $\dim_k I^n < \binom{n+1}{1} = n + 1$. Hence $y_1 I^{n-1} = I^n$, a contradiction. If $r > 1$, let $\overline{R} = R/y_1 I^{n-1}$. We claim that $\overline{R}$ and $\overline{I}$ satisfy $S(r-1, n)$:

- $m\bar{I}^n = 0$ still.
- $\dim_k \bar{I}^n = \dim_k I^n - \dim_k y_1 I^{n-1} < \binom{n+r}{r} - \binom{n+r-1}{r} = \binom{n+r-1}{r-1}$.
- $\{\bar{y}_i\}_{i=2}^{\infty}$ suffices.
- Let $G = \{\sigma \in S_\omega : \sigma(y_1) = y_1\}$. G fixes $y_1 I^{n-1}$, and hence acts on $\bar{R}$ in the appropriate way.

So we may apply the lemma and conclude that

$$\bar{I}^n = (\bar{y}_2, \ldots, \bar{y}_r)\bar{I}^{n-1}, \text{ or}$$
$$I^n = (y_2, \ldots, y_r)I^{n-1} + y_1 I^{n-1}$$
$$= (y_1, \ldots, y_r)I^{n-1},$$

a contradiction.

Case 2. $d < \binom{n+r-1}{r}$. Let $K = [0 : y_1]_{m-1}$. Let $\bar{R} = R/K$. The map $I^{n-1} \xrightarrow{y_1} y_1 I^{n-1}$ is surjective and has kernel K. Hence $\bar{I}^{n-1} \cong y_1 I^{n-1}$. We claim that $\bar{R}$ and $\bar{I}$ satisfy $S(r, n-1)$:

- $m\bar{I}^{n-1} \cong m y_1 I^{n-1} \subseteq m I^n = 0$.
- $\dim_k \bar{I}^{n-1} = \dim_k y_1 I^{n-1} < \binom{n-1+r}{r}$.
- $\{\bar{y}_i\}_{i=2}^{\infty}$ suffices.
- Each $\sigma \in S_\omega$ fixes I^{n-1}, hence $G = \{\sigma \in S_\omega : \sigma(y_1) = y_1\}$ fixes K. So G acts on $\bar{R}$ in the appropriate way.

So we may apply the lemma again and conclude that

$$\bar{I}^{n-1} = (\bar{y}_2, \ldots, \bar{y}_{r+1})\bar{I}^{n-1}$$

or,

$$I^{n-1} = (y_2, \ldots, y_{r+1})I^{n-2} + K.$$

By hypothesis,

$$(y_1, \ldots, y_{r+1})I^{n-1} = I^n,$$
$$y_1 I^{n-1} + (y_2, \ldots, y_{r+1})I^{n-1} = I^n,$$
$$y_1(y_2, \ldots, y_{r+1})I^{n-2} + y_1 K + (y_2, \ldots, y_{r+1})I^{n-1} = I^n,$$
$$(y_2, \ldots, y_{r+1})I^{n-1} = I^n.$$

The action of an appropriate σ yields the desired result. $\qquad\square$

Proof of Theorem 9.39. We first show how the condition $S(r, n)$ is brought in. Let $\{x_1, \ldots, x_p\}$ be a set of elements in I such that $(x_1, \ldots, x_p)^n = I^n$. Let $\mathbf{U} = \{u_{j=1, i=1}^{p, \infty}\}$ be a set of distinct indeterminates over R and set $S = R[\mathbf{U}]_{mR[\mathbf{U}]}$, and let S_ω act on S by $\sigma(u_{i,j}) = u_{\sigma(i),j}$. Finally, define $y_j = \sum_{i=1}^{p} u_{ij}x_i$. Note every set $\{y_s, \ldots, y_{s+p-1}\}$ in S generates the same ideal as $(x_1, \ldots, x_p)$.

Now we pass to S and assume $mI^n = 0$. We already have $(y_1,\ldots,y_p)I^{n-1} = I^n$, so if $p \le r$ we are done. If $p-1 \ge r$, we have $S(r,n)$, and hence $S(p-1,n)$, so the lemma yields $(y_1,\ldots,y_{p-1})I^{n-1} = I^n$, and we proceed by induction.

It is clear how a specialization of the u_{ij} can bring us back to R. $\square$

In our case, with $r = d$,

$$A_n = (z_1,\ldots,z_d)A_{n-1}$$

means that $r(A) < n$. It follows that if we take $n > a(A)$ and satisfying

$$\binom{n+d}{d} > P_A(n),$$

then $r(A) < n$.

Proposition 9.43. *Let A be a standard graded algebra. It is always possible to bound $r(A)$ given the Hilbert series of A.*

Example 9.44. Let us consider two examples. If $\dim A = 1$, $P_A(n) = e_0$, so that taking $\binom{n+1}{1} > e_0$, we have

$$r(A) \le \sup\{a(A), e_0 - 1\}.$$

If $\dim A = 2$, its Hilbert polynomial is

$$P_A(n) = e_0(n+1) - e_1.$$

According to the observation above, if we pick n such that

$$\binom{n+2}{2} > e_0(n+1) - e_1,$$

that is,

$$n^2 + (3 - 2e_0)n + (2 + 2e_0 - 2e_1) > 0,$$

then

$$r(A) \le \sup\{a(A), n - 1\}.$$

Remark 9.45. It would be interesting to find similar estimations for both $\operatorname{reg}(A)$ and $\operatorname{adeg}(A)$: From the Hilbert function to predict estimates for $\operatorname{reg}(A)$ and $\operatorname{adeg}(A)$.

Remark 9.46. Bounds for n can be roughly estimated as follows, according to [DST88, p. 108] (see also [Ost66]): Let $f(t) = a_d t^d + \cdots + a_1 t + a_0$ be a polynomial with real coefficients. For any root α of $f(t)$,

$$|\alpha| \le \begin{cases} 1 + \max\left\{\left|\dfrac{a_{d-1}}{a_d}\right|, \left|\dfrac{a_{d-2}}{a_d}\right|, \ldots, \left|\dfrac{a_0}{a_d}\right|\right\} & \text{(Cauchy)} \\[3mm] \max\left\{\left|\dfrac{da_{d-1}}{a_d}\right|, \left|\dfrac{da_{d-2}}{a_d}\right|^{\frac{1}{2}}, \ldots, \left|\dfrac{da_0}{a_d}\right|^{\frac{1}{d}}\right\} & \text{(Cauchy)} \\[3mm] 2\max\left\{\left|\dfrac{a_{d-1}}{a_d}\right|, \left|\dfrac{a_{d-2}}{a_d}\right|^{\frac{1}{2}}, \ldots, \left|\dfrac{a_0}{a_d}\right|^{\frac{1}{d}}\right\} & \text{(Knuth)} \end{cases}$$

Of course Hilbert polynomials are not random ones, so one should expect better bounds in these cases.

The Relation Type of an Algebra

Definition 9.47. *Let $A = k[x_1, \ldots, x_n]/I$ be a standard algebra over a field k. Suppose that $I \subset (x_1, \ldots, x_n)^2$. The relation type of A is the least integer s such that $I = (I_1, \ldots, I_s)$, where I_i is the ith graded component of I. We will denote it by $\mathrm{rt}(A)$.*

This integer $\mathrm{rt}(A)$ is independent of the presentation. The notion can be extended to the cases of graded algebras over a commutative ring k.

Proposition 9.48. *Let k be an infinite field and let A be a standard Cohen–Macaulay k–algebra. Then $\mathrm{rt}(A) \leq \mathrm{r}(A) + 1$.*

Proof. Let
$$R = k[z_1, \ldots, z_d] \hookrightarrow A = k[x_1, \ldots, x_n]/I$$
be a standard Noether normalization ($\deg z_i = 1$). Since A is Cohen–Macaulay, $A = \bigoplus_q R b_q$ and $\mathrm{r}(A) = \max_q \{\deg(b_q)\}$. Computing the Castelnuovo regularity $\mathrm{reg}(A)$ with respect to the ring R gives $\mathrm{reg}(A) = \mathrm{r}(A)$, while computing it with respect to the ring $k[x_1, \ldots, x_n]$ gives $\mathrm{reg}(A) \geq \mathrm{rt}(A) - 1$. $\qquad\square$

Remark 9.49. It is usually the case that $\mathrm{rt}(A)$ is much smaller than $\mathrm{r}(A)$. For example, let $\mathbf{T}$ be a $m \times n$ matrix of distinct indeterminates, and let $A = k[T_{ij}\text{'s}]/I$, where I is the ideal generated by all minors of order 2 of $\mathbf{T}$. It can be shown that $\mathrm{r}(A) = \min\{m, n\}$, while $\mathrm{rt}(A) = 2$ by definition.

More general relationships between $\mathrm{rt}(A)$ and $\mathrm{r}(A)$ are not known, one difficulty being that they behave differently with respect to operations such as taking hyperplane section.

Cayley–Hamilton Theorem

Let $\varphi : E \mapsto E$ be an endomorphism of a finitely generated module over a commutative ring A. φ will satisfy a monic equation
$$\varphi^n + a_{n-1}\varphi^{n-1} + \cdots + a_0 I = 0, \quad a_i \in A,$$
where I is the identity endomorphism of E. By a Cayley–Hamilton theorem we understand any method to get such equations in a deliberate fashion.

From now on A is a standard graded ring and $R = k[\mathbf{z}] \hookrightarrow A$ is a fixed Noether normalization. To determine $\mathrm{r}(A)$, we look for equations of integral dependence of the elements of A with respect to R.

A simple approach is to find graded R–modules on which A acts as endomorphisms (e.g. A itself). The most naive path to the equation is through the Cayley–Hamilton theorem developed by G. Almkivist ([Al78]), working as follows. Let E be a finitely generated R–module and let
$$f : E \mapsto E$$

be an endomorphism. Map a free graded module over E and lift f:

$$
\begin{array}{ccc}
F & \xrightarrow{\;\pi\;} & E \\
{\scriptstyle\varphi}\big\downarrow & & \big\downarrow{\scriptstyle f} \\
F & \xrightarrow{\;\pi\;} & E
\end{array}
$$

Let

$$P_\varphi(t) = \det(tI + \varphi) = t^n + \cdots + a_n$$

be the characteristic polynomial of φ, $n = \mathrm{rank}(F)$. By the usual Cayley–Hamilton theorem, we have that $P_\varphi(f) = 0$. The drawback is that n, which is at least the minimal number of generators of E, may be too large. One should do much better using a trick of [Al78]. Lift f to a mapping from a projective resolution of E into itself,

$$
\begin{array}{ccccccccccc}
0 & \longrightarrow & F_s & \longrightarrow & \cdots & \longrightarrow & F_1 & \longrightarrow & F_0 & \longrightarrow & E & \longrightarrow & 0 \\
& & \big\downarrow{\scriptstyle\varphi_s} & & & & \big\downarrow{\scriptstyle\varphi_1} & & \big\downarrow{\scriptstyle\varphi_0} & & \big\downarrow{\scriptstyle f} & & \\
0 & \longrightarrow & F_s & \longrightarrow & \cdots & \longrightarrow & F_1 & \longrightarrow & F_0 & \longrightarrow & E & \longrightarrow & 0
\end{array}
$$

and define

$$P_f(t) = \prod_{i=0}^{s} (P_{\varphi_i}(t))^{(-1)^i}.$$

This rational function is actually a polynomial in $R[t]$ (see [Al78]). Indeed

$$P_f(t) = \frac{a(t)}{b(t)} = \frac{t^m + a_1 t^{m-1} + \cdots + a_m}{t^n + b_1 t^{n-1} + \cdots + b_n}$$

reduces to the characteristic polynomial of f acting on the vector space $E \otimes_R K$, where K is the field of fractions of R. It follows that

$$P_f(t) = c(t) = t^e + c_1 t^{e-1} + \cdots + c_e \in K[t].$$

Multiplying out we obtain $a(t) = b(t) \cdot c(t)$, from which it will follow that all the coefficients of $c(t)$ lie in R.

If E is a graded module and f is homogeneous, then $P_f(t)$ is a homogeneous polynomial, $\deg E = \deg P_f(t)$.

Proposition 9.50 (Cayley–Hamilton Theorem). *Let E be a graded module over a ring of polynomials and let f be an endomorphism of E. If the rank of E over R is e, $P_f(t)$ is a monic polynomial of degree e. Moreover, if E is torsion–free over R then $P_f(f) \cdot E = 0$. Furthermore, if E is a faithful A–module and $f \in A_1$ then $f^e \in (\mathbf{z})A_+$.*

Proof. Most of these properties are proved in [Al78]. Passing over to the field of fractions of R, the characteristic polynomial of the vector space mapping

$$f \otimes K : E \otimes K \longrightarrow E \otimes K$$

is precisely $P_f(t)$.

The existence of an equation

$$f^e + c_1 f^{e-1} + \cdots + c_e = 0, \; c_i \in R,$$

and the fact that $\mathrm{Hom}_R(E,E)$ is graded implies that there is a similar equation where $c_i \in (\mathbf{z})^i$. □

Remark 9.51. Without the torsion-free hypothesis the assertions may fail. For instance, if $A = k[x,y]/(xy,y^2)$, f is multiplication by y on A, then $P_f(t) = t$, but $P_f(f) \neq 0$.

Remark 9.52. A question of independent interest is to find R–modules of small multiplicity that afford embeddings

$$A \hookrightarrow \mathrm{Hom}_R(E,E).$$

For example, the relationship between these multiplicities may be as large as $\deg(E) = n$ and $\deg(R) = [\frac{n^2}{2}] + 1$ ([Sch05]). There are however certain restrictions to be overcome: If the Cohen–Macaulay type of the localization of A at its minimal primes is at most 3, then $\deg(E) \geq \deg(A)$ (see [Gu72]).

Remark 9.53. In case the module E is A itself, we do not need the device of Proposition 9.50, as we can argue directly as follows. For $u \in A_1$, $R[u] \simeq R[t]/I$, where $I = f \cdot J$, $\mathrm{height}(J) \geq 2$. But if A is torsion-free over R, $R[u]$ will have the same property and necessarily $J = (1)$. This means that the rank of $R[u]$, which is the degree of f, is at most $\deg(A)$.

The case when A is Cohen–Macaulay has an added feature. If A is a hypersurface ring, that is $A = R[u]$ for some $u \in A$, then $\deg f = \deg(A) - 1 = \mathrm{r}(A) - 1$. In all other cases however $\mathrm{r}(A) < \deg(A) - 1$.

Arithmetic Degree of an Algebra Versus its Reduction Number

Proposition 9.50 had a very restrictive condition on the module: all associated primes have the same dimension. To be able to overcome this, we proceed as follows. Given

$$\varphi : E \longmapsto E,$$

we show that there is a filtration by characteristic submodules

$$E = E_1 \supset E_2 \supset \cdots \supset E_n = 0$$

such that

$$\mathrm{adeg}(E) = \sum \deg(E_i/E_{i+1}),$$

where the factors are torsion free over appropriate subrings of a Noether normalization of $A/\mathrm{ann}(E)$. We then combine the various characteristic polynomials.

Theorem 9.54. *Let A be an affine algebra over an infinite field k, let $k[\mathbf{z}]$ be a Noether normalization of A where the z_i are forms of degree 1, and let M be a finitely generated graded, faithful A–module. Then every element of A satisfies a monic equation over $k[\mathbf{z}]$ of degree at most $\mathrm{adeg}(M)$.*

Proof. Let

$$(0) = L_1 \cap L_2 \cap \cdots \cap L_n$$

be an equidimensional decomposition of the trivial submodule of M, derived from an indecomposable primary decomposition by collecting the components of the same dimension. If $I_i = $ annihilator (M/L_i), then each ring A/I_i is unmixed, equidimensional and

$$\dim A/I_i > \dim A/I_{i+1}.$$

Since k is infinite, there exists a Noether normalization $k[z_1, \ldots, z_d]$ of A such that for each ideal I_i, a subset of the $\{z_1, \ldots, z_d\}$ generates a Noether normalization for A/I_i.

First, we are going to check that $\mathrm{adeg}(M)$ can be determined by adding the arithmetic degrees of the factors of the filtration

$$M \supset L_1 \supset L_1 \cap L_2 \supset \cdots \supset L_1 \cap L_2 \cap \cdots \cap L_n = (0),$$

at the same time that we use the Cayley–Hamilton theorem.

We write the arithmetic degree of M as

$$\mathrm{adeg}(M) = a_1(M) + a_2(M) + \cdots + a_n(M),$$

where $a_i(M)$ is the contribution of the prime ideals minimal over I_i. (Warning: This does not mean that $a_i(M) = \mathrm{adeg}(M/L_i)$.) We first claim that (set $L_0 = M$)

$$\mathrm{adeg}(M) = \sum_{i=1}^{n} \mathrm{adeg}(L_1 \cap \cdots \cap L_{i-1}/L_1 \cap \cdots \cap L_i).$$

Indeed, there is an embedding

$$F_i = L_1 \cap \cdots \cap L_{i-1}/L_1 \cap \cdots \cap L_i \hookrightarrow M_i = M/L_i,$$

showing that F_i is equidimensional of the same dimension as M_i. If $\mathfrak{p}$ is an associated prime of I_i, localizing we get $(L_1 \cap \cdots \cap L_i)_{\mathfrak{p}} = (0)$ which shows that

$$\Gamma_{\mathfrak{p}}(M_{\mathfrak{p}}) \subset \Gamma_{\mathfrak{p}}((L_1 \cap \cdots \cap L_{i-1})_{\mathfrak{p}}),$$

while the converse is clear. This shows that the geometric degree of the module F_i is exactly the contribution of $e_i = a_i(M)$ to $\mathrm{adeg}(M)$.

We are now ready to use Proposition 9.50 on the modules F_i. Let $f \in A$ act on each F_i. For each integer i, we have a polynomial

$$P_i(t) = t^{e_i} + c_1 t^{e_i - 1} + \cdots + c_{e_i},$$

with $c_j \in (\mathbf{z})^j$, and such that $P_i(f) \cdot F_i = (0)$. Consider the polynomial

$$P_f(t) = \prod_{i=1}^{n} P_i(t),$$

and evaluate it on f from left to right. As

$$P_i(f) \cdot F_i = 0,$$

meaning that $P_i(f)$ maps $(L_1 \cap \cdots \cap L_{i-1})$ into $(L_1 \cap \cdots \cap L_i)$, a simple inspection shows that $P_f(f) = 0$, since M is a faithful module. $\square$

Observe that if A is a standard algebra and f is an element of A_1, then $P_f(t)$ gives an equation of integrality of f relative to the ideal generated by $\mathbf{z}$.

Corollary 9.55. *Let A be a standard graded algebra and denote*

$$\mathrm{edeg}(A) = \inf\{\ \mathrm{adeg}(M) \mid M \text{ faithful graded module }\}.$$

For any standard Noether normalization R of A, every element of A satisfies an equation of degree $\mathrm{edeg}(A)$ over R.

Reduction Equations from Integrality Equations

If A is an standard algebra, for a given element $u \in A_1$, a typical equation of reduction looks like

$$u^e \in (\mathbf{z}){A_1}^{e-1},$$

which is less restrictive than an equation of integrality. One should therefore expect these equations to have lower degrees. Unfortunately we do not yet see how to approach it.

The following argument shows how to pass from integrality equation to some reduction equations, but unfortunately injects the issue of characteristic into the fray.

Proposition 9.56. *Let $A = k[A_1]$ be a standard algebra over a field k of characteristic zero. Let $R = k[\mathbf{z}] \hookrightarrow A$ be a Noether normalization, and suppose that every element of A_1 satisfies a monic equation of degree e over $k[\mathbf{z}]$. Then $\mathrm{r}(A) \leq e - 1$.*

Proof. Let $u_1, \ldots, u_n$ be a set of generators of A_1 over $k[\mathbf{z}]$, and consider the integrality equation of

$$u = x_1 u_1 + \cdots + x_n u_n,$$

where the x_i are elements of k. By assumption, we have

$$u^e = (x_1 u_1 + \cdots + x_n u_n)^e = a_1 u^{e-1} + \cdots + a_e,$$

where $a_i \in (\mathbf{z})^i$. Expanding u^e we obtain

$$\sum_{\alpha} a_{\alpha} m_{\alpha} u^{\alpha} \in (\mathbf{z}) A_1^{e-1},$$

where $\alpha = (\alpha_1, \ldots, \alpha_n)$ is an exponent of total degree e, a_α is the multinomial coefficient $\binom{e}{\alpha}$, and m_α is the corresponding 'monomial' in the x_i. We must show

$$u^\alpha \in (\mathbf{z})A_1^{e-1}$$

for each α.

To prove the assertion, it suffices to show that the span of the vectors $(a_\alpha m_\alpha)$, indexed by the set of all monomials of degree e in n variables, has the dimension of the space of all such monomials. Indeed, if these vectors lie on a hyperplane

$$\sum_\alpha c_\alpha T_\alpha = 0,$$

we would have a homogeneous polynomial

$$f(X_1, \ldots, X_n) = \sum_\alpha c_\alpha m_\alpha X^\alpha$$

which vanishes on k^n. This means that all the coefficients $c_\alpha a_\alpha$ are zero, and therefore each c_α is zero since the m_α do not vanish in characteristic zero. $\qquad\square$

Example 9.57 (Doering–Gunston). Let Δ be a 2-dimensional simplicial complex, and let

$$\Delta = \Delta_1 \cup \cdots \cup \Delta_c$$

be its decomposition into connected components. We assume that they all have dimension 2. Then the diagonal embedding

$$0 \to A = k[\Delta] \longrightarrow B = k[\Delta_1] \times \cdots \times k[\Delta_c] \longrightarrow C \to 0$$

is the S_2–ification of A. A computation of Hilbert functions shows that C is a vector space (of dimension $c - 1$), concentrated in degree 1. Since each $k[\Delta_i]$ is Cohen–Macaulay, its regularity is at most 2 and therefore $\mathrm{reg}(B) \leq 2$. By [Ei95, Corollary 20.19], it follows that $\mathrm{reg}(A) \leq 2$. As a consequence $\mathrm{r}(A) \leq 2$, which is lower than the general estimate given above.

Arithmetic Degree Versus Number of Generators

There are, as expected, relationships between numbers of generators of modules and ideals and their degrees. We exploit one of these now (see [Be77]). More general relationships are discussed in the next section.

Theorem 9.58. *Let A be an standard graded algebra with a presentation $A = S/I$, where $S = k[x_1, \ldots, x_n]$, such that I does not contain linear forms. Let $d = \dim A$ and $a = \mathrm{adeg}(A)$. If $\mathrm{depth}\, A = \dim A - 1$, then*

$$\nu(I) \leq a^{n-d} + (n - d) - \deg(A). \tag{9.26}$$

Proof. We may assume that k is an infinite field, so that without loss of generality we may further assume that $R = k[x_1, \ldots, x_d]$ is a Noether normalization of A. According to Theorem 9.54, the image of each x_i satisfies a monic, homogeneous equation over R, in other words, for $i = d+1, \ldots, n$, there exist

$$f_i = x_i^a + c_{1i}x_i^{a-1} + \cdots + c_{ai} \in I,$$

where c_{ji} are homogeneous polynomials in R of degree $a - j$.

Denote by J the ideal generated by the f_i's and consider the exact sequence of R–modules,

$$0 \to I/J \longrightarrow S/J \longrightarrow S/I \to 0.$$

Note that R is also a Noether normalization of S/J. The hypothesis on the depth of A implies that I/J is a free module over R, of rank

$$\deg(S/J) - \deg(S/I) = a^{n-d} - \deg(A).$$

The rest of the assertion follows from $v(I) \leq v(I/J) + v(J)$.   □

Remark 9.59. If A is Cohen–Macaulay one has

$$\boxed{v(I) \leq \deg(A)^{n-d-1} + n - d - 1.} \tag{9.27}$$

This follows by choosing J to be the ideal generated by f_i, for $i = d+1, \ldots, n-1$. One now has that I/J is a Cohen–Macaulay module of the ring S/J, which has multiplicity $\deg(A)^{n-d-1}$.

Finding larger Cohen–Macaulay ideals L, $J \subset L \subset I$, would lead to better estimates.

Problem 9.60. Let $\triangle$ be a simplicial complex of dimension $\dim \triangle$. If the ring $k[\triangle]$ is Cohen–Macaulay, is there a combinatorial description of $r(k[\triangle])$? One has that $r(k[\triangle]) \leq \dim k[\triangle]$, and the inequality is strict if and only if $\chi(\triangle) = 1$, since this forces the last component of the h–vector of $\triangle$ to vanish.

Problem 9.61. Let I be a monomial ideal of $R = k[x_1, \ldots, x_n]$ and let J be its radical, $J = \sqrt{I}$. Find a general bound for $r(R/I)$ in terms of $r(R/J)$ and the degrees of the generators of I.

Problem 9.62. Let ψ be an endomorphism of a graded module E. Find a method that constructs the characteristic polynomial of $P_\psi(t)$ discussed in Proposition 9.50.

Exercise 9.63. Let $A = k[x_1, \ldots, x_n]/I$ be a standard graded algebra over the infinite field k. If $\dim A = d$, show that for all generic $d \times n$ matrices (a_{ij}), with entries in k, the linear forms

$$z_i = \sum_{j=1}^n a_{ij}x_j, \quad i = 1, \ldots, d,$$

generate a Noether normalization $R = k[z_1, \ldots, z_d]$ of A such that $r_R(A) = r(A)$.

Exercise 9.64. Suppose A and B are standard algebras over the field k. Prove the product formula

$$\mathrm{adeg}(A \otimes_k B) = \mathrm{adeg}(A) \cdot \mathrm{adeg}(B).$$

Exercise 9.65. Let $(R, \mathfrak{m})$ be a Cohen–Macaulay local ring and let $G = \mathrm{gr}_{\mathfrak{m}}(R)$ be the associated graded ring of $\mathfrak{m}$. Prove that if depth $G \geq \dim R - 1$ then the reduction number of G is independent of the chosen Noether normalization (by forms of degree 1).

9.4 Homological Degrees of a Module

We can refine the definition of the arithmetic degree of a module into several notions of homological degrees. Let M be a graded module over the ring of polynomials $R = k[x_1, \ldots, x_d]$. Without loss of generality, we are going to assume that $\dim M = d$. Denote

$$M_i = \mathrm{Ext}_R^i(M, R).$$

M_i is a graded module of codimension $\geq i$, so has its own arithmetic degree:

$$\mathrm{adeg}(M_i) = \square_i + \triangle_i + \bullet + \cdots,$$

where $\square_i$ is the contribution of the associated primes of codimension i, $\triangle_i$ the contribution of the primes of codimension $i + 1$, with $\bullet$ denoting contributions of higher codimension.

The following table displays these definitions, and note that each entry can be expressed as the *degree* of an appropriate Ext. (This obscure sentence will be clarified shortly.)

	0	1	2	3	4	$\cdots$	d
M_0	$\square_0$	$\triangle_0$	0	0	0	$\cdots$	0
M_1	0	$\square_1$	$\triangle_1$	$\bullet$	$\bullet$	$\cdots$	$\bullet$
M_2	0	0	$\square_2$	$\triangle_2$	$\bullet$	$\cdots$	$\bullet$
M_3	0	0	0	$\square_3$	$\triangle_3$	$\cdots$	$\bullet$
$\vdots$	$\vdots$	$\vdots$	$\vdots$	$\vdots$	$\square_4$	$\cdots$	$\vdots$
$\vdots$	$\vdots$	$\vdots$	$\vdots$	$\vdots$	$\vdots$	$\ddots$	$\vdots$
M_d	0	0	0	0	0	$\cdots$	$\square_d$

Observe the information carried by the diagonal:

$$\mathrm{adeg}(M) = \square_0 + \square_1 + \cdots + \square_d \tag{9.28}$$

It is not clear that adding all the entries of the table will amount to anything significant. In the case of (9.28) the summands refer to different objects so each carries a different aspect of adeg(M). Perhaps we should try the same with the other diagonals, such as, what is the meaning of the integer

$$\triangle_1 + \triangle_2 + \cdots + \triangle_{d-1}$$

(note that $\triangle_0 = \triangle_d = 0$)? A larger issue is which weighted combination of these numbers has interesting properties? For example, a candidate worthy of consideration is an *extended* arithmetic degree:

$$\text{Adeg}(M) = \sum_{i \geq 0} \text{adeg}(\text{Ext}_R^i(M, R)).$$

It has several of the properties of adeg($\cdot$) and captures contributions of prime ideals in the whole cohomology of M.

The actual choice we shall make will dig deeper into the elements of this table to obtain derived tables.

We point out that the first nonzero of the modules $\text{Ext}_R^i(M, R)$ has a special feature:

Proposition 9.66. *Let R be a (locally) Gorenstein ring and let M be a finitely generated R–module. Suppose the annihilator of M has codimension r. Then the module $\text{Ext}_R^r(M, R)$ is nonzero, and all of its associated primes are of codimension r.*

Proof. This is a consequence of the following: If $\mathbf{x}$ is an R–regular sequence of length r, contained in the annihilator of M, then

$$\text{Ext}_R^r(M, R) \simeq \text{Hom}_{R/(\mathbf{x})}(M, R/(\mathbf{x})),$$

which is the dual of a module over the Cohen–Macaulay ring $R/(\mathbf{x})$. In particular the associated prime ideals of $\text{Ext}_R^r(M, R)$ are contained in the set of associated primes of $R/(\mathbf{x})$. (More strongly, $\text{Ext}_R^r(M, R)$ satisfies the condition S_2 of Serre as an $R/(\mathbf{x})$–module.) $\qquad\square$

Homological Degree of a Module

To give a proper generality to the next degree to be introduced, let $(R, \mathfrak{m})$ be an Artinian local ring and let A be a finitely generated, graded R algebra, $A = A_0[A_1]$, where A_0 is a finite R–algebra. Such algebras are homomorphic images of polynomial rings $S = R'[x_1, \ldots, x_n]$ where R' is a Gorenstein local ring which is finite over R.

The largest of the degrees attached to a module that we will consider is:

Definition 9.67. *Let M be a finitely generated graded module over the graded algebra A and let S be a Gorenstein graded algebra mapping onto A. Assume that $\dim S = r$, $\dim M = d$. The homological degree of M is the integer*

$$\text{hdeg}(M) = \deg(M) + \sum_{i=r-d+1}^{r} \binom{d-1}{i-r+d-1} \cdot \text{hdeg}(\text{Ext}_S^i(M,S)).$$

It becomes more compact when $\dim M = \dim S = d > 0$:

$$\text{hdeg}(M) = \deg(M) + \sum_{i=1}^{d} \binom{d-1}{i-1} \cdot \text{hdeg}(\text{Ext}_S^i(M,S)). \tag{9.29}$$

Note that $\text{hdeg}(\cdot)$ has been defined recursively on the dimension of the support of the module. The explanation for the binomial coefficients will appear later when we explore this notion under the effect of hyperplane sections.

The low dimension cases are (assuming $\dim S = \dim M$):

(i) If $\dim M = 0$, $\text{hdeg}(M) = \ell(M)$.
(ii) If $\dim M = 1$, $\text{hdeg}(M) = \text{adeg}(M) = \deg(M) + \ell(\text{Ext}_S^1(M,S))$.
(iii) If $\dim M = 2$, $\text{hdeg}(M) = \text{adeg}(M) + \ell(\text{Ext}_S^2(\text{Ext}_S^1(M,S),S))$.

Example 9.68. To see how the various degrees compare, let $S = k[x,y,z,w]$, and J an unmixed, non-Cohen–Macaulay ideal of codimension 2. Define $I = (x,y,z,w)J$ and $A = S/I$. From the sequence

$$0 \to J/I \simeq k^{\nu(J)} \longrightarrow S/I \longrightarrow S/J \to 0,$$

the long sequence of Ext gives

$$\text{adeg}(A) = \text{gdeg}(A) + \nu(J),$$
$$\text{hdeg}(A) = \text{adeg}(A) + \ell(\text{Ext}_S^3(S/J,S)).$$

Thus $\text{hdeg}(A)$ and $\text{adeg}(A)$ differ by the length of the Hartshorne-Rao module of A.

Thus, if $J = (x,y) \cap (z,w)$, $\nu(J) = 4$ and $\text{Ext}_S^3(S/J,S) \simeq k$, which gives $\deg(A) = \text{gdeg}(A) = 2$, $\text{adeg}(A) = 6$ and $\text{hdeg}(A) = 7$.

Remark 9.69. The relationship between $\text{hdeg}(A)$ and $\text{reg}(A)$ is a source of interest. Example 9.35(b) satisfies

$$r(A) = 1 < \text{adeg}(A) = 2 < \text{reg}(A) = t < \text{hdeg}(A) = 2 + t^2.$$

In general we do not know whether, for a standard graded algebra A of dimension at least 2, $\text{reg}(A) \leq \text{adeg}(A)$ always holds.

Let us begin listing elementary properties of this degree.

Proposition 9.70. *Let M be a graded module over the graded algebra A and let S be a Gorenstein graded algebra such that $S/I = A$.*

(a) *$\mathrm{hdeg}(M)$ is independent of S.*
(b) *$\deg(M) \leq \mathrm{gdeg}(M) \leq \mathrm{adeg}(M) \leq \mathrm{hdeg}(M)$, and equality holds throughout if and only if M is Cohen–Macaulay.*

Proof. (a) Given two Gorenstein graded algebras over R, S_1 and S_2, mapping onto A, we can find another Gorenstein graded algebra S mapping onto S_1 and onto S_2 such that the diagram

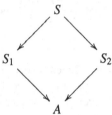

is commutative. This means that to prove (a) we may consider the case of algebras S and S' such that $S = S'/I$.

We are going to show that up to shifts,

$$\mathrm{Ext}_S^i(M,S) = \mathrm{Ext}_{S'}^{i+r}(M,S'),$$

where $r = \dim S' - \dim S$.

There exists a regular sequence $\mathbf{x}$ consisting of r homogeneous elements of S' contained in the ideal I, and therefore annihilating the module M.

Before we go on with the proof, we recall a useful tool for calculation of Exts (see [Kap74, p. 155]).

Proposition 9.71 (Rees Lemma). *Let R be a commutative ring, E and F R–modules and $x \in R$. If x is a regular element on R and F and $xE = 0$, then*

$$\mathrm{Ext}_R^n(E,F) \simeq \mathrm{Ext}_{R/(x)}^{n-1}(E,F/xF), \quad \forall n \geq 1. \tag{9.30}$$

In our case, this means that

$$\mathrm{Ext}_{S'/(\mathbf{x})}^i(M,S'/(\mathbf{x})) = \mathrm{Ext}_{S'}^{i+r}(M,S'),$$

so that we may assume that S and S' have the same dimension.

We may further assume that $\dim S = \dim S' = \dim M$. Let us show that in this case,

$$\mathrm{Ext}_S^i(M,S) = \mathrm{Ext}_{S'}^i(M,S').$$

Let us begin with the right hand side of this equation. Let

$$\mathbb{E}_\bullet : \qquad 0 \to S' \longrightarrow E_0 \longrightarrow \cdots \longrightarrow E_d \to 0,$$

be an injective resolution of S'. Applying $\text{Hom}_{S'}(M, \cdot)$, we get a complex

$$\text{Hom}_{S'}(M, \mathbb{E}_\bullet) = \text{Hom}_S(M, \text{Hom}_{S'}(S, \mathbb{E}_\bullet)) \qquad (9.31)$$

of S–modules, in which the cohomology of $\text{Hom}_{S'}(S, \mathbb{E}_\bullet)$ is $\text{Ext}^i_{S'}(S, S')$, that vanishes in all dimensions but 0, since S is a maximal Cohen–Macaulay module for S'. It follows that $\text{Hom}_{S'}(S, \mathbb{E}_\bullet)$ is an S–injective resolution of $\text{Hom}_{S'}(S, S') = S$, since S is a Gorenstein ring. The cohomology of both sides of (9.31) gives the desired assertion.

(b) The inequalities are clear and the Cohen–Macaulayness of M is, by local duality, expressed by the vanishing of all $\text{Ext}^i_S(M, S)$ for $i > \text{codim } M$. $\qquad\qquad \square$

Definition 9.72. *The difference* $\text{hdeg}(M) - \text{deg}(M)$ *will be called the* homological Cohen–Macaulay deviation *of* M.

Remark 9.73. By the theorem of local duality, $\text{hdeg}(M)$ can be equally defined in the following setting. Let $R \mapsto A$ be a finite homomorphism, where R is a Gorenstein algebra. Then

$$\text{Ext}^i_R(M, R) \simeq \text{Ext}^i_S(M, S),$$

where S is as above.

The next properties, some of which have attached restrictions, begin to explore the means to obtain *a priori* estimates for these degrees.

Proposition 9.74. *Let M be a graded module over the graded algebra A and let S be a Gorenstein graded algebra such that $S/I = A$.*

(a) *Let $L = H^0_\mathfrak{m}(M)$ be the submodule of M of elements with finite support. Then*

$$\text{hdeg}(M) = \text{hdeg}(M/L) + \ell(L).$$

In particular, if $\dim M = 1$ then $\text{adeg}(M) = \text{hdeg}(M)$.
(b) *If $\dim M = 2$ and h is a regular hyperplane section on M, then*

$$\text{hdeg}(M) \geq \text{hdeg}(M/hM).$$

Proof. We may again assume that $\dim S = \dim M = d$. Part (a) is clear since

$$\text{Ext}^i_S(M, S) = \text{Ext}^i_S(M/L, S), \quad 1 \leq i < d$$
$$\text{Ext}^d_S(M/L, S) = 0$$
$$\ell(\text{Ext}^d_S(M, S)) = \ell(L).$$

To prove (b), starting from the exact sequence

$$0 \to M \xrightarrow{h} M \longrightarrow \overline{M} \to 0,$$

we obtain the long exact sequence of Ext,

$$0 \to \mathrm{Hom}_S(M,S) \longrightarrow \mathrm{Hom}_S(M,S) \longrightarrow \mathrm{Ext}^1_S(\overline{M},S) \longrightarrow$$
$$\mathrm{Ext}^1_S(M,S) \longrightarrow \mathrm{Ext}^1_S(M,S) \longrightarrow \mathrm{Ext}^2_S(\overline{M},S) \to 0,$$

since $\mathrm{Hom}_S(\overline{M},S) = \mathrm{Ext}^2_S(M,S) = 0$. As for the other modules of this sequence, both $\mathrm{Hom}_S(M,S)$ and $\mathrm{Ext}^1_S(\overline{M},S)$ have no embedded primes, and $F = \mathrm{Ext}^2_S(\overline{M},S)$ is a module of finite length which is the cokernel of the endomorphism induced by multiplication by h on the module $E = \mathrm{Ext}^1_S(M,S)$, of Krull dimension at most 1.

In arbitrary dimensions, we have by standard properties,

$$\deg(\mathrm{Ext}^1_S(\overline{M},S)) = \deg(\overline{M})$$
$$= \deg(M)$$
$$= \mathrm{adeg}(\mathrm{Hom}_S(M,S)).$$

If $\dim E = 0$, since

$$\ell(F) \leq \ell(E),$$

by adding we get the assertion.

Suppose $\dim E = 1$, and let L denote its submodule of finite support and put $G = E/L$. One has $\mathrm{adeg}(E) = \mathrm{adeg}(G) + \ell(L)$. Consider the exact sequence, spawned by the snake lemma, obtained by multiplication on E by h; note that it induces endomorphisms on L and on G:

$$0 \to {}_hL \longrightarrow {}_hE \longrightarrow {}_hG \longrightarrow L/hL \longrightarrow E/hE \longrightarrow G/hG \to 0.$$

From above, we have that $E/hE = \mathrm{Ext}^2_S(\overline{M},S)$, which is a module of finite length, and thus G/hG has finite length as well. But G is a module of dimension 1 without embedded primes, which means that h is a system of parameters for it and thus must be a regular element on it; this means that ${}_hG = 0$, and therefore $\ell({}_hE) < \infty$, which we take into the exact sequence

$$0 \to \mathrm{Hom}_S(M,S)/h \cdot \mathrm{Hom}_S(M,S) \longrightarrow \mathrm{Ext}^1_S(\overline{M},S) \longrightarrow {}_hE \to 0,$$

and get

$$\deg(\mathrm{Hom}_S(M,S)) = \deg(\mathrm{Ext}^1_S(\overline{M},S)),$$

as in the previous case. Since we also have

$$\mathrm{adeg}(E) = \ell(L) + \deg(G)$$
$$= \ell(L) + \deg(G/hG)$$
$$\geq \ell(L/hL) + \ell(G/hG)$$
$$\geq \ell(E/hE)$$
$$= \ell(F),$$

we add as earlier to prove the assertion. □

For reasons that will become clearer later, it would be important that Proposition 9.74(b) be valid without restriction on the dimension.

Conjecture 9.75. If M is a graded module and h is a regular hyperplane section then

$$\text{hdeg}(M) \geq \text{hdeg}(M/hM). \qquad (9.32)$$

Generalized Cohen–Macaulay Modules

For some special modules it is possible to have a more explicit expression of its homological degree.

Theorem 9.76. *Let S be as above and assume* $\dim M = \dim S = d$. *If M is a Cohen–Macaulay module on the punctured spectrum then*

$$\text{hdeg}(M) = \deg(M) + \sum_{i=0}^{d-1} \binom{d-1}{i} \cdot \ell(H_{\mathfrak{m}}^i(M)).$$

Proof. It is straightforward:

$$\text{hdeg}(M) = \deg(M) + \sum_{i=1}^{d} \binom{d-1}{i-1} \cdot \text{hdeg}(\text{Ext}_S^i(M,S))$$

$$= \deg(M) + \sum_{i=1}^{d} \binom{d-1}{i-1} \cdot \ell(\text{Ext}_R^i(M,R))$$

$$= \deg(M) + \sum_{i=1}^{d} \binom{d-1}{i-1} \cdot \ell(H_{\mathfrak{m}}^{d-i}(M))$$

$$= \deg(M) + \sum_{i=0}^{d-1} \binom{d-1}{i} \cdot \ell(H_{\mathfrak{m}}^i(M)). \qquad \square$$

Note that the binomial term is the invariant of Buchsbaum in the theory of Buchsbaum modules ([SVo86, Chapter 1, Proposition 2.6]).

These modules also behave nicely under regular hyperplane sections. Indeed suppose

$$0 \to M \xrightarrow{h} M \longrightarrow \overline{M} \to 0$$

is exact and consider the long exact sequence of cohomology

$$0 \to M_0 \xrightarrow{h} M_0 \longrightarrow \overline{M}_0 \longrightarrow M_1 \xrightarrow{h} M_1 \longrightarrow \overline{M}_1 \longrightarrow \cdots$$

$$M_{d-2} \xrightarrow{h} M_{d-2} \longrightarrow \overline{M}_{d-2} \longrightarrow M_{d-1} \xrightarrow{h} M_{d-1} \longrightarrow \overline{M}_{d-1} \to 0,$$

where we may assume $\dim M = \dim S = d$ and set

$$M_i = \text{Ext}_S^i(M,S),$$
$$\overline{M}_i = \text{Ext}_S^{i+1}(\overline{M},S).$$

We break up this sequence into two families of shorter sequences

$$0 \to L_i \longrightarrow M_i \overset{h}{\longrightarrow} M_i \longrightarrow N_i \to 0$$

$$0 \to N_i \longrightarrow \overline{M}_i \longrightarrow L_{i+1} \to 0$$

for $0 \le i \le d-1$.

If $i = 0$, $L_0 = 0$ and $\deg(M) = \deg(\overline{M})$ For $i \ge 1$, the modules of both sequences have finite lengths and thus

$$\ell(N_i) = \ell(L_i)$$
$$\ell(\overline{M}_i) = \ell(L_i) + \ell(L_{i+1}).$$

As $\ell(M_i) \ge \ell(L_i)$, we get that

$$\ell(\overline{M}_i) \le \ell(M_i) + \ell(M_{i+1}),$$

so multiplying by $\binom{d-2}{i-1}$ and adding we get

$$\mathrm{hdeg}(M) \ge \mathrm{hdeg}(M/hM).$$

Strict equality will result if $h \cdot M_i = 0$, as in the case of a Buchsbaum module.

Homologically Associated Primes of a Module

One way to look at the associated primes of a module M is as a set of visible obstructions to carrying out on M constructions inspired from linear algebra (e.g. building duals). Some other obstructions may only become visible when we look at objects derived from a projective resolution of M.

Motivated by the notion of the homological degree of a graded module, we introduce the following definition.

Definition 9.77. *Let R be a Noetherian local ring and let M be a finitely generated R–module. Suppose S is a Gorenstein local ring with a surjective morphism $S \mapsto R$. The* homologically associated primes *of M are the prime ideals of R*

$$\text{h-Ass}(M) = \bigcup_{i \ge 0} \mathrm{Ass}(\mathrm{Ext}_S^i(M,S)). \tag{9.33}$$

There remains to prove that this definition is independent of the Gorenstein ring S. In the main case of interest, when R is a graded algebra, the independence is assured if S is also taken graded as earlier in this section.

The set

$$\text{h–Ass}(M) \setminus \mathrm{Ass}(M),$$

might be called the *hidden associated primes* of M.

Let S be a Gorenstein standard graded ring with infinite residue field and M be a finitely generated graded module over S. We recall that a *superficial* element of order r for M is an element $z \in S_r$ such that $0 :_M z$ is a submodule of M of finite length.

To establish the Bertini's property of $\operatorname{hdeg}(\cdot)$ predicted in Conjecture 9.75 for arbitrary regular hyperplane sections we follow [Vas98].

Almost every instance of a 'Bertini's theorem' requires an appropriate notion of generic hyperplane. Here the following will be adequate.

Definition 9.78. *A generic hyperplane section of M is an element $h \in S_1$ that is superficial for all the iterated Exts*

$$M_{i_1, i_2, \ldots, i_p} = \operatorname{Ext}_S^{i_1}(\operatorname{Ext}_S^{i_2}(\cdots(\operatorname{Ext}_S^{i_{p-1}}(\operatorname{Ext}_S^{i_p}(M, S), S), \cdots, S))),$$

and all sequences of integers $i_1 \geq i_2 \geq \cdots \geq i_p \geq 0$.

By local duality it follows that, up to shifts in grading, there are only finitely many such modules. Actually, it is enough to consider those sequences in which $i_1 \leq \dim S$ and $p \leq 2 \dim S$, which ensures the existence of such 1–forms as h.

Theorem 9.79. *Let S be a standard Gorenstein graded algebra and let M be a finitely generated graded module. If $h \in S$ is a regular, generic hyperplane section then*

$$\operatorname{hdeg}(M) \geq \operatorname{hdeg}(M/hM).$$

Its proof has the following consequence:

Corollary 9.80. *Let M and h be as above and let r be a positive integer. Then*

$$\operatorname{hdeg}(M/h^r M) \leq r \cdot \operatorname{hdeg}(M).$$

Remark 9.81. Two general lines of investigation yet to be exploited of these degrees are their behavior under linkage and the role played by Koszul homology.

Some Conjectures and Questions

There are several conjectures that purport to connect the various measures of complexity of an algebra A. These questions arise from numerous examples and special cases when one seeks to link the degree data of the algebra with multiplicity data.

One of long-standing is ([EiG84]):

Conjecture 9.82 (Eisenbud-Goto). If A is a standard graded domain over an algebraically closed field k then

$$\boxed{\operatorname{reg}(A) \leq \deg(A) - \operatorname{codim} A + 2.} \tag{9.34}$$

A weaker question asks whether $\operatorname{r}(A) \leq \deg(A) - \operatorname{codim} A + 2$.

A key relationship between $\mathrm{reg}(A)$ and $\mathrm{hdeg}(A)$ is given by ([DGV98]):

Theorem 9.83. *Let A be a standard graded algebra over a field k. Then*

$$\mathrm{reg}(A) < \mathrm{hdeg}(A). \tag{9.35}$$

The relationship $\mathrm{hdeg}(A) > \mathrm{reg}(A)$ is too one–sided. It would be of interest to know whether $\mathrm{hdeg}(A)$ is bounded by a polynomial function of $\mathrm{reg}(A)$.

Conjecture 9.84. Let I be a homogeneous ideal of $S = k[x_1, \ldots, x_n]$ and $in(I)$ its initial ideal with respect to some term order. Then

$$r(S/I) \leq r(S/in(I)). \tag{9.36}$$

One simple instance is when $in(I)$ is a Cohen–Macaulay ideal. In such case I will also be Cohen–Macaulay and the Hilbert functions, which coincide, provide all the information that is needed, $r(S/I) = r(S/in(I))$. [1]

Let us display some of these questions and others in a diagram. The notation employed is: $A = S/I$, $A' = S/in(I)$, a solid arrow denotes an established inequality (some only in characteristic zero), while a broken arrow signifies a conjectural one (sometimes corrected by $\pm$).

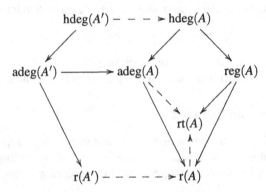

Remark 9.85. Most likely the relationship between $\mathrm{hdeg}(A)$ and $\mathrm{reg}(A)$ is not so uniform as 'predicted' by the diagram.

Problem 9.86. Let R be a Noetherian local ring and let M be a finitely generated R–module. Is the set

$$\bigcup_{i \geq 0} \mathrm{Ass}(\mathrm{Ext}_R^i(M, R))$$

always finite?

---

[1] A. Conca [Con03] and N. V. Trung [Tru03] have settled this question affirmatively.

Exercise 9.87. Let $A = S/I$, with S a Gorenstein local ring. Suppose that A is unmixed and $\dim A = 2$. Prove that

$$\mathrm{hdeg}(A) = \deg(A) + \ell(\widetilde{A}/A),$$

where $\widetilde{A}$ is the S_2–ification of A.

9.5 Complexity Bounds in Local Rings

Let $(R, \mathfrak{m})$ be a Noetherian local ring and let $I \subset R$ be an ideal. The study of certain algebras constructed with I, such as its Rees algebra $R[It]$ with its special fiber $F(I) = R[It] \otimes (R/\mathfrak{m})$, provide a setting for application of the methods of the two precedent sections. Here we focus on $F(I)$, which is an ordinary standard graded algebra and attempt to connect arithmetical properties of I to the complexity bounds of $F(I)$.

Homological Multiplicity

We begin by observing that it still makes sense to define the arithmetic degree of the local ring R: In the formula (9.5), the multiplicity $\deg(R/\mathfrak{p})$ of the local ring $R/\mathfrak{p}$ replaces $\deg(A/\mathfrak{p})$. The geometric degree is defined similarly, but $\mathrm{reg}(A)$ has no obvious extension.

In order to define $\mathrm{hdeg}(R)$ we must assume that R is the homomorphic image of a Gorenstein ring. Since the definition of this degree seems to depend on the presentation $S \mapsto R$, we simply take the minimum of such values. In some cases, for example, if R is complete, $\mathrm{hdeg}(R)$ is by local duality independent of S.

It would be of interest to construct degree functions $\mathrm{Deg}(\cdot)$, sharper than $\mathrm{hdeg}(\cdot)$, but still satisfying the conditions:

(i) If $L = \Gamma_{\mathfrak{m}}(M)$ is the submodule of elements of finite support of M and $\overline{M} = M/L$, then
$$\mathrm{Deg}(M) = \mathrm{Deg}(\overline{M}) + \ell(L).$$

(ii) If $h \in S$ is a regular hyperplane section on M, then
$$\mathrm{Deg}(M) \geq \mathrm{Deg}(M/hM).$$

(iii) If M is a Cohen–Macaulay module, then
$$\mathrm{Deg}(M) = \deg(M).$$

In the previous section, we saw that $\mathrm{hdeg}(M)$ does have this attribute, but it may be overstated when it deals with condition (ii). Getting equality would be considerably better.

Proposition 9.88. *Let* $\text{Deg}(\cdot)$ *be a degree function satisfying* (i) *and* (ii) *above. Then for any finitely generated module M with minimal number of generators* $\nu_R(M)$,

$$\nu_R(M) \leq \text{Deg}(M).$$

Proof. We induct on $d = \dim(M)$. If $d = 0$, $\text{Deg}(M) = \ell(M) \geq \nu_R(M)$. In case $d > 0$, if $L = \Gamma_{\mathfrak{m}}(M) \neq 0$, the condition (i) implies that it will be enough to prove the assertion for $N = M/L$, a module of depth > 0. Let h be a special hyperplane section of M; from (ii) we have

$$\text{Deg}(N) \geq \text{Deg}(N/hN) \geq \nu_R(N/hN) = \nu_R(N),$$

the second inequality following the induction hypothesis and the last equality by Nakayama lemma. $\qquad\qquad\square$

Definition 9.89. *Let R be a local ring that is the homomorphic image of a Gorenstein ring. The* homological multiplicity *is the integer $e_h(R) = \text{hdeg}(R)$. A similar degree can be defined for any $\text{Deg}(\cdot)$ function by $e_D(R) = \text{Deg}(R)$.*

The standard characterization of regular local rings becomes:

Theorem 9.90. *Let R be a local ring that is a homomorphic image of a Gorenstein ring. Then R is a regular local ring if and only if $e_h(R) = 1$.*

Corollary 9.91. *Let S be a Gorenstein local ring (or a standard Gorenstein graded algebra) and let M be a finitely generated (resp. finitely generated, graded) S–module. Then*

$$\boxed{\text{hdeg}(M) \geq \nu_S(M).} \qquad\qquad (9.37)$$

Bounding the Number of Generators of Ideals with Multiplicities

The notion of homological degrees can be employed to estimate the number of generators of an ideal in terms of various degrees of R and of R/I.

Example 9.92. Let R be a Cohen–Macaulay local ring of dimension d and let I be a Cohen–Macaulay ideal of height g. Suppose $S \rightarrow R$ is a Gorenstein presentation and $\dim S = d$.

The exact sequence

$$0 \rightarrow I \longrightarrow R \longrightarrow R/I \rightarrow 0 \qquad\qquad (9.38)$$

permits a comparison between $\text{hdeg}(R/I) = \deg(R/I)$ and $\text{hdeg}(I)$.

If $g \leq 1$, I is Cohen–Macaulay and

$$hdeg(I) = deg(I) = \begin{cases} deg(R) - deg(R/I), & \text{if } g = 0, \\ deg(R), & \text{if } g = 1. \end{cases}$$

If $g \geq 2$ then I is not a Cohen–Macaulay module and

$$Ext_S^{g-1}(I,S) \simeq Ext_S^g(R/I,S).$$

We then have

$$hdeg(I) = deg(R) + \binom{d-1}{g-2} deg(R/I).$$

Since in general we have $v(I) \leq hdeg(I)$, this gives the estimate

$$v(I) \leq deg(R) + \binom{d-1}{g-2} deg(R/I).$$

For the purpose of finding $v(I)$ this can be considerably improved. This takes place as follows. First we make a reduction to the case of an infinite residue field in the usual manner. If $d > g$, we may reduce the exact sequence (9.38) modulo a superficial element h for both R and R/I. The sequence (9.38) gives rise to

$$0 \rightarrow I' \longrightarrow R' \longrightarrow R'/I' \rightarrow 0,$$

in which $I' = I/hI$ (and therefore $v(I) = v(I')$) and the multiplicities are retained. Repeating we end up in the case $d = g$,

$$v(I) \leq deg(R) + \binom{g-1}{g-2} deg(R/I) = deg(R) + (g-1) deg(R/I).$$

This is however not yet as sharp as the estimate of [Val81].

It is not difficult to see that if R is a local ring which is the homomorphic image of a Gorenstein ring and I is an ideal of R, of dimension $d - g$, and R/I is Cohen–Macaulay, then the similar estimate for the number of generators of I holds,

$$v(I) \leq hdeg(R) + (g-1) deg(R/I).$$

Bounding the Hilbert Function

Several bounds on the multiplicity of graded algebras such as $F(I)$, arising from Cohen–Macaulay rings, were obtained by J.D. Sally and others ([Sal78]). We give two adapted proofs where $e_h(R)$ replaces the ordinary multiplicity in the case of Cohen–Macaulay ring.

Theorem 9.93. *Let $(R, \mathfrak{m})$ be a local ring of dimension $d > 0$, which is the homomorphic image of a Gorenstein ring. Let I be an $\mathfrak{m}$–primary ideal and s the index of nilpotency of R/I. Then*

$$v(I) \leq s^{d-1} \cdot e_h(R) + d - 1. \tag{9.39}$$

Proof. We may assume that the residue field of R is infinite as the passage from say R to $R[T]_{\mathfrak{m}R[T]}$ can only improve $e_h(R)$.

The proof is by induction on d. Suppose $(R, \mathfrak{m})$ is a local ring of dimension 1, and I is $\mathfrak{m}$–primary. If $L = \Gamma_{\mathfrak{m}}(R)$, the exact sequence

$$0 \to I \cap L \longrightarrow I \longrightarrow I\overline{R} \to 0,$$

where $\overline{R} = R/L$, shows that

$$v(I) \leq \ell(I \cap L) + v(I\overline{R}),$$

and the claim follows since $\overline{R}$ is Cohen–Macaulay and the bound $v(I\overline{R}) \leq \deg(\overline{R})$ holds for all ideals. It follows that

$$v(I) \leq \ell(L) + \deg(\overline{R}) = \mathrm{hdeg}(R) = e_h(R).$$

For completeness, here is the argument. Pick a superficial element $x \in \mathfrak{m}$ of order 1, that yields $\deg(\overline{R}) = \ell(\overline{R}/x\overline{R}) = \ell(I\overline{R}/xI\overline{R})$, while the surjection

$$I\overline{R}/xI\overline{R} \longrightarrow I\overline{R}/\mathfrak{m}I\overline{R} \to 0$$

gives $v(I\overline{R}) = \ell(I\overline{R}/\mathfrak{m}I\overline{R}) \leq \ell(I\overline{R}/xI\overline{R})$.

The argument also shows that $e_h(R) \geq e_h(\overline{R}) + \ell(\Gamma_{\mathfrak{m}}(R))$, we may assume that depth $(R) \geq 1$ in all cases.

If $d > 1$, pass to the $d - 1$ dimensional ring $R/(x^s)$. The ideal $I/(x^s)$ is $\mathfrak{m}/(x^s)$–primary so, by induction,

$$v(I/(x^s)) \leq s^{d-2} \cdot e_h(R/(x^s)) + d - 2.$$

Hence

$$v(I) \leq v(I/(x^s)) + 1 \leq s^{d-2} \cdot s \cdot e_h(R) + d - 1,$$

as $e_h(R/(x^s)) \leq s \cdot e_h(R)$ by Corollary 9.80. $\qquad\square$

We recall that if I is an ideal, a reduction is another ideal $J \subset I$ such that $I^{n+1} = JI^n$ and the least such integer n, denoted $r_J(I)$, is called the reduction number of I relative to J. They are controlling elements in the theory of Rees algebras (see [Vas94] for more details). If $(R, \mathfrak{m})$ is a Noetherian local ring with infinite residue field, the lifts of Noether normalizations of $F(I)$ give minimal reductions for I; the reduction number $r(I)$ of I is the minimum of all such $r_J(I)$.

As in [Sal76], and with the same proof, we have:

Corollary 9.94. *Let $(R, \mathfrak{m})$ be a local ring of dimension $d > 0$, with infinite residue field, which is the homomorphic image of a Gorenstein ring. Then*

$$r(\mathfrak{m}) \leq d! \cdot e_h(R) - 1.$$

Proof. Theorem 9.93 gives the bound

$$\nu(\mathfrak{m}^n) \le n^{d-1} \cdot e_h(R) + d - 1,$$

so that to obtain a minimal reduction of $\mathfrak{m}$ of reduction number $d! \cdot e_h(R) - 1$, making use of Theorem 9.39, it suffices to note that

$$(d! \cdot e_h(R))^{d-1} \cdot e_h(R) - 1 < \binom{d! \cdot e_h(R) + d}{d}.$$

The method permits to estimate $r(I)$ for ideals primary for the maximal ideal. A similar calculation gives that if I is an $\mathfrak{m}$–primary ideal

$$r(I) \le d! \cdot \ell(R/I)^{d-1} \cdot e_h(R) - 1.$$

In the Cohen–Macaulay case the estimates above and in the literature for the reduction number of $\mathfrak{m}$ are too pessimistic. One has ([DGV98]):

Theorem 9.95. *Let* $(R, \mathfrak{m})$ *be a Cohen–Macaulay local ring of dimension* $d > 0$. *Let* I *be an* $\mathfrak{m}$*–primary ideal and* s *the index of nilpotency of* R/I. *Then*

$$\nu(I) \le e(R) \binom{s+d-2}{d-1} + \binom{s+d-2}{d-2}. \tag{9.40}$$

Proof. We may assume that the residue field of R is infinite. Let $J = (a_1, \dots, a_d)$ be a minimal reduction of $\mathfrak{m}$. By assumption, $J^s \subset I$.

Set $J_0 = (a_1, \dots, a_{d-1})^s$. This is a Cohen–Macaulay ideal of height $d - 1$, and the multiplicity of R/J_0 is (easy exercise)

$$e(R/J_0) = e(R) \binom{s+d-2}{d-1}.$$

Consider the exact sequence

$$0 \to I/J_0 \longrightarrow R/J_0 \longrightarrow R/I \to 0.$$

We have that I/J_0 is a Cohen–Macaulay ideal of the one–dimensional Cohen–Macaulay ring R/J_0. This implies that

$$\nu(I/J_0) \le e(R/J_0).$$

One the other hand, we have

$$\nu(I) \le \nu(J_0) + \nu(I/J_0) \le \binom{s+d-2}{d-2} + e(R/J_0),$$

to establish the claim.   □

Corollary 9.96. *Let $(R, \mathfrak{m})$ be a Cohen–Macaulay local ring of dimension $d > 0$, with infinite residue field and multiplicity $e(R)$. If R is not a regular local ring then the reduction number of $\mathfrak{m}$ is bounded by*

$$r(\mathfrak{m}) \leq d \cdot e(R) - 2d + 1. \qquad (9.41)$$

Proof. We apply Theorem 9.95 to the powers of the ideal $\mathfrak{m}$. One has

$$v(\mathfrak{m}^n) \leq e(R) \binom{n+d-2}{d-1} + \binom{n+d-2}{d-2}.$$

According to Theorem 9.39, it suffices to find n such that

$$v(\mathfrak{m}^n) < \binom{n+d}{d},$$

as $r(\mathfrak{m}) \leq n$. It will be enough to choose n so that

$$e(R) \binom{n+d-2}{d-1} + \binom{n+d-2}{d-2} < \binom{n+d}{d},$$

which is equivalent with

$$e(R) \frac{n}{n+d-1} + \frac{d-1}{n+d-1} < \frac{n+d}{d}.$$

This inequality will be satisfied for $n = d \cdot e(R) - 2d + 2$, as desired. $\qquad \square$

Remark 9.97. These estimates can be extended to $\mathfrak{m}$–primary ideals of arbitrary Cohen–Macaulay rings ([DGV98]). If the index of nilpotency of R/I is s and the multiplicity of I is $e(I)$, then

$$r(I) \leq \min\{ d \cdot s^{d-1} e(R) - d - 1, d \cdot e(I) - 2d + 1 \}.$$

Special cases work better. For instance, if $R = k[x_1, \ldots, x_n]$, where k is a field of characteristic zero, and I is an ideal of dimension 0 generated by forms of degree r, a calculation with Alberto Corso showed that $e(I) = r^n$ but $r(I) \leq r^{n-1} - 1$.

Tangent Cones

The presence of good depth properties in the Rees algebra of I or on its special fiber $F(I)$ cuts down these estimates for the reduction of I.

Theorem 9.98. *Let $(R, \mathfrak{m})$ be an analytically equidimensional local ring with residue field of characteristic zero. Let I be an $\mathfrak{m}$–primary ideal. Suppose that the Rees algebra $R[It]$ of I satisfies the condition S_2 of Serre (e.g. $R[It]$ is normal). Then $r(I) \leq e(I; R) - 1$. ($e(I; R)$ is the multiplicity of the ideal I.)*

Proof. The hypothesis implies that the associated graded ring of I,

$$\mathrm{gr}_I(R) = R[It] \otimes_R (R/I),$$

has the condition S_1 of Serre and is equidimensional (see [NVa93] for details).

Since the characteristic of the residue field $R/\mathfrak{m}$ is zero, the Artinian ring R/I will contain a field k that maps onto the residue field $R/\mathfrak{m}$. Let $k[t_1,\ldots,t_d]$ be a standard Noether normalization of the special fiber $F(I) = \mathrm{gr}_I(R) \otimes (R/\mathfrak{m})$. We can lift it to $\mathrm{gr}_I(R)$ and therefore assume that $\mathrm{gr}_I(R)$ is a finitely generated, graded module over $A = k[t_1,\ldots,t_d]$ of rank equal to the multiplicity $e(I;R)$. As all associated primes of $\mathrm{gr}_I(R)$ have the same dimension we can bound reduction number by arithmetic degree according to Theorem 9.54

$$\mathrm{r}(I) = \mathrm{r}(\mathrm{gr}_I(R)) < \mathrm{adeg}(\mathrm{gr}_I(R)) = \deg(\mathrm{gr}_I(R)) = e(I;R),$$

giving the estimate. $\qquad\qquad\qquad\qquad\qquad\qquad\qquad\qquad\qquad\qquad\square$

The following (see [Tru87]) places a different kind of constraint on $\mathrm{gr}_I(R)$:

Theorem 9.99. *Let R be a Buchsbaum (resp. Cohen–Macaulay) local ring of dimension $d \geq 1$, let I be an $\mathfrak{m}$–primary ideal and suppose that depth $\mathrm{gr}_I(R) \geq d-1$. Then $\mathrm{r}(I) \leq \deg(R)t^{d-1}$ (resp. $\mathrm{r}(I) \leq \deg(R)t^{d-1} - 1$), where $t = \mathrm{nil}(R/I)$.*

Question 9.100. Let $(R,\mathfrak{m})$ be a local ring that is the homomorphic image of a Gorenstein ring and let I be an ideal of R. How are $e_h(R[It])$ and $e_h(\mathrm{gr}_I(R))$ related? (At least in case I is $\mathfrak{m}$–primary.)

Exercise 9.101. Let A be an equidimensional standard graded algebra of dimension at most 4. If $\widetilde{A}$ is the S_2–ification of A, prove that as an A–module

$$v(\widetilde{A}) \leq 1 + \mathrm{hdeg}(A) - \deg(A).$$

Exercise 9.102. Let R be a local ring and M a finitely generated R–module. Describe the conditions under which the following equality holds:

$$v(M) = \mathrm{hdeg}(M).$$

A

A Primer on Commutative Algebra

We collect here several cornerstones and some of the basic navigational tools of Commutative Algebra, with proofs assembled from many sources. The results treated tend to be named results and/or proofs with ingredients that one can weave into algorithms. They make up far less than what is needed to roam freely in this part of algebra, and are not intended as replacement for a basic textbook (e.g. [Kap74], [Mat80], [ZS60], and particularly [Ei95]), although an effort was made to provide many full proofs.

A.1 Noetherian Rings

Several results discussed are valid in broader generality, but we will not have the need to exploit this fact.

- Chain conditions
- Hilbert basis theorem, Cohen theorem, power series
- Artin–Rees lemma
- Filtrations and Rees algebras

Chain Conditions

Throughout rings are commutative with an identity. For most notions used in the study of such rings (zerodivisor, unit, idempotent, nilpotent, prime ideal, multiplicative set, etc) we refer to any of the textbooks mentioned above.

Key to commutative algebra are the various chain conditions on ideals and modules. They permit the building of structures and comparisons.

Definition A.1. *Let R be a ring and let M be an R-module.*

(a) *M is* Noetherian[1] *if it satisfies the ascending chain condition on submodules, or equivalently, if every submodule of M is finitely generated.*

(b) *M is Artinian if it satisfies the descending chain condition on submodules.*

(c) *M has a composition series if the there a filtration of submodules,*

$$0 \subset M_1 \subset \cdots \subset M_n = M,$$

whose factors M_i/M_{i-1} are simple modules. The integer n is independent of the filtration, and it is called the length of M (notation: $\ell(M)$).

These notions make manipulations of ideals and modules possible. For example, if

$$0 \to N \longrightarrow M \longrightarrow P \to 0$$

is an exact sequence of R–modules, then M is Noetherian (resp. Artinian, or has a composition series) if and only if N and P are Noetherian (resp. Artinian, or have composition series).

Despite the wording, there is considerable lack of symmetry between the ascending and the descending conditions. The reason lies with the intimate role that the ring R plays in the former: If I is the annihilator of M, then R/I is a Noetherian ring. This means that M is Noetherian if and only if M is a finitely generated R-module and R/I is a Noetherian ring. As for condition (c), it is the combination of the other two, (a)+(b)=(c).

Prime Ideals

To a major extent, the foci of attention of the study of a commutative ring R are its prime ideals. The following summarizes some general properties and introduces terminology.

Proposition A.2. *Let R be a commutative ring. The following assertions hold:*

(a) *Every proper ideal I is contained in a prime ideal.*

(b) *Let I be an ideal of a ring R and suppose that it is contained in the set theoretic union of a finite collection of ideals I_i*

$$I \subset \bigcup_{i=1}^{s} I_i.$$

Then I is contained in one of these ideals in the following two cases: (i) the I_i are prime ideals for all but at most 2 i's; (ii) R contains an infinite field k.

---

[1] It is not an unreasonable delusion to think of a Noetherian ring as very large rather than the other way around: The finite generation of its ideals can also be achieved in the presence of plenty of coefficients!

(c) *The nilradical of an ideal I is the set*

$$\sqrt{I} = \{x \in R \mid x^n \in I \text{ for some } n\}$$

and

$$\sqrt{I} = \bigcap \mathfrak{p}, \text{ for all prime ideals } \mathfrak{p} \supset I.$$

(d) *The Jacobson radical of R is the ideal $J = \bigcap \mathfrak{m}$, for all maximal ideals of R. It is the largest ideal with the property: $\forall x \in J$ and $\forall r \in R$, $1 + rx$ is a unit of R.*

Proof. For practice, let us consider the proof of (b).

(i): We may assume that I is not contained in the union of fewer than s of the I_i. We argue by contradiction.

If $s = 2$, picking $a \in I \setminus I_1 \subset I_2$ and $b \in I \setminus I_2 \subset I_1$ then $a + b$ cannot be contained in either I_1 or I_2. Assume $s > 2$ and I_s is prime. Let $a \in I \setminus I_1 \cup \cdots \cup I_{s-1} \subset I_s$ and pick $b \in I \cdot I_1 \cdots I_{s-1} \setminus I_s$. Then $a + b$ cannot lie in any I_i.

(ii) The case $s = 2$ being trivial, suppose $s > 2$ and assume I is not contained in the union of fewer I_i. This means that for every $t = 1, \ldots, s$, we can find

$$a_t \in I_t \setminus \bigcup_{i \neq t} I_i.$$

Consider, for each $c \in k$, a linear combination

$$b = a_1 + ca_2 + \cdots + c^{s-1}a_s = \sum_{t=1}^{s} c^{t-1}a_t.$$

Since k is infinite, we can find s different elements $c_1, \ldots, c_s \in k$ such that the corresponding $b_1, \ldots, b_s$ belong to the same subset, say, I_1. But I_1 is a k–vector space so that any linear combination of the b_t will belong to it; moreover the matrix

$$\begin{bmatrix} 1 & c_1 & c_1^2 & \cdots & c_1^{s-1} \\ 1 & c_2 & c_2^2 & \cdots & c_2^{s-1} \\ \vdots & \vdots & \vdots & \ddots & \vdots \\ 1 & c_s & c_s^2 & \cdots & c_s^{s-1} \end{bmatrix}$$

has determinant different from zero (it is a Vandermonde determinant). Thus I_1 contains $a_1, \ldots, a_s$, contradicting the choice of the a_t's. $\square$

Hilbert Basis Theorem, Cohen Theorem, Power Series Rings

The key that unlocks the door to the realm of Commutative Algebra is:

Theorem A.3 (Hilbert Basis Theorem). *If R is a Noetherian ring and x is an indeterminate then $R[x]$ is Noetherian.*

Proof. The reader should write down her/his favorite proof! $\square$

Theorem A.4 (Cohen Theorem). *A ring R is Noetherian if and only if every prime ideal is finitely generated.*

Corollary A.5. *If R is a Noetherian ring and x is an indeterminate over R then the ring $R[[x]]$ of formal power series is Noetherian.*

Proof. Let $\mathfrak{p}$ be a prime ideal of $R[[x]]$. If $x \in \mathfrak{p}$, $\mathfrak{p} = (x, \mathfrak{q})$ where $\mathfrak{q}$ is a prime ideal of R.

Assume that $x \notin \mathfrak{p}$. Let $I \subset R$ be the ideal generated by the constant terms of all elements of $\mathfrak{p}$. Since I is finitely generated, there are finitely many power series in $\mathfrak{p}$, $f_1, \ldots, f_m$, whose constant terms generate I. We claim that $\mathfrak{p} = (f_1, \ldots, f_m)$. If $g \in \mathfrak{p}$, write $f = x^r \cdot g$, $g(0) \neq 0$. Since $\mathfrak{p}$ is prime, $g \in \mathfrak{p}$. We also have

$$g - \sum r_{i0} f_i = x g_1, \; r_{i0} \in R.$$

Iterating we build all the coefficients of the power series required to express g as a linear combination of the f_i's.   □

Theorem A.6. *A commutative ring R is Artinian if and only if R is Noetherian and every prime ideal is maximal.*

Noetherian Graded Rings

One of the first 'practical' uses of the Hilbert basis theorem was:

Proposition A.7. *Let $R = \sum_{n \geq 0} R_n$ be a positively graded commutative ring and set $R_+ = \sum_{n > 0} R_n$. Then R is Noetherian if and only if R_0 is Noetherian and R_+/R_+^2 is a finitely generated R_0–module.*

Proof. Suppose the conditions on R_0 and R_+ hold. Since R_+/R_+^2 is a direct sum of R_0–modules, there exists $r \in \mathbb{N}$ such that $R_n = \sum_{0 < i < r} R_i R_{n-i}$ for $n \geq r$. Pick a finite set $\{b_1, \ldots, b_s\}$ of elements in $\bigcup_{0 < i < r} R_i$ that generate R_+/R_+^2. The claim is that $R = R_0[b_1, \ldots, b_s]$. Since R_1 is a direct summand of R_+/R_+^2, it is finitely generated by the b_j of degree 1. The next summand, R_2/R_1^2 is generated by the images of the b_j of degree 2, while R_1^2 is generated by the 2-products of the earlier b_j's. From the exact sequence

$$0 \to R_1^2 \longrightarrow R_2 \longrightarrow R_2/R_1^2 \to 0,$$

it follows that R_2 consists of the degree 2 elements of $R_0[b_1, \ldots, b_s]$. We proceed in this fashion until $n = r$, when no new generators are needed.

For the converse, it suffices to note that $R_0 = R/R_+$ and that R_+/R_+^2 is an R–module annihilated by R_+.   □

Semisimple Rings

A particularly useful class of Artinian ring are the semisimple algebras. We recall that an algebra R is semisimple if every ideal I is a direct summand, $R = I \oplus J$, which means that I is generated by one idempotent.

Theorem A.8. *Let R be a finitely generated algebra over a field k. Then R is semisimple if and only if R is a finite-dimensional k-algebra without non-trivial nilpotent elements. In this case,*

$$R = K_1 \times \cdots \times K_m$$

where the K_i are finite field extensions of k.

A major class of semisimple rings is obtained as follows.

Theorem A.9. *Let k be a perfect field and let $K_1, \ldots, K_m$ be finite extensions of k. Then*

$$K_1 \otimes_k \cdots \otimes_k K_m$$

is a semisimple ring.

Nakayama Lemma

The following is the best known general purpose comparison device to test equality of two modules.

Lemma A.10 (Nakayama Lemma). *Let R be a commutative ring and let J denote its Jacobson radical. If M is a finitely generated R–module and N is a submodule such that $M = N + JM$, then $M = N$. In particular, if $M = JM$ then $M = 0$.*

Filtrations and Rees Algebras

A *filtration* of a ring R is a family F of subgroups F_i of R indexed by some set S. The most useful kinds are indexed by an ordered monoid S and are multiplicative

$$F_i \cdot F_j \subset F_{i+j}, \ i, j \in S.$$

They tend to be either *increasing* or *decreasing*, that is $F_i \subset F_j$ if $i < j$ or conversely.

The Rees algebra of F is the graded ring

$$R(F) := \bigoplus_{i \in S} F_i,$$

with natural addition and multiplication. If the filtration is decreasing, there is another algebra attached to it, the associated graded ring

$$\mathrm{gr}_F(R) := \bigoplus_{i \in S} F_i / F_{>i},$$

with $F_{>i} = \bigcup_{j>i} F_j$. If the filtration is increasing, the associated graded ring is defined similarly by changing the sign of i. (Appendix B contains a broad study of filtrations suited for computation.)

Some algebras of interest arise from special filtrations of a commutative ring, multiplicative decreasing $\mathbb{N}$–filtrations $F = \{R_n, \ n \in \mathbb{N}\}$ of R where each R_n is an ideal of R, and

$$R_m \cdot R_n \subset R_{m+n}.$$

Its Rees algebra can be coded as a subring of the polynomial ring

$$R(F) = \sum_{n \in \mathbb{N}} R_n t^n \subset R[t].$$

In addition to the associated graded ring as above, we also have the extended Rees algebra

$$R_e(F) = R(F)[t^{-1}] = \sum_{n \in \mathbb{Z}} R_n t^n \subset R[t, t^{-1}].$$

These representations are useful when computing Krull dimensions. Very important is the isomorphism

$$R_e(F)/(t^{-1}) \simeq \mathrm{gr}_F(R) = \bigoplus_{n=0}^{\infty} R_n/R_{n+1}.$$

It provides a mechanism to pass properties from $\mathrm{gr}_F(R)$ to R itself (see Appendix B).

A major example is the I–adic filtration of an ideal I, $R_n = I^n$, $n \geq 0$. Its *Rees algebra*, which will be denoted by $R[It]$, has its significance centered on the fact that it provides an algebraic realization for the classical notion of blowing–up a variety along a subvariety, and plays an important role in the birational study of algebraic varieties, particularly in the study of desingularization.

Another filtration is the one associated to the symbolic powers $I^{(n)}$ of the ideal I. If I is a prime ideal, its nth symbolic power is the I–primary component of I^n. (There is a more general definition if I is not prime.) Its Rees algebra

$$R(I) := \sum_{n \geq 0} I^{(n)} t^n,$$

the *symbolic Rees algebra* of I, also represents a blowup, inherits more readily the divisorial properties of R, but has its usefulness limited because it is not always Noetherian. The presence of Noetherianess in $R(I)$ is loosely linked to the number of equations necessary to define set–theoretically the subvariety $V(I)$. In turn, the lack of Noetherianess of certain cases has been used to construct counterexamples to Hilbert's 14th Problem.

Artin–Rees Lemma

This is a backbone of commutative algebra of nearly the same pedigree as Hilbert results in the 1870's papers.

Theorem A.11 (Artin-Rees Lemma). *Let R be a Noetherian ring and let I and J be two ideals. There exists an integer c such that for all $n \geq c$ the following equality holds*

$$J \cap I^n = I^{n-c}(J \cap I^c).$$

(A.1)

Proof. Let $a_1, \ldots, a_n$ be a generating set of the ideal I and consider the R-subalgebra of the ring of polynomials $A = R[t]$,

$$B = R[a_1 t, \ldots, a_n t].$$

Since R is Noetherian and B is finitely generated, B is also Noetherian.

Grading A in the usual fashion, B is a graded subalgebra, the Rees algebra of I:

$$B = R + It + I^2 t^2 + \cdots + I^n t^n + \cdots.$$

Define $L_n = J \cap I^n$ and set

$$L = L_0 + L_1 t + L_2 t^2 + \cdots + L_n t^n + \cdots.$$

L is clearly a homogeneous ideal of B, so there is a finite set of forms that generates it,

$$L = (b_1 t^{d_1}, \ldots, b_s t^{d_s}).$$

Let $c = \sup\{d_1, \ldots, d_s\}$; for $n \geq c$, we must have

$$L_n = \sum_{i=1}^{s} I^{n-d_i} b_i,$$

from which the assertion follows. $\qquad\square$

Theorem A.12 (Krull Intersection Theorem). *Let R be a Noetherian ring and let I be an ideal of R. If*

$$L = \bigcap_{n \geq 1} I^n,$$

then $L = I \cdot L$. In particular, if I is contained in the Jacobson radical of R, then

$$\bigcap_{n \geq 1} I^n = 0.$$

Proof. It suffices to put $J = L$ in the Artin-Rees Lemma. The second assertion follows from Nakayama lemma. $\qquad\square$

Remark A.13. Actually, using the Nakayama lemma one can give another description of L. Consider the multiplicative set $S = \{1 + a, \ a \in I\}$. In the ring $S^{-1}R$ the ideal $S^{-1}I$ is contained in the Jacobson radical. Thus the equality $S^{-1}L = S^{-1}I \cdot S^{-1}L$ implies (by Nakayama lemma) that $S^{-1}L = 0$. This means that there is $x \in I$ such that $(1+x)L = 0$.

The theorem above applies equally to modules; more precisely, if M is a finitely generated R–module, then

$$L = \bigcap_{n \geq 1} I^n M,$$

satisfies $L = I \cdot L$. This can be readily seen by making use of the *idealization trick*, consisting in giving the direct sum $S = R \oplus M$ a ring structure by decreeing

$$(a, x) \cdot (b, y) = (a \cdot b, a \cdot y + b \cdot x).$$

Now one applies the theorem to the ring S and its ideal $I \oplus M$.

Another important use of the Artin–Rees lemma is to the identification of two topologies defined by the powers of an ideal I. If M is a finitely generated module over a Noetherian ring R, then the family of submodules $\{I^n M \mid \forall n \geq 0\}$ defines a system of neighborhoods of $0 \in M$. If $N \subset M$ is a submodule, there are two topologies defined on N, the induced one, $\{I^n M \cap N\}$, and its own I-adic topology. The Artin–Rees lemma identifies them.

Exercise A.14 (Bass). Prove that a ring R is Noetherian if and only if the direct sum of any set of injective modules is an injective module.

Exercise A.15. If E is a finitely generated module over the Noetherian ring R, prove that there exists a filtration of submodules,

$$0 = E_0 \subset E_1 \subset \cdots \subset E_n = E,$$

such that $E_i/E_{i-1} = R/\mathfrak{p}_i$, where $\mathfrak{p}_i$ are prime ideals.

Exercise A.16. Prove that if I is an ideal of an Artinian ring R, then some power of I is generated by one idempotent.

Exercise A.17. Explain how to set up a computation to determine the integer c in Theorem A.11.

A.2 Krull Dimension

The notion of dimension of a ring R is totally influenced by geometry, being given by lengths of chains of closed irreducible sets of its prime spectrum $\mathrm{Spec}(R)$. The great advantage here lies in the fact that such sets are each determined by a unique prime ideal,

$$\mathfrak{p}_0 \subset \mathfrak{p}_1 \subset \cdots \subset \mathfrak{p}_n.$$

These are ideals that can be manipulated nicely, which will provide for many numerical estimates of lengths of chains of prime ideals.

- Codimension of ideals
- Systems of parameters
- Determinantal ideals

Codimension of Ideals

We introduce a measure of size for Noetherian rings and its ideals. Its justification lies on Theorem A.20. A more numerical approach passes through the theory of Hilbert functions (see Appendix B). There exists a method based on ordinary field theory that is good enough for most of our purposes that will be treated later. Some far-fetched arcane of homological algebra also serves this need to size up rings.

Definition A.18. *Let R be a Noetherian ring and let $\mathfrak{p}$ be a prime ideal. The codimension or height of $\mathfrak{p}$ is the supremum of the lengths of the chains of prime ideals contained in $\mathfrak{p}$. The height of an ideal I is the infimum of the heights of its minimal primes. The Krull dimension of R is the supremum of the heights of its prime ideals.*

This definition can also be extended to modules. Given a finitely generated module M over a Noetherian R, its *dimension* $\dim M$ is the supremum of all $\dim R/\mathfrak{p}$, where $\mathfrak{p}$ runs through the set of associated primes of M. This dimension equals the dimension of the ring R/I, where I is the annihilator of M. The height codim M of this ideal is called the *codimension* of M.

There is also a notion of *codimension of an algebra*, $A = k[x_1, \ldots, x_n]/I$: it is the codimension of the ideal I provided it does not contain any form of degree 1.

Krull Principal Ideal Theorem

The following result is central to the abstract theory of Noetherian rings. It is the bedrock on which many of its concepts are built. For its reach, its proofs tend to be surprisingly short. We will follow the treatment of [Ei95], [ZS60] and particularly [Kap74]. The reader is advised to read the lively discussion on dimension in [Ei95].

Theorem A.19. *Let R be a Noetherian ring, x a non-unit of R and let $\mathfrak{m}$ be a minimal prime ideal over (x). Then height $\mathfrak{m} \leq 1$.*

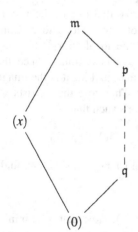

Proof. Our assumption says that there are no prime ideals between (x) and $\mathfrak{m}$.

We will want to argue that if $\mathfrak{p}$ and $\mathfrak{q}$ are prime ideals, the situation described in the diagram cannot occur. By localizing R at $\mathfrak{m}$, we can assume R to be a local ring with $\mathfrak{m}$ as its maximal ideal. Consider the chain $(x) + \mathfrak{p}^{(n)}$, where $\mathfrak{p}$ is the prime ideal as in this diagram, and $\mathfrak{p}^{(n)}$ is the nth symbolic power of $\mathfrak{p}$. This is a descending chain of ideals, all of which contain (x). Since $R/(x)$ is a Noetherian ring with a single prime ideal, it must be Artinian.

This means that for some $n \in \mathbb{Z}^+$ we have that

$$\mathfrak{p}^{(n)} \subset (x) + \mathfrak{p}^{(n+1)}.$$

Therefore any $a \in \mathfrak{p}^{(n)}$ can be written as $a = rx + b$ where $r \in R$ and $b \in \mathfrak{p}^{(n+1)}$; since $\mathfrak{p}^{(n+1)} \subset \mathfrak{p}^{(n)}$ we have that $a - b = rx \in \mathfrak{p}^{(n)}$. However $x \notin \mathfrak{p}$ since $\mathfrak{m}$ is minimal over (x) and r is in $\mathfrak{p}^{(n)}$, and thus we have

$$\mathfrak{p}^{(n)} \subset x\mathfrak{p}^{(n)} + \mathfrak{p}^{(n+1)}. \tag{A.2}$$

Since R is a local ring, we can apply Nakayama lemma to (A.2) and conclude that

$$\mathfrak{p}^{(n)} = \mathfrak{p}^{(n+1)}. \tag{A.3}$$

Localizing at $\mathfrak{p}$, this equality becomes

$$\mathfrak{p}_\mathfrak{p}^n = \mathfrak{p}_\mathfrak{p}^{n+1}. \tag{A.4}$$

We can apply Nakayama lemma to (A.4), and get that the maximal ideal $\mathfrak{p}R_\mathfrak{p}$ of $R_\mathfrak{p}$ is nilpotent, and therefore $\mathfrak{p}$ cannot properly contain another prime. $\qquad\square$

This theorem has the following fuller form.

Theorem A.20. *Let R be a Noetherian ring and let $\mathfrak{m}$ be a minimal prime ideal over $(x_1, \ldots, x_n)$. Then* height $\mathfrak{m} \leq n$.

A consequence of this result is that prime ideals of Noetherian rings have finite height and thus $\mathrm{Spec}(R)$ satisfies the descending chain condition. In particular if R is a local ring then $\dim R < \infty$.

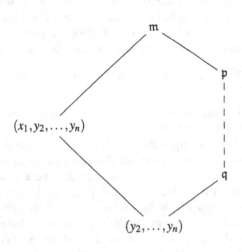

Proof. We may assume that R is a local ring with $\mathfrak{m}$ as its maximal ideal. Let $\mathfrak{p}$ be any prime ideal with no other prime between itself and $\mathfrak{m}$. It will be enough to show that $\mathfrak{p}$ is a minimal prime over an ideal generated by $n - 1$ elements.

We have that one of the x_i, say x_1, does not belong to $\mathfrak{p}$. Consider the ideal $(\mathfrak{p}, x_1)$; the ring $R/(\mathfrak{p}, x_1)$ is Artinian, since the only maximal ideal is the image of $\mathfrak{m}$. Therefore there exists an integer s such that

$$\mathfrak{m}^s \subset (\mathfrak{p}, x_1).$$

In particular all the s-th powers of the x_i are contained in $(\mathfrak{p}, x_1)$; thus we can find $y_i \in \mathfrak{p}$ and $a_i \in R$ such that

$$x_i^s = y_i + a_i x_1,$$

for $i = 2, \ldots, n$. We claim that $\mathfrak{p}$ is minimal over $(y_2, \ldots, y_n)$. Note that $\mathfrak{m}$ is minimal over $(x_1, y_2, \ldots, y_n)$. Suppose there exists a prime ideal, say $\mathfrak{q}$, between $\mathfrak{p}$ and

$(y_2, \ldots, y_n)$ and view this diagram in the ring $R/(y_2, \ldots, y_n)$, we have a situation that would contradict Theorem A.19. □

The connection between Krull dimension defined in this fashion and the one defined for graded algebras in terms of the degrees of the Hilbert polynomial is as follows ([Mat80]).

Theorem A.21. *Let* $(R, \mathfrak{m})$ *be a Noetherian local ring of dimension* $d \geq 1$ *and let*

$$F(\mathfrak{m}) = R/\mathfrak{m} \oplus \mathfrak{m}/\mathfrak{m}^2 \oplus \cdots,$$

be the tangent cone of $\mathfrak{m}$. *The Hilbert polynomial of* $F(\mathfrak{m})$ *has degree* $d - 1$.

Systems of Parameters

From the proof of Krull's theorem we restate what essentially is a converse:

Theorem A.22. *Let* R *be a Noetherian ring and let* $\mathfrak{p}$ *be a prime ideal of height* r. *Then* $\mathfrak{p}$ *is a minimal prime over an ideal* $(x_1, \ldots, x_r)$ *generated by* r *elements.*

The set $\{x_1, \ldots, x_r\}$ is called a *system of parameters* for $\mathfrak{p}$. In the case of a local ring R, a system of parameters for its maximal ideal is also called a system of parameters of R.

Rings of Polynomials

Proposition A.23. *Let* R *be a Noetherian ring and let* $\mathfrak{P}$ *be a prime ideal of the polynomial ring* $R[x]$, *and set* $\mathfrak{p} = R \cap \mathfrak{P}$. *Then*

$$\text{height } \mathfrak{P} = \begin{cases} \text{height } \mathfrak{p}, & \text{if } \mathfrak{P} = \mathfrak{p}R[x], \\ \text{height } \mathfrak{p} + 1, & \text{otherwise} \end{cases}$$

Proof. It is clear that if $\mathfrak{p}$ is a prime ideal of R, then $\mathfrak{p}R[x]$ is the kernel of the canonical homomorphism

$$\psi : R[x] \mapsto (R/\mathfrak{p})[x],$$

and is therefore a prime ideal of $R[x]$. If follows that if

$$\mathfrak{p}_0 \subset \mathfrak{p}_1 \subset \cdots \subset \mathfrak{p}_s$$

is a chain of primes of R. Then

$$\mathfrak{p}_0 R[x] \subset \mathfrak{p}_1 R[x] \subset \cdots \subset \mathfrak{p}_s R[x]$$

is a chain of primes of $R[x]$ of the same length, which shows that

$$\text{height } \mathfrak{p} \leq \text{height } \mathfrak{p}R[x].$$

To prove the reverse inequality, we may localize at $\mathfrak{p}$ without affecting the height of the prime ideals $\mathfrak{p}$ and $\mathfrak{p}R[x]$. Let $(x_1,\ldots,x_s)$ be a system of parameters for $\mathfrak{p}$. By definition there exists an integer m such that $\mathfrak{p}^m R_\mathfrak{p} \subset (x_1,\ldots,x_s)R_\mathfrak{p}$, from which it is clear that $\mathfrak{p}^m R[x]_\mathfrak{p} \subset (x_1,\ldots,x_s)R[x]_\mathfrak{p}$, and thus by Theorem A.20, height $\mathfrak{p}R[x]_\mathfrak{p} \leq s$, which takes care of the first assertion.

Suppose now $\mathfrak{P}$ is not an extended prime of $R[x]$, that is $\mathfrak{p} = \mathfrak{P} \cap R$, $\mathfrak{P} \neq \mathfrak{p}R[x]$. This means that height $\mathfrak{P} \geq$ height $\mathfrak{p} + 1$. We may again localize at $\mathfrak{p}$, so that for simplicity of notation, assume that $(R,\mathfrak{p})$ is a local ring. We then have the embedding

$$\mathfrak{P}/\mathfrak{p}R[x] \hookrightarrow R[x]/\mathfrak{p}R[x] = (R/\mathfrak{p})[x]$$

into a principal ideal domain. Therefore $\mathfrak{P}/\mathfrak{p}R[x]$ is going to be generated by a single element, or equivalently

$$\mathfrak{P} = (f, \mathfrak{p}R[x]).$$

Clearly $\mathfrak{P}$ is minimal over $(x_1,\ldots,x_s,f)$, hence by Theorem A.20

$$\text{height}_{R[x]} \, \mathfrak{P} \leq 1 + s = 1 + \text{height}_R \, \mathfrak{p},$$

as desired.   □

Theorem A.24. *Let R be a Noetherian ring and let $x_1,\ldots,x_n$ be a set of independent indeterminates over R. Then*

$$\dim R[x_1,\ldots,x_n] = \dim R + n.$$

In particular if R is a field k, then

$$\dim k[x_1,\ldots,x_n] = n.$$

Sums versus Products

A well-known trick gives very broad generalization of Theorem A.20 to general ideals in rings of polynomials (generalized by Serre to all regular local rings).

Theorem A.25. *Let I and J be ideals of a polynomial ring $R = k[x_1,\ldots,x_n]$ over a field k. If $I + J \neq R$ then $\text{height}(I,J) \leq \text{height } I + \text{height } J$.*

Proof. Let J' be a copy of the ideal J in the ring of polynomials $R' = k[y_1,\ldots,y_n]$. Observe that

$$R/(I,J) = R \otimes_k R'/(I,J',\Delta),$$

where $\Delta = (x_1 - y_1,\ldots,x_n - y_n)$. From the dimension formula for the direct product of affine varieties,

$$\dim R \otimes_k R'/(I,J') = \dim R/I + \dim R'/J',$$

and the usual equality in rings of polynomials $\dim R = \text{height } I + \dim R/I$, a final application of Theorem A.20 gives the assertion.   □

Codimension of Determinantal Ideals

Let R be a Noetherian ring and let φ be a $m \times n$ matrix with entries in R:

$$\varphi = \begin{bmatrix} a_{11} & \cdots & a_{1n} \\ \vdots & \ddots & \vdots \\ a_{m1} & \cdots & a_{mn} \end{bmatrix}.$$

Estimating the sizes of the ideal $I_t(\varphi)$ generated by all $t \times t$ minors of φ is important for many of our constructions, in particular those that involve Jacobian ideals.

The classical bound for the sizes $EN(m,n;t)$ of these ideals is the theorem of Eagon and Northcott [EN62] (for more details and generalizations, see [BVe88]):

Theorem A.26. *The ideals* $I_t(\varphi)$ *satisfy*

$$\text{height } I_t(\varphi) \leq EN(m,n;t) = (m-t+1)(n-t+1), \qquad \text{(A.5)}$$

where equality is reached when φ *is a generic matrix in* $m \cdot n$ *indeterminates.*

Proof. We only prove the first assertion, leaving the rest to the reader. The case $t = 1$ is Theorem A.20, so that we may assume $t \geq 2$. Denote $I = I_t(\varphi)$ and let $\mathfrak{p}$ be a minimal prime of I. Localizing at $\mathfrak{p}$ we may assume that R is a local ring and denote still by $\mathfrak{p}$ its maximal ideal; it is enough to show that $\dim R \leq EN(m,n;t)$.

If one of the entries of φ is a unit, say a_{11} is an invertible element of R, through a series of elementary row and column operations the matrix φ can be transformed into a matrix

$$\begin{bmatrix} 1 & 0 \cdots 0 \\ 0 & \\ \vdots & \varphi' \\ 0 & \end{bmatrix}.$$

Since $t > 1$, it is clear that $I_t(\varphi) = I_{t-1}(\varphi')$. We induct on t, which means that all the entries of φ may be assumed to lie in the maximal ideal of R.

Consider now the matrix

$$\psi = \begin{bmatrix} a_{11} + x & \cdots & a_{1n} \\ \vdots & \ddots & \vdots \\ a_{m1} & \cdots & a_{mn} \end{bmatrix},$$

where x is an indeterminate over R and let $L = I_t(\psi)$. Since $t > 1$, note that $L \subset \mathfrak{p}R[x]$, so in particular height $L \leq \dim R$. On the other hand, $(L,x) = (I,x)$, from which we claim that L is $\mathfrak{p}R[x]$–primary. Otherwise there would exist a minimal prime $\mathfrak{Q}$ of L properly contained in $\mathfrak{p}R[x]$. But then, in the ring $R[x]/\mathfrak{Q}$, the image of $(\mathfrak{p},x)$ would be a maximal ideal of codimension at least two, but minimal over the principal ideal generated by the image of x; this would contradict Theorem A.19. We may now localize at $\mathfrak{p}R[x]$ and decrement t. □

Valuative Dimension

There are other notions of *dimension* of commutative rings, from transcendence degrees for classical rings to various dimensions with a homological flavor. We recall one dimension which is useful in the study of non–Noetherian domains, the *valuative dimension* of a ring.

If A is an integral domain of field of quotients K, its valuative dimension is the supremum of the Krull dimension of all valuation rings of K that contain A ([Gil72]). If A is a Noetherian domain this is the same as its Krull dimension.

We use it here to obtain the Krull dimension of the ordinary Rees algebra (see [Val79]).

Theorem A.27. *Let R be a Noetherian ring and I be an ideal of R. For each minimal prime $\wp$ of R set $c(\wp) = 1$ if $I \not\subset \wp$, and $c(\wp) = 0$ otherwise. Then*

$$\dim R[It] = \sup\{\dim R/\wp + c(\wp)\}.$$

In particular, $\dim R[It] \leq \dim R + 1$, and equality will hold when I contains regular elements.

Proof. We may assume that $\dim R < \infty$. Let N be the nilradical of R. Since $N[t]$ is the nilradical of $R[t]$, it follows that $N[t] \cap R[It]$ is the nilradical of $R[It]$, so that we may assume that R is reduced and the minimal primes of $R[It]$ have the form $P_i = \wp_i R[t] \cap R[It]$. In turn, this implies that to determine $\dim R[It]$ we may assume that R is a domain. In case $I = 0$, then $R[It] = R$. If $I \neq 0$, let V be any valuation ring of R. Since I is finitely generated, $V \cdot R[It] = V[at] \simeq V[t]$, which is a ring of Krull dimension $\dim V + 1$. $\square$

Exercise A.28. Let R be a Noetherian ring of dimension n. Prove that any radical ideal I is the radical of an ideal generated by $n + 1$ elements.

Exercise A.29 (Valla). Let $(R, \mathfrak{m})$ be a Noetherian local ring and let $x_1, \ldots, x_d$ be a system of parameters for the maximal ideal of R. Set $A = R[Jt]$ for the Rees algebra of the ideal J generated by the x_i's. Show that the elements

$$\{x_1, x_1 t - x_2, x_2 t - x_3, \ldots, x_{d-1} t - x_d, x_d t\}$$

form a system of parameters for the maximal ideal $(\mathfrak{m}, Jt)$ of A.

Exercise A.30. Let R be a Noetherian ring and let φ be a $n \times n$ symmetric matrix with entries in R. Prove that the ideal I generated by all $t \times t$ minors of φ, $t \leq n$, has codimension at most

$$\binom{n-t+2}{2},$$

and that this bound is reached if φ is a generic matrix.

Exercise A.31. Let R and S be two commutative Noetherian rings, and let E be an abelian group which is finitely generated as an R–module and as an S–module. Suppose that the actions of R and S on E commute, that is, for $r \in R$, $s \in S$ and $e \in E$, $r \cdot se = s \cdot re$. Prove that the Krull dimension of E over either ring is the same.

Exercise A.32. A commutative ring R is a Hilbert ring if every prime ideal is the intersection of the maximal ideals containing it. Prove that if $(R, \mathfrak{m})$ is a Noetherian local ring and $f \in \mathfrak{m}$, then the localization R_f is a Hilbert ring.

Exercise A.33. Let A be a Noetherian domain of Krull dimension 1. Show that any ring $A \subset B \subset K$, where K is the field of fractions of A, is a Noetherian ring of Krull dimension at most 1.

A.3 Graded Algebras

These algebras are the setting for a strong interaction between algebraic and combinatorial constructions. Most of these aspects are discussed in Appendix B, limiting ourselves here to some algebraic notions.

- Chains of homogeneous prime ideals
- Analytic spread and reduction number
- Superficial elements

Dimension of Graded Algebras

Let A be a $\mathbb{Z}$–graded Noetherian ring. We may now form chains of homogeneous prime ideals

$$P_0 \subset P_1 \subset \cdots \subset P_r$$

and define the *homogeneous* Krull dimension $\dim_h A$ as the supremum of the lengths of all such chains.

Proposition A.34. *For any $\mathbb{Z}$–graded Noetherian ring A,*

$$\dim A = \dim_h A.$$

Proof. See [BH93, Section 1.5].

Dimension Formula

Proposition A.35. *Let B be a Noetherian integral domain that is finitely generated as an algebra over a subring A. Suppose there exists a prime ideal Q of B such that $B = A + Q$, $A \cap Q = 0$. Then*

$$\dim(B) = \dim(A) + \text{height}(Q) = \dim(A) + \text{tr.deg.}_A(B).$$

Proof. We may assume that $\dim A$ is finite; $\dim(B) \geq \dim(A) + \text{height}(Q)$ by our assumption. On the other hand, by the standard dimension formula of [Mat80, Theorem 23], for any prime ideal P of B, $\mathfrak{p} = P \cap A$, we have

$$\text{height}(P) \leq \text{height}(\mathfrak{p}) + \text{tr.deg.}_A(B) - \text{tr.deg.}_{k(\mathfrak{p})}k(P).$$

The inequality

$$\dim(B) \leq \dim(A) + \text{height}(Q)$$

follows from this formula and reduction to the affine algebra obtained by localizing B at the zero ideal of A. $\qquad\qquad\qquad\qquad\qquad\qquad\qquad\qquad\qquad\qquad\square$

Corollary A.36. *Let B be a $\mathbb{N}$–graded Noetherian ring and let A denote its degree 0 component. Then for any homogeneous prime ideal P, with $\mathfrak{p} = P \cap A$,*

$$\dim(B/P) = \dim(A/\mathfrak{p}) + \text{tr.deg.}_{k(\mathfrak{p})}k(P)$$

and

$$\dim(B_\mathfrak{p}) = \dim(A_\mathfrak{p}) + \text{tr.deg.}_A(B).$$

Corollary A.37. *Let $A = A_0[A_1]$ be a $\mathbb{N}$–graded Noetherian algebra where A_0 is a local ring. Then*

$$\dim A = \dim A_0 + \text{height}(A_1).$$

Proof. The proofs are immediate after reduction to the case where A_0 is a domain, which is left to the reader.

Superficial Element and Analytic Spread

Let $A = A_0[A_1]$ be a $\mathbb{N}$–graded Noetherian algebra over the local ring $(A_0, \mathfrak{m})$. The *special fiber* of A is the ring $F(A) = A \otimes_{A_0} (A_0/\mathfrak{m})$.

Definition A.38. *An element $h \in A_n$ is called a* superficial *element of degree n if $(0 : h)_m = 0$ for $m \gg 0$. In the case of an arbitrary Noetherian ring R and of an ideal I, then $h \in I$ is said to be a superficial element of degree n if $I^{m+n} : h = I^m$ for $m \gg 0$.*

When A_0 is an Artin ring, and $h \in A_1$, such elements have the property that the Hilbert polynomials of A and of A/hA coincide.

Definition A.39. *The* analytic spread *of A is the dimension of the ring $F(A)$,*

$$\ell(A) = \dim F(A).$$

Proposition A.40. *Let $A = A_0[A_1]$ be a $\mathbb{N}$–graded Noetherian algebra over the local ring $(A_0, \mathfrak{m})$ with infinite residue field. If $\ell(A) = \ell$ there exist $x_1, \ldots, x_\ell \in A_1$ such that $A_{r+1} = (x_1, \ldots, x_\ell)A_r$.*

The proof uses the following version of prime avoidance lemmas that is useful in the study of graded algebras.

Proposition A.41. *Let $A = A_0[A_1]$ be a Noetherian graded algebra over the local ring $(A_0, \mathfrak{m})$ with infinite residue field, generated by elements of degree 1. Let $P_1, \ldots, P_r$ be a family of homogeneous prime ideals that do not contain A_1. Then there exists $h \in A$ satisfying*

$$h \in A_1 \setminus \mathfrak{m}A_1 \cup \bigcup_{i=1}^{r} P_i.$$

Furthermore, if the P_i's include all the associated primes of A that do not contain A_1 then $0: (0:hA)$ contains some power of A_+.

Proof. Denote by $C_1, \ldots, C_r$ the components of degree 1 of these homogeneous ideals. Consider the vector space $V = A_1/\mathfrak{m}A_1$ over the residue field $A_0/\mathfrak{m}$ and its subspaces $V_i = (C_i + \mathfrak{m}A_1)/\mathfrak{m}A_1$. By Nakayama lemma and the choices of the P_i's, $V_i \neq V$, $\forall i$. Since $A_0/\mathfrak{m}$ is infinite, it follows that

$$V \neq \bigcup_{i=1}^{r} V_i,$$

by any of the usual tricks. (For example, let $e_1, \ldots, e_n$ be a basis V. Since each V_i is a proper subspace, it is contained in the locus of the linear polynomial $f_i(\mathbf{x}) = a_{i1}x_1 + \cdots + a_{in}x_n$. Then any point where

$$f(\mathbf{x}) = \prod_{i=1}^{r} f_i(\mathbf{x})$$

does not vanish provides a vector that does not lie in any V_i.)

For the second assertion, choose then $h \in A_1$ whose image in V does not lie in any V_i. Let $\mathfrak{P}$ be a minimal prime of $0: (0:hA)$; it suffices to show that $A_+ \subset \mathfrak{P}$. (If h is a regular element, $0: (0:hA) = A$.) Note that $\mathfrak{P}$ consists of zero divisors and contains h. This means that $\mathfrak{P}$ is an associated prime of A but distinct from any of the P_i, and therefore must contain A_+. $\qquad\square$

Proof of Proposition A.40. Let $\{P_1, \ldots, P_r\}$ consist of all minimal primes of $\mathfrak{m}A$, and let $x_1 \in A_1$ be chosen as above. The ring $A' = A/(x_1) = A_0[A_1/A_0x_1)]$ has for special fiber $F(A') = F(A)/(\overline{x_1})$, and therefore $\ell(A') = \ell - 1$, from the choice of x_1. The sequence $\mathbf{x} = \{x_1, \ldots, x_\ell\}$ is produced by iteration, and $\ell(A/(\mathbf{x})) = 0$. This means that the ring $F(A/(\mathbf{x}))$ is Artinian and therefore for all $r \gg 0$, $A_{r+1} = (\mathbf{x})A_r + \mathfrak{m}A_{r+1}$ and $A_{r+1} = (\mathbf{x})A_r$ by Nakayama lemma. $\qquad\square$

Reduction Number

We already discussed in Chapter 9 this notion but it may be worthwhile to recall it here in its simplest context.

Definition A.42. *Let $\mathbf{x} = x_1, \ldots, x_\ell$ be a set of elements as above. The smallest integer r for which $A_{r+1} = (\mathbf{x})A_r$ is called the* reduction number *of A with respect to $\mathbf{x}$. The infimum of all these integers is called the* reduction number *of A.*

A.4 Integral Extensions

Integral homomorphisms express a very close geometric connection between commutative rings. Here we give an overview of this basic relationship, with some proofs.

- Integral elements
- Dimension formulas
- Zariski main theorem

Integral Elements

Let A and B be two rings and assume that there exists a map from A into B

$$\varphi: A \longmapsto B. \tag{A.6}$$

Let b be an element in B; it makes good sense to ask whether this element is transcendental, algebraic, integral etc. Once you have a map like (A.6), we also have the following map

$$\Phi: A[T] \longmapsto B, \quad T \longmapsto b \quad \text{and} \quad \Phi_{|A} = \varphi. \tag{A.7}$$

If $\ker \Phi = (0)$, then we say that b is transcendental, otherwise it is algebraic.

Definition A.43. *The element $b \in B$ is integral over A if there exists a relation of the form*

$$b^n + a_{n-1} b^{n-1} + \cdots + a_1 b + a_0 = 0,$$

for $a_i \in A$ and some $n \in \mathbb{N}$. In other words, b is integral over A if there exists a monic polynomial in the kernel of (A.7). We say that B is integral over A (or that B is an integral extension of A) if every element of B is integral over A.

Let us give now a criterion of integrality (due, more or less, to Weierstrass) that will be used repeatedly in what will follow.

Proposition A.44. *Let $A \hookrightarrow B$ two rings. An element $b \in B$ is integral over A if and only if there exists a finitely generated A-module L with $A \subset L \subset B$ and $bL \subset L$.*

Proof. Consider $L = A[1, b, b^2, \ldots, b^{n-1}] = \left\{ \sum_{i=0}^{n-1} \alpha_i b^i \mid \alpha_i \in A \right\} \subset B$. It is easy to check that L is a finitely generated A-module which contains all the powers of b, e.g.

$$b^n = -a_{n-1} b^{n-1} - a_{n-2} b^{n-2} - \cdots - a_1 b - a_0$$
$$b^{n+1} = bb^n = b(-a_{n-1} b^{n-1} - a_{n-2} b^{n-2} - \cdots - a_1 b - a_0) =$$
$$= -a_{n-1} b^n - a_{n-2} b^{n-1} - \cdots - a_1 b^2 - a_0 b.$$

$L = A[b]$ is a finitely generated and faithful (since it contains 1) A-module such that $bL \subset L$.

Conversely, let L be a finitely generated A-module contained in B such that $bL \subset L$. We want to show that b is integral over A. Let us assume that $L = (c_1, \ldots, c_n)$. We have the following set of equations

$$
\begin{cases}
bc_1 = \sum_{j=1}^{n} a_{1j}c_j \\
bc_2 = \sum_{j=2}^{n} a_{2j}c_j \\
\vdots \\
bc_n = \sum_{j=1}^{n} a_{nj}c_j
\end{cases}
$$

where $a_{ij} \in A$. Those relations can be translated in the following matrix equation

$$
\begin{bmatrix}
b - a_{11} & -a_{12} & \cdots & -a_{1n} \\
-a_{21} & b - a_{22} & \cdots & -a_{2n} \\
\vdots & \vdots & \ddots & \vdots \\
-a_{n1} & -a_{n2} & \cdots & b - a_{nn}
\end{bmatrix}
\begin{bmatrix}
c_1 \\ c_2 \\ \vdots \\ c_n
\end{bmatrix}
=
\begin{bmatrix}
0 \\ 0 \\ \vdots \\ 0
\end{bmatrix}.
\tag{A.8}
$$

Let $\Psi = (b\delta_{ij} - a_{ij})_{1 \le i,j \le n}$ be that matrix, and let us multiply (A.8) by the adjoint of Ψ, $\mathrm{Adj}(\Psi)$; using the fact that $\mathrm{Adj}(\Psi)\Psi = \Psi\mathrm{Adj}(\Psi) = \det(\Psi)I_n$, where I_n denotes the identity $n \times n$ matrix, we get

$$
\det(\Psi)
\begin{bmatrix}
c_1 \\ c_2 \\ \vdots \\ c_n
\end{bmatrix}
=
\begin{bmatrix}
0 \\ 0 \\ \vdots \\ 0
\end{bmatrix},
$$

hence $\det(\Psi)$ belongs to the annihilator of L; since L contains 1 (being a unitary, hence, faithful A-module) we must have $\det(\Psi) = 0$. Expanding the determinant we get the desired relation of integrality for b. $\square$

Corollary A.45. *Let $A \hookrightarrow B$ be two rings, and let $\overline{A}$ be the set of all the elements of B that are integral over A. Then $\overline{A}$ is a subring of B containing A.*

Proof. We have to show that if $a, b \in B$ are integral over A then $a \pm b$ and ab are integral over A. We know that there exist A–submodules of B, L and M, such that $aL \subset L$ and $bM \subset M$. Thus

$$
abLM \subset aLbM \subset LM
$$
$$
(a \pm b)LM \subset aLM \pm bLM \subset LM.
$$

The assertion will follow by Proposition A.44. $\square$

Proposition A.46. *Let $A \hookrightarrow B \hookrightarrow C$ be a chain of rings; if $c \in C$ is integral over B and B is integral over A then c is integral over A. In particular integral extensions of integral extensions are integral extensions (over the smaller ring). Also, if $B = \overline{A}$ is the integral closure of A in C, then $\overline{A}$ is integrally closed in C.*

Proof. Indeed, if $c^n + b_{n-1}c^{n-1} + \cdots + b_1c + b_0 = 0$ with $b_i \in B$ then

$$
A[b_0, b_1, \ldots, b_{n-1}, c] = \sum_{i=0}^{n-1} A[b_0, b_1, \ldots, b_{n-1}]c^i,
$$

and since $A[b_0, \ldots, b_{n-1}]$ is a finite A-module, so is $A[b_0, \ldots, b_{n-1}, c]$. $\square$

Dimension Theory

Let A and B be two rings. Assume that there exists a map from A into B,

$$\varphi: A \longmapsto B.$$

The mapping of the corresponding prime spectra

$$\varphi_a: \operatorname{Spec}(B) \longmapsto \operatorname{Spec}(A), \quad \mathfrak{P} \mapsto \varphi_a(\mathfrak{P}) = \mathfrak{P} \cap A = \mathfrak{p}. \tag{A.9}$$

has particularly nice features in case B is integral over A.

Proposition A.47. *If B is integral over A then the contraction to A of any maximal ideal of B is a maximal ideal of A and conversely. In other words, the restriction of φ_a to the set of maximal ideals of B induces a mapping onto he set of maximal ideals of A.*

Proof. Let $\mathfrak{p} = \mathfrak{P} \cap A$. Clearly, $\mathfrak{p}$ is the kernel of the map obtained by composing the inclusion of A into B with the natural surjection π of B onto $B/\mathfrak{P}$

$$A \hookrightarrow B \to B/\mathfrak{P}.$$

Therefore the "first theorem of homomorphism" guarantees the existence of an inclusion from $A/\mathfrak{p}$ into $B/\mathfrak{P}$

$$A/\mathfrak{p} \hookrightarrow B/\mathfrak{P}; \tag{A.10}$$

moreover, we have that $B/\mathfrak{P}$ is integral over $A/\mathfrak{p}$. Indeed, if $\bar{b} = b + \mathfrak{P} \in B/\mathfrak{P}$, where b satisfies a relation of the form

$$b^n + a_{n-1}b^{n-1} + \cdots + a_1 b + a_0 = 0,$$

then we also have

$$\pi(b^n + a_{n-1}b^{n-1} + \cdots + a_1 b + a_0) = 0. \tag{A.11}$$

If we expand (A.11), we get a monic relation for $\bar{b}$ with coefficients in $A/\mathfrak{p}$

$$\bar{b}^n + \bar{a}_{n-1}\bar{b}^{n-1} + \cdots + \bar{a}_1 \bar{b} + \bar{a}_0 = \bar{0}.$$

Now we claim that $\mathfrak{p}$ is maximal if and only if $\mathfrak{P}$ is maximal. Let's suppose $B/\mathfrak{P}$ to be a field, i.e. $\mathfrak{P}$ is a maximal ideal of B; we want to argue that $A/\mathfrak{p}$ is also a field. Pick $a \in A/\mathfrak{p}$; we want to show that its inverse exists in $A/\mathfrak{p}$. Certainly the inverse of a exists in $B/\mathfrak{P}$, call it b. Since b is integral over $A/\mathfrak{p}$ its satisfies a relation of the form

$$b^n + a_{n-1}b^{n-1} + \cdots + a_1 b + a_0 = 0, \tag{A.12}$$

where $a_i \in A/\mathfrak{p}$.

Pick n to be the least positive integer for which we have a relation such as (A.12). We want to show that $n = 1$. Assume $n > 1$. This implies that $a_0 \neq 0$; otherwise, we could factor b out from (A.12)

$$b(b^{n-1} + a_{n-1}b^{n-2} + \cdots + a_1) = 0$$

and since $B/\mathfrak{P}$ is an integral domain (being a field) and $b \neq 0$, we would have a relation for b of lower degree: a contradiction.

Rewrite (A.12) as follows

$$b(b^{n-1} + a_{n-1}b^{n-2} + \cdots + a_2 b + a_1) = -a_0, \qquad (A.13)$$

and multiply both sides of (A.13) by a; since $ab = 1$ we get

$$b^{n-1} + a_{n-1}b^{n-2} + \cdots + a_2 b + a_1 + a a_0 = 0,$$

which gives us again a relation for b of lower degree. The contradiction implies that $n = 1$, hence $b \in A/\mathfrak{p}$.

Conversely, let $b \in B/\mathfrak{P}$; then there exists a relation of the form $b^n + a_{n-1}b^{n-1} + a_{n-2}b^{n-2} + \cdots + a_1 b + a_0$ with $a_i \in A/\mathfrak{p}$. As we did before, we can take n to be the least positive integer for which that happens; hence we may assume $a_0 \neq 0$, i.e. a_0 is invertible. Then we have the following equation

$$b(b^{n-1} + a_{n-1}b^{n-2} + \cdots + a_1) = -a_0, \qquad (A.14)$$

and so if we multiply both sides of (A.14) by $-a_0^{-1}$ we get

$$b[-a_0^{-1}(b^{n-1} + a_{n-1}b^{n-2} + \cdots + a_1)] = 1,$$

which gives us the desired inverse of b, and clearly this inverse lies in B.   □

Lemma A.48. *Let $A \hookrightarrow B$ be two rings and assume that B is integral over A. Let S be a multiplicatively closed subset of A. Then there is a ring monomorphism $A_S \hookrightarrow B_S = B \otimes_A A_S$ and B_S is integral over A_S.*

Theorem A.49 (Lying Over Theorem). *Let $A \hookrightarrow B$ be two rings; if B is integral over A then the map φ_a defined as in (A.9) is surjective, i.e. given $\mathfrak{p} \in \mathrm{Spec}(A)$ we can find $\mathfrak{P} \in \mathrm{Spec}(B)$ such that $\varphi_a(\mathfrak{P}) = \mathfrak{P} \cap A = \mathfrak{p}$.*

Proof. Pick $\mathfrak{p} \in \mathrm{Spec}(A)$. We want to find a prime $\mathfrak{P} \in \mathrm{Spec}(B)$ which contracts to $\mathfrak{p}$.

Clearly $A_{\mathfrak{p}} \hookrightarrow B_{\mathfrak{p}}$ is still an integral extension, and $\mathfrak{p}A_{\mathfrak{p}}$ is the maximal ideal of $A_{\mathfrak{p}}$. Consider $\mathfrak{p}B_{\mathfrak{p}}$; if it is the maximal ideal you are done. Otherwise you can find a maximal ideal containing it. In both cases, the contraction is going to be $\mathfrak{p}A_{\mathfrak{p}}$, the maximal ideal of $A_{\mathfrak{p}}$. Hence we want to show that $\mathfrak{p}B_{\mathfrak{p}}$ is a proper ideal, namely

$$\mathfrak{p}B_{\mathfrak{p}} \neq B_{\mathfrak{p}}.$$

Assume $\mathfrak{p}B_{\mathfrak{p}} = B_{\mathfrak{p}}$; then you can write

$$1 = \sum_{i=1}^n r_i b_i,$$

where $r_i \in \mathfrak{p}A_{\mathfrak{p}}$ and $b_i \in B_{\mathfrak{p}}$. Consider $C = A_{\mathfrak{p}}[b_1, \ldots, b_n]$; it is a finitely generated subring of $B_{\mathfrak{p}}$. As each b_i is integral over $A_{\mathfrak{p}}$, C is finitely generated as a submodule of $B_{\mathfrak{p}}$. But this says that $C = \mathfrak{p}C$, which is impossible by Nakayama's lemma. Thus $\mathfrak{p}B_{\mathfrak{p}} \neq B_{\mathfrak{p}}$. Finally, take any prime in $B_{\mathfrak{p}}$ that contracts to $\mathfrak{p}A_{\mathfrak{p}}$. Lift it in B and contract it in A.   □

Theorem A.50 (Going Up Theorem). *Let $A \hookrightarrow B$ be two rings and assume that B is integral over A. Let $\mathfrak{p} \subset \mathfrak{q}$ be two prime ideals of A and assume that $\mathfrak{p}$ is the contraction to A of a prime ideal of B, i.e. $\mathfrak{p} = \mathfrak{P} \cap A$ for some $\mathfrak{P} \in \mathrm{Spec}(B)$. Then there exists $\mathfrak{Q} \in \mathrm{Spec}(B)$ such that $\mathfrak{P} \subset \mathfrak{Q}$ and $\mathfrak{q} = \mathfrak{Q} \cap A$.*

Proof. Since $\mathfrak{p} = \varphi_a(\mathfrak{P}) = \mathfrak{P} \cap A$ we can think of $B/\mathfrak{P}$ as an integral extension of $A/\mathfrak{p}$ (in fact the condition that an element of B is integral over A is preserved by the homomorphism $\pi \colon B \longrightarrow B/\mathfrak{P}$)

$$A/\mathfrak{p} \hookrightarrow B/\mathfrak{P} \quad \text{integral extension.}$$

Since $\mathfrak{q}/\mathfrak{p}$ is a prime ideal in $A/\mathfrak{p}$, there exists, by Theorem A.49, a prime ideal of $B/\mathfrak{P}$ lying over $\mathfrak{q}/\mathfrak{p}$. But every ideal of $B/\mathfrak{P}$ is the quotient of an ideal in B, in other words

$$A/\mathfrak{p} \xrightarrow{\ \text{inj.}\ } B/\mathfrak{P}$$

$$\mathfrak{q}/\mathfrak{p} \longleftarrow\!\!\shortmid \ \mathfrak{Q}/\mathfrak{P}$$

such that $\mathfrak{Q}/\mathfrak{P} \cap A/\mathfrak{p} = \mathfrak{q}/\mathfrak{p}$, where $\mathfrak{Q} \in \mathrm{Spec}(B)$. Thus $\mathfrak{Q} \cap A = \mathfrak{q}$. $\square$

Theorem A.49 and Theorem A.50 enable us to make several assertions about the dimensions of A and B whenever $A \hookrightarrow B$ and B is an integral extension of A. Let us see the first one of these results.

Corollary A.51. *Let $A \hookrightarrow B$ be two rings and assume that B is integral over A. Then $\dim(A) = \dim(B)$.*

Proof. Given any chain of primes in A

$$\mathfrak{p}_0 \subset \mathfrak{p}_1 \subset \cdots \subset \mathfrak{p}_n \subset A,$$

we can lift it, using the lying over theorem, to a chain of primes in B

$$\mathfrak{P}_0 \subset \mathfrak{P}_1 \subset \cdots \subset \mathfrak{P}_n \subset B,$$

one has $\dim(A) \leq \dim(B)$.

Conversely, given a chain of primes in B

$$\mathfrak{P}_0 \subset \mathfrak{P}_1 \subset \cdots \subset \mathfrak{P}_n \subset B,$$

we can contract it to a chain of primes in A

$$\mathfrak{p}_0 \subset \mathfrak{p}_1 \subset \cdots \subset \mathfrak{p}_n \subset A, \tag{A.15}$$

If we prove that two distinct primes in B cannot have the same contraction in A, then we are done because (A.15) says that $\dim(A) \geq \dim(B)$. This last fact is guaranteed by the next lemma. $\square$

Lemma A.52. *Let $A \hookrightarrow B$ be two rings and assume that B is integral over A. If $\mathfrak{P} \subset \mathfrak{Q}$ are two distinct prime ideals of B, then their restrictions to A will be two different prime ideals $\mathfrak{p} \subset \mathfrak{q}$ of A.*

Proof. Let $\mathfrak{p} = \mathfrak{P} \cap A$ and let's consider, as usual, the following integral extension

$$A/\mathfrak{p} \hookrightarrow B/\mathfrak{P}.$$

Notice that both $A/\mathfrak{p}$ and $B/\mathfrak{P}$ are integral domains and that $\mathfrak{P}$ and $\mathfrak{Q}$ corresponds respectively to the zero ideal and to $\mathfrak{Q}/\mathfrak{P}$ of $B/\mathfrak{P}$. Therefore we are in the following general situation. Let $R \hookrightarrow S$ be two domains and assume that S is integral over R. If $\mathfrak{a}$ is a prime ideal of S then $\mathfrak{a} \cap R \neq (0)$. Indeed, let $s \in \mathfrak{a}$; since s is integral over R it satisfies a relation of the form

$$s^n + r_{n-1}s^{n-1} + \cdots + r_1 s + r_0 = 0 \quad r_i \in R.$$

Pick n to be the least positive integer for which we have such a relation. Thus $R \ni r_0 \neq 0$ and of course $r_0 = -s^n - r_{n-1}s^{n-1} - \cdots - r_1 s \in \mathfrak{a}$. □

The last of the triad (whose proof is left for look up) is:

Theorem A.53 (Going Down Theorem). *Let A be normal domain and let $A \hookrightarrow B$ be an integral extension which is torsionfree as an A–module. Let $\mathfrak{p} \subset \mathfrak{q}$ be two prime ideals of A and assume that $\mathfrak{q}$ is the contraction to A of a prime ideal of B, i.e. $\mathfrak{q} = \mathfrak{Q} \cap A$ for some $\mathfrak{Q} \in \mathrm{Spec}(B)$. Then there exists $\mathfrak{P} \in \mathrm{Spec}(B)$ such that $\mathfrak{P} \subset \mathfrak{Q}$ and $\mathfrak{p} = \mathfrak{P} \cap A$.*

Integrally Closed Domains

There are many important classes of integrally closed domains besides factorial domains. We recall two.

Definition A.54. *A integral domain R is a valuation domain if for any nonzero elements $x, y \in R$, then $\frac{x}{y}$ or $\frac{y}{x}$ lie in R.*

This means that the ideals of R are linearly ordered. It follows that R is an integrally closed local ring. By abuse of terminology, one says that a *valuation* of R is a valuation domain $R \subset V \subset K$, where K is the field of fraction of R. The *valuative dimension* of R is the supremum of the Krull dimension of all its valuations.

Their usefulness lies in part because of the following set of properties.

Theorem A.55. *Let A be an integral domain. Then*

(a) *The integral closure of A is the intersection of all valuations of R.*
(b) *For any prime ideal $\mathfrak{p}$ of R, there exists a valuation V of R such that its maximal ideal $\mathfrak{m}$ satisfies $\mathfrak{p} = A \cap \mathfrak{m}$.*
(c) *If A is a Noetherian domain, its Krull dimension equals its valuative dimension.*

Definition A.56. *A commutative ring D is a* Dedekind domain *if it is a Noetherian integrally closed domain of dimension* ≤ 1.

For convenience we include fields among these rings. There are literally hundreds of characterizations of these rings, from which we single out:

Theorem A.57. *Let D be an integral domain. The following conditions are equivalent:*

(a) *D is a Dedekind domain.*
(b) *Each ideal of D is a projective module.*
(c) *Each nonzero ideal of D is invertible.*
(d) *D is Noetherian, of dimension* ≤ 1, *and for each nonzero ideal I*

$$(I \cdot I^{-1})^{-1} = D.$$

The engine of proof is Proposition 6.11, along with the interpretation of (b) and (c) in terms of a presentation of the ideal I. We leave the proof to the reader as an exercise.

An important technique to create Dedekind domains lies in taking the integral closure of a base Dedekind domain in a finite extension of its field of fractions.

Zariski Main Theorem

Integral extensions occur in many important results of algebraic geometry, of which we describe one.

Let $\varphi : A \mapsto B$ be a homomorphism of commutative rings and $\mathfrak{P} \in \mathrm{Spec}(B)$, and denote $\mathfrak{p} = \varphi_a(\mathfrak{P})$. The set of all primes of B contracting to $\mathfrak{p}$ define the fiber of $\mathfrak{p}$ or of $\mathfrak{P}$; it is given as $X = \mathrm{Spec}(B \otimes_A (A/\mathfrak{p})_{\mathfrak{p}})$. A prime ideal $\mathfrak{Q}$ of a ring R is said to be *isolated* if it is both maximal and a minimal prime of R.

Theorem A.58 (Zariski Main Theorem). *Let A be a commutative ring and B an algebra of finite type over A. Let $\overline{A}$ be the integral closure of A in B and $\mathfrak{P} \in \mathrm{Spec}(B)$. If $\mathfrak{P}$ is isolated in its fiber, there exists $f \in \overline{A}$, $f \notin \mathfrak{P}$ such that $\overline{A}_f = B_f$.*

Proof. See [Ray70, p. 42]. □

Exercise A.59. Let $\psi : A \mapsto B$ be an integral extension where A is an integral domain. For $b \in B$, let $I(b)$ the kernel of the natural mapping

$$A[T] \mapsto B, \ T \mapsto b.$$

Show that if A is integrally closed and B is torsionfree as an A–module, then $I(b)$ is a principal ideal. Prove also that if A is a domain which is not integrally closed and $b \in \overline{A} \setminus A$, then $I(b)$ is not a principal ideal.

A.5 Finitely Generated Algebras over Fields

Our aim will be the study of fundamental aspects of the theory of finitely generated algebras. The terminology *affine ring* refers to a finitely generated algebra over a field k.

- Noether normalization
- Nullstellensatz
- Integral closure

Noether Normalization

Let $B = k[x_1, \ldots, x_n]$ be a finitely generated algebra over a field k and assume that the x_i's are algebraically dependent. Our goal is to find a new set of generators $y_1, \ldots, y_n$ for B such that

$$k[y_2, \ldots, y_n] \hookrightarrow B = k[y_1, \ldots, y_n]$$

is an integral extension.

Let $k[X_1, \ldots, X_n]$ be the ring of polynomials over k in n variables; to say that the x_i's are algebraically dependent means that the map

$$\pi \colon k[X_1, \ldots, X_n] \to B, \quad X_i \mapsto x_i$$

has non trivial kernel, call it I. Assume that f is a nonzero polynomial in I,

$$f(X_1, \ldots, X_n) = \sum_\alpha a_\alpha X_1^{\alpha_1} X_2^{\alpha_2} \cdots X_n^{\alpha_n},$$

where $0 \neq a_\alpha \in k$ and all the multi-indices $\alpha = (\alpha_1, \ldots, \alpha_n)$ are distinct.

Our goal will be fulfilled if we can change the X_i's into a new set of variables, Y_i's, such that f can be written as a monic (up to a scalar multiple) polynomial in Y_1 and with coefficients in the remaining variables, i.e.

$$f = aY_1^m + b_{m-1}Y_1^{m-1} + \cdots + b_1 Y_1 + b_0 \tag{A.16}$$

where $0 \neq a \in k$ and $b_i \in k[Y_2, \ldots, Y_n]$.

We are going to consider two changes of variables that work for our purposes: the first one, a clever idea of Nagata, does not assume anything about k; the second one assumes k to be infinite and has certain efficiencies attached to it.

The first change of variables replaces the X_i by Y_i given by

$$Y_1 = X_1, \ Y_i = X_i - X_1^{p^{i-1}} \text{ for } i \geq 2,$$

where p is some integer yet to be chosen. If we rewrite f using the Y_i's instead of the X_i's, it becomes

$$f = \sum_\alpha a_\alpha Y_1^{\alpha_1} (Y_2 + Y_1^p)^{\alpha_2} \cdots (Y_n + Y_1^{p^{n-1}})^{\alpha_n} \tag{A.17}$$

Expanding each term of this sum, there will be only one term pure in Y_1, namely

$$a_\alpha Y_1^{\alpha_1 + \alpha_2 p + \cdots + \alpha_n p^{n-1}}.$$

Furthermore, from each term in (A.17) we are going to get one and only one such a power of Y_1. Such monomials have higher degree in Y_1 than any other monomial in which Y_1 also occurs. If we choose $p > \sup\{\alpha_i \mid a_\alpha \neq 0\}$, the exponents $\alpha_1 + \alpha_2 p + \cdots + \alpha_n p^{n-1}$ are distinct since they have different p–adic expansions. This provides for the required equation.

If k is an infinite field, we consider another change of variables that preserves degrees. It will have the form

$$Y_1 = X_1, \; Y_i = X_i - c_i X_1 \text{ for } i \geq 2,$$

where the c_i are to be properly chosen. Using this change of variables in the polynomial f, we obtain

$$f = \sum_\alpha a_\alpha Y_1^{\alpha_1} (Y_2 + c_2 Y_1)^{\alpha_2} \cdots (Y_n + c_n Y_1)^{\alpha_n}. \tag{A.18}$$

We want to make choices of c_i in such a way that when we expand (A.18) we achieve the same goal as before, i.e. a form like in (A.16). For that, it is enough to work on the homogeneous component f_d of f of highest degree, in other words, we can deal with f_d alone. But

$$f_d(Y_1, \ldots, Y_n) = h_0(1, c_2, \ldots, c_n) Y_1^d + h_1 Y_1^{d-1} + \cdots + h_d,$$

where h_i are homogeneous polynomials of $k[Y_2, \ldots, Y_n]$, with $\deg h_i = i$, and we can view $h_0(1, c_2, \ldots, c_n)$ as a nontrivial polynomial function in the c_i. Since k is infinite, we can pick the c_i, so that $0 \neq h_0(1, c_2, \ldots, c_n) \in k$.

Theorem A.60 (Noether Normalization Theorem). *Let k be a field and let $B = k[x_1, \ldots, x_n]$ be a finitely generated k-algebra; then there exist $z_1, \ldots, z_d$ algebraically independent elements of B such that B is integral over the polynomial ring $A = k[z_1, \ldots, z_d]$.*

Proof. We may assume that the x_i's are algebraically dependent. From the preceding, we can find $y_1, \ldots, y_n$ in B such that

$$k[y_2, \ldots, y_n] \hookrightarrow k[y_1, \ldots, y_n] = B$$

is an integral extension, and if necessary we iterate.   □

Corollary A.61. *Let k be a field and $\psi : A \mapsto B$ a k–homomorphism of finitely generated k–algebras. If $\mathfrak{P}$ is a maximal ideal of B then $\mathfrak{p} = \psi^{-1}(\mathfrak{P})$ is a maximal ideal of A.*

Proof. Consider the embedding of k-algebras

$$A/\mathfrak{p} \hookrightarrow B/\mathfrak{P},$$

where by the preceding $B/\mathfrak{P}$ is a finite dimensional k–algebra. It follows that the integral domain $A/\mathfrak{p}$ must be a field.   □

Degree of a Prime Ideal

The following is the notion of degree for ideals which are not necessarily graded.

Remark A.62. Let $\mathfrak{p}$ be a prime ideal of the polynomial ring $A = k[x_1, \ldots, x_n]$. We can use Noether normalization to define the *degree* of $\mathfrak{p}$ or of $A/\mathfrak{p}$.

(a) If $\mathfrak{p}$ is a homogeneous ideal, its degree $\deg(\mathfrak{p})$ is the multiplicity of the graded algebra $A/\mathfrak{p}$ (see Appendix B). This means that if

$$C = k[z_1, \ldots, z_d] \hookrightarrow A/\mathfrak{p}$$

is a Noether normalization in which the z_i's are forms of degree 1, then $\deg(\mathfrak{p})$ is the torsionfree rank of $A/\mathfrak{p}$ as a C–module.

(b) If $\mathfrak{p}$ is not homogeneous, $\deg(\mathfrak{p})$ will be the minimum rank of $A/\mathfrak{p}$ relative to all possible Noether normalizations.

(c) The two definitions agree when $\mathfrak{p}$ is graded. There is also a notion of degree for prime ideals in local rings, which is discussed in Chapter 9.

Hilbert Nullstellensatz

One of the many uses of Noether normalization is to give a proof of the celebrated:

Theorem A.63 (Hilbert Nullstellensatz). *Let k be a field with algebraic closure $\bar{k}$ and let A be the polynomial ring $k[X_1, \ldots, X_n]$, and let I be an ideal of A. If the polynomial $f \in A$ vanishes at all the zeros of I in $\bar{k}^n$ then $f \in \sqrt{I}$:*

$$f \in \sqrt{I} \Leftrightarrow V(I) \subset V(f). \qquad (A.19)$$

Proof. It is enough to show that if an ideal $J \subset A$ is proper, then it must have a zero. First, it is clear that the ideal generated by J in $\bar{k}[x_1, \ldots, x_n]$ is also proper, so we may assume k is algebraically closed.

Without loss of generality, we may assume that J is a maximal ideal. By Theorem A.60, A/J is a finite algebraic extension of k and therefore $A/J \simeq k$ as k–algebras. This means that for each X_i there exist $c_i \in k$ such that $X_i - c_i \in J$, which shows that $J = (X_1 - c_1, \ldots, X_n - c_n)$.

To prove the theorem, we apply the preceding to the ideal

$$L = (I, 1 - ft) \subset k[X_1, \ldots, X_n, t].$$

From the assumption on f, it follows that L must be the unit ideal,

$$1 = g(\mathbf{X}, t)(1 - ft) + \sum_j h_j(\mathbf{X}, t) f_j(\mathbf{X}), \quad f_j \in I.$$

Evaluating at $t \to 1/f$, and clearing denominators, yields the assertion. $\qquad \square$

The Ideal of Leading Coefficients

The following simulacrum of an initial ideal is useful in theoretical arguments and in Noether normalization.

Theorem A.64 (Suslin). *Let S be a Noetherian ring and I be an ideal of the polynomial ring $S[x]$. Denote by $c(I)$ the ideal of S generated by the leading coefficients of all the elements of I. Then*

$$\text{height } c(I) \geq \text{height } I.$$

Proof. Let $P_1, \ldots, P_n$ be the minimal primes of I. From the equality $\cap P_i = \sqrt{I}$ we have that for some integer N,

$$\left(\prod_{i=1}^{n} P_i \right)^N \subset I.$$

Now for any two ideals, I, J, one has $c(I) \cdot c(J) \subset c(IJ)$ since if a is the leading coefficient of $f \in I$ and b is that of $g \in J$ then either $ab = 0$ or ab is the leading coefficient of fg. Therefore if $c(I) \subset \mathfrak{q}$ for a prime $\mathfrak{q} \subset S$ we have $c(P_i) \subset \mathfrak{q}$ for some i. This means that if the assertion of the theorem is true for prime ideals, then

$$\text{height } \mathfrak{q} \geq \text{height } c(P_i) \geq \text{height } P_i \geq \text{height } I$$

and we are done since height $c(I) = \inf\{\text{height } \mathfrak{q}\}$.

Assume then that $I = P$ is prime and let $\mathfrak{p} = P \cap S$. If $P = \mathfrak{p}[x]$ then $c(P) = \mathfrak{p}$ and the statement follows from Theorem A.20. If $P \neq \mathfrak{p}[x]$ then $c(P) \neq \mathfrak{p}$; indeed otherwise for

$$f = a_n x^n + \cdots + a_0 \in P,$$

$a_n \in \mathfrak{p} \subset P$. so $f - a_n x^n \in P$ and $a_{n-1} \in \mathfrak{p}$ by an easy induction. It follows that

$$\text{height } c(P) > \text{height } \mathfrak{p} = \text{height } P - 1,$$

to establish the claim.   □

Integral Closure of Affine Domains

Proposition A.65. *Let A be an integrally closed domain of field of fractions K, and let F be a finite separable field extension of K. The integral closure B of A in F is contained in a finitely generated A–module. In particular, if A is Noetherian then B is also Noetherian. Furthermore, if A is an affine algebra over a field k, then B is an affine algebra over k.*

Proof. Let $y_1, \ldots, y_m$ be a basis of F/K; by premultiplying the elements of a basis by elements of A we may assume $y_i \in B$. Denote by Tr the trace function of F/K; by assumption Tr is a non–degenerate quadratic form on the K–vector space F. Denote by $z_1, \ldots, z_m$ a basis dual to the y_i's,

$$\text{Tr}(y_i z_j) = \delta_{ij}.$$

Let $z = \sum_j a_j z_j$, $a_j \in K$. If z is integral over A then each $z y_i \in B$ and its trace,

$$\text{Tr}(z y_i) = a_i \in K,$$

is also integral over A. It follows that $B \subset \sum_j A z_j$. □

Remark A.66. In this situation, there exists a basis of F/K of the form $y_i = y^i$, $y \in B$, $i \leq m$. If $f(t)$ is the minimal polynomial of y and D is its discriminant, the argument above can be re-arranged to show that $D \cdot B \subset \sum_i A y^i$.

Theorem A.67. *The integral closure of an affine domain over a field k is an affine domain over k.*

Proof. We follow the proof in [Ser65]. Let A be an affine domain over k with field of fractions K, denote by F a finite field extension of K, and let B be the integral closure of A in F. By Noether normalization, A is integral over a subring $A_0 = k[x_1, \ldots, x_n]$ generated by algebraically independent elements. From the transitivity of integrality, B is the integral closure of A_0 in F, so we may take $A_0 = A$. We claim that B is a finitely generated A–module. Without loss of generality, we can assume that the extension F/K is a normal extension.

Let E be the subfield of elements of F that are purely inseparable over K. If char $K = 0$, we simply apply Proposition A.65; otherwise it will suffice to prove that the integral closure of A in E is as asserted. Since $E = k(z_1, \ldots, z_m)$ is a finite extension of $k(x_1, \ldots, x_n)$, there exists an exponent $q = p^s$ such that $z^q \in k(x_1, \ldots, x_n)$, $\forall z \in E$. Let k' be the finite extension of k obtained by adjoining to k the qth roots of all the coefficients that occur when z_i^q is written as a rational function of the x_j; note that $E \subset E' = k_0(x_1^{\frac{1}{q}}, \ldots, x_n^{\frac{1}{q}})$. But the polynomial ring $B' = k_0[x_1^{\frac{1}{q}}, \ldots, x_n^{\frac{1}{q}}]$ is clearly the integral closure of A in E', and B is contained in it. □

A.6 The Method of Syzygies

Syzygies are devices[2] to study modules over general rings by examining the linear relations amongst its generators. As such it is part of Homological Algebra, but its focused concern with the details of matrices with entries in commutative rings give the subject its own character. It has another character, alluded to in section 1.3, as a method to study polynomial relations. We shall take it here simply as the study of free resolutions.

- Hilbert syzygy theorem
- Koszul complex and depth
- Finite free resolutions

[2] The word is derived from the Greek *syzykos*, loosely meaning yoked together, related, and had been used in astronomy to denote bodies in conjunction, or opposition: in an eclipse, with Sun, Earth and Moon aligned, they are said to be in syzygy. The notion of syzygy was introduced in algebra by Sylvester (Cambridge & Dublin Math. J. **276** (1850); more explicitly, Phil. Trans. **143** (1853)) and finally clarified by Hilbert (Math. Annalem **36** (1890)).

Hilbert Syzygy Theorem

Throughout R will denote a commutative ring, and modules are defined over R, although the treatment extends to general rings; we will be influenced by the treatment in [Kap66].

Given a module M, a projective presentation of M is a surjective homomorphism

$$0 \to K \longrightarrow P_0 \overset{\varphi}{\longrightarrow} M \to 0,$$

where P_0 is a projective R–module. When P is a free module, on a basis e_α, the module K is generated by the relations among the elements $\varphi(e_\alpha)$. K is called the *module of syzygies* of M. Iterating the process gives rise to a complex,

$$\cdots \longrightarrow P_n \overset{\varphi_n}{\longrightarrow} P_{n-1} \longrightarrow \cdots \longrightarrow P_1 \overset{\varphi_1}{\longrightarrow} P_0 \longrightarrow M \to 0,$$

called a *projective resolution* of M. If the complex has finite length, M is said to have finite projective resolution and the infimum among such integers is the *projective dimension* of M, in notation proj $\dim_R M$.[3]

When the P_n are chosen to be free modules, the maps φ_n are represented by matrices. If R is a *PID*, a high point of this construction is that all modules have projective dimension at most one and if M is finitely generated, after basis changes in P_1 and P_0, the matrix φ_1 has zeros in the off–diagonal positions. (This being, of course, the basic theorem for modules over PID's.)

The following elementary result, essentially due to Fitting, shows that the homological character of the higher modules of syzygies of a given module is independent of the chosen resolution.

Proposition A.68 (Schanuel Lemma). *Let M be an R–module and*

$$0 \to 0 \longrightarrow K \longrightarrow P \overset{\varphi}{\longrightarrow} M \to 0,$$
$$0 \to 0 \longrightarrow K' \longrightarrow P' \overset{\varphi'}{\longrightarrow} M \to 0$$

be two projective presentations of M. Then $K \oplus P' \simeq K' \oplus P$.

Proof. Set

$$S = \{(x,x') \in P \oplus P' \mid \varphi(x) = \varphi'(x')\},$$

and note that projections into the first and second components are surjections onto P and P', with kernels respectively isomorphic to K' and K. Since P and P' are projective, the mappings split to give the desired assertion.   □

Corollary A.69. *Let M be an R–module and suppose*

$$0 \to K \longrightarrow P_n \longrightarrow P_{n-1} \longrightarrow \cdots \longrightarrow P_1 \longrightarrow P_0 \longrightarrow M \to 0$$
$$0 \to L \longrightarrow Q_n \longrightarrow Q_{n-1} \longrightarrow \cdots \longrightarrow Q_1 \longrightarrow Q_0 \longrightarrow M \to 0$$

---

[3] Syzygies, unlike diamonds, are better when they do not go on forever.

are two exact complexes where the P_i's and the Q_i's are projective, then

$$K \oplus Q_n \oplus P_{n-1} \oplus \cdots \simeq L \oplus P_n \oplus Q_{n-1} \oplus \cdots.$$

In particular, the projective dimension of a module can be defined by any of its projective resolutions.

Definition A.70. *The* global dimension *of the ring R is*

$$\text{global dim } R = d(R) = \max\{ \text{ proj dim}_R M, \text{ for all } R\text{–modules}\}.$$

If $d(R)$ is finite, we say that R is *regular*. As a measure of size, $d(R)$ is too strict. For most rings, $d(R) = \infty$ simply because some module has infinite projective dimension. For this reason, it is often necessary to consider in the definition above only those modules with finite projective resolutions.

Theorem A.71 (Hilbert Syzygy Theorem). *Let $R[x]$ denote the ring of polynomials in one indeterminate over R. Then*

$$\boxed{d(R[x]) = d(R) + 1.} \tag{A.20}$$

In particular, for a field k, the ring of polynomials $k[x_1, \ldots, x_n]$ has global dimension n, while the ring $\mathbb{Z}[x_1, \ldots, x_n]$ has global dimension $n + 1$.

Proof. We begin with a useful observation. For a given $R[x]$–module M consider the sequence

$$0 \to R[x] \otimes_R M \xrightarrow{\psi} R[x] \otimes_R M \xrightarrow{\varphi} M \to 0, \tag{A.21}$$

where

$$\psi(x^n \otimes e) = x^n \otimes xe - x^{n+1} \otimes e,$$
$$\varphi(x^n \otimes e) = x^n \cdot e.$$

It is a straightforward verification that this sequence of $R[x]$–modules and homomorphisms is exact.

Let M be an R–module and let

$$0 \to P_r \longrightarrow \cdots \longrightarrow P_1 \longrightarrow P_0 \longrightarrow M \to 0$$

be a projective resolution. Since $R[x]$ is R-free, tensoring the complex with $R[x]$ yields an $R[x]$–projective resolution of $R[x] \otimes_R M$, and proj $\dim_{R[x]} (R[x] \otimes_R M) \leq$ proj $\dim_R M$.

Suppose now that M is an $R[x]$–module, view it as an R–module and use it in the sequence (A.21): by elementary considerations we obtain,

$$\text{proj dim}_{R[x]} M \le 1 + \text{proj dim}_{R[x]} (R[x] \otimes_R M) \le 1 + \text{proj dim}_R M,$$

which shows that

$$d(R[x]) \le d(R) + 1.$$

For the reverse inequality, we argue as follows (see [Kap66] for additional details). Any R–module M can be made into an $R[x]$–module by defining $f(x)e = f(0)e$, for $e \in M$. With this structure, we claim that

$$\text{proj dim}_{R[x]} M = \text{proj dim}_R M + 1.$$

From (A.21) we already have that the left hand side cannot exceed the right hand side of the expression. To prove equality, we use induction on $n = \text{proj dim}_R M$.

If $n = 0$, that is, if M is R–projective, then M cannot be $R[x]$–projective, since it is annihilated by x, which is a regular element of $R[x]$.

If $n > 0$, map a free R–module F onto M,

$$0 \to K \longrightarrow F \longrightarrow M \to 0,$$

$\text{proj dim}_R K = n - 1$ and by induction $\text{proj dim}_{R[x]} K = n$. Since $\text{proj dim}_{R[x]} F = 1$, by the preceding case, $\text{proj dim}_{R[x]} M = n + 1$, unless, possibly, $n = 1$.

To deal with this last case, map a free $R[x]$–module G over M with kernel L. The assumption to be contradicted is that L is $R[x]$–projective. Since $xM = 0$, $xG \subset L$, and the exact sequence

$$0 \to L/xG \to G/xG \longrightarrow M \to 0$$

says that L/xG is R-projective. But we also have the exact sequence

$$0 \to xG/xL \longrightarrow L/xL \longrightarrow L/xG \to 0,$$

and therefore xG/xL is R–projective. Since $xG/xL \simeq G/L \simeq M$, we get the desired contradiction. $\square$

Regular Sequences

We recall the notion that embodies the most efficient manner to generate certain ideals.

Definition A.72. *Let R be a ring and let E be an R-module. A regular sequence on E is an ordered set $\{x_1, \dots, x_n\}$ of elements of R with the following properties:*

(i) $E \ne (x_1, \dots, x_n)E,$

(ii) $(x_1, \dots, x_{i-1})E : x_i = (x_1, \dots, x_{i-1})E,$ *for $i = 1, \dots, n$.*

When $E = R$ we say that $\mathbf{x}$ is a regular sequence. An ideal generated by a regular sequence is called a complete intersection.

If R is a Noetherian local ring and the x_i's are taken in its maximal ideal, then for any finitely generated R–module E (i) is automatically satisfied by Nakayama lemma, and if (ii) holds then any permutation of the regular sequence is also a regular sequence. For more detail, see [Kap74].

The following describes some of the basic properties of regular sequences.

Proposition A.73. *Let* $\mathbf{x} = x_1, \ldots, x_n$ *be a sequence of elements of* R *and let* E *be a finitely generated module for which* $E \neq (x_1, \ldots, x_n)E$.

(a) *If* $\mathbf{x}$ *is a regular sequence on* E *then* $\mathrm{height}(\mathbf{x}) = n$.
(b) *If* R *is a graded* k–*algebra for some field* k *and the* x_i *are homogeneous of positive degree, then* $\mathbf{x}$ *is a regular sequence if and only if the subring* $B = k[x_1, \ldots, x_n]$ *is a ring of polynomials in* n *variables and* R *is flat over* B.

Koszul Complex and Depth

We give here a discussion of what is likely the most useful complex in commutative algebra. It permits the introduction of various measures of size for ideals and modules.

Let E be an R–module and denote by $\bigwedge(E)$ the exterior algebra of E. Given an element $\varphi \in \mathrm{Hom}_R(E, R)$, one defines a mapping ∂ on $\bigwedge(E)$, given in degree r by

$$\partial(e_1 \wedge \cdots \wedge e_r) = \sum_{i=1}^{r} (-1)^{i-1} \varphi(e_i)(e_1 \wedge \cdots \wedge \widehat{e_i} \wedge \cdots \wedge e_r).$$

∂ sends $\wedge^r E$ to $\wedge^{r-1} E$, and it is easy to see that $\partial^2 = 0$.
We will refer to the complex

$$\mathbb{K} = \mathbb{K}(E, \varphi) = \{\textstyle\bigwedge(E), \partial\}$$

as the *Koszul complex* associated to E and φ. For an R–module M, we can attach coefficients to $\mathbb{K}(E, \varphi)$ by forming the chain complex $\mathbb{K}(E, \varphi; M) = \mathbb{K}(E, \varphi) \otimes_R M$.

A consequence of the definition of ∂ is that, if ω and ω' are homogeneous elements of $\bigwedge(E)$, of degrees p and q, respectively, then

$$\partial(\omega \wedge \omega') = \partial(\omega) \wedge \omega' + (-1)^p \omega \wedge \partial(\omega'). \tag{A.22}$$

This implies that the cycles $Z(\mathbb{K})$ form a subalgebra of $\mathbb{K}$, and that the boundaries $B(\mathbb{K})$ form a two–sided ideal of $Z(\mathbb{K})$. Thus the homology of the complex, $H(\mathbb{K})$, inherits a structure of R–algebra.

Proposition A.74. $H(\mathbb{K}(E, \varphi; M))$ *is annihilated by the ideal* $\varphi(E)$.

Proof. If $M \simeq R$, it suffices to note that if $e \in E$ and $\omega \in Z_r(\mathbb{K})$, then from (A.22) $\partial(e \wedge \omega) = \varphi(e)\omega$. The same argument holds when coefficients are attached. $\square$

Since $H_0(\mathbb{K}(E,\varphi;M)) = M/\varphi(E)M$, the main problem of the elementary theory of these complexes is to find criteria for the vanishing of the higher homology modules. The most satisfying setting is the case when E is a free R–module, $E \simeq R^n = Re_1 \oplus \cdots \oplus Re_n$. In view of Proposition A.74, such complexes are more interesting when $\varphi(R^n)$ is a proper ideal of R. It is convenient to consider the elements $x_i = \varphi(e_i)$ and view $\mathbb{K}(R^n,\varphi)$ as the (graded) tensor product of n Koszul complexes associated to maps of the kind $R \xrightarrow{x} R$. That is, if we denote such a complex by $\mathbb{K}(x)$, we have

$$\mathbb{K}(R^n,\varphi) = \mathbb{K}(x_1) \otimes \cdots \otimes \mathbb{K}(x_n).$$

We will denote such complex by $\mathbb{K}(x_1,\ldots,x_n)$, or $\mathbb{K}(\mathbf{x})$, with $\mathbf{x} = \{x_1,\ldots,x_n\}$.

The vanishing of the homology modules of $\mathbb{K}(\mathbf{x};M)$ has the following module theoretic explanation.

Proposition A.75 (Grade–Sensitivity of Koszul Complexes). *Let R be a Noetherian ring, and let $\mathbf{x} = \{x_1,\ldots,x_n\}$ be a sequence of elements generating the ideal I. Let M be a finitely generated R–module with $M \neq IM$, $\mathbb{K}(\mathbf{x};M)$ be the corresponding Koszul complex and let q be the largest integer for which $H_q(\mathbb{K}(\mathbf{x};M)) \neq 0$. Then all maximal M–regular sequences in I have length equal to $n-q$.*

Proof. Note that since $H_0(\mathbb{K}(\mathbf{x};M)) = M/IM$, and the complex $\mathbb{K}(\mathbf{x};M)$ has length n, $0 \leq q \leq n$.

We use descending induction on q. From the definition of $\mathbb{K}(\mathbf{x};M)$, $H_n(\mathbb{K}(\mathbf{x};M))$ consists of the elements of M which are annihilated by I. If this module is nonzero we are done. If not, that is $q < n$, the ideal I is not contained in any associated prime of M and therefore there is $a \in I$ which is a regular element on M. Consider the short exact sequence induced by multiplication by a,

$$0 \to M \xrightarrow{a} M \longrightarrow M/aM \to 0.$$

Tensoring it with the complex of free modules $\mathbb{K}(\mathbf{x})$, we get the exact sequence of Koszul complexes,

$$0 \to \mathbb{K}(\mathbf{x};M) \xrightarrow{a} \mathbb{K}(\mathbf{x};M) \longrightarrow \mathbb{K}(\mathbf{x};M/aM) \to 0.$$

In homology we get the long exact sequence,

$$H_{q+1}(\mathbb{K}(\mathbf{x};M)) \to H_{q+1}(\mathbb{K}(\mathbf{x};M/aM)) \to H_q(\mathbb{K}(\mathbf{x};M)) \xrightarrow{a} H_q(\mathbb{K}(\mathbf{x};M)).$$

From the definition of q, we obtain $H_i(\mathbb{K}(\mathbf{x};M/aM)) = 0$ for $i > q+1$. On the other hand, by Proposition A.74, $H_q(\mathbb{K}(\mathbf{x};M))$ is annihilated by I, and thus $aH_q(\mathbb{K}(\mathbf{x};M)) = 0$. Taken together we have

$$H_{q+1}(\mathbb{K}(\mathbf{x};M/aM)) \simeq H_q(\mathbb{K}(\mathbf{x};M)),$$

from which an easy induction suffices to complete the proof. □

The last equality in the proof shows also:

Corollary A.76. *If* $\mathbf{a} = a_1, \ldots, a_{n-q}$ *is a maximal regular sequence on M contained in I, then*

$$H_q(\mathbb{K}(\mathbf{x};M)) = (\mathbf{a}M:I)/\mathbf{a}M.$$

Definition A.77. *Let I be an ideal of a Noetherian ring R, and let M be a finitely generated R–module. The I–depth of M is the length of a maximal regular sequence on M contained in I. If R is a local ring and I is the maximal ideal (in which case the condition* $M/IM \neq 0$ *is automatically satisfied by Nakayama lemma), the I–depth of M is called the depth of M. If* $M = R$, *the I–depth of R is called the grade of I, and denoted* grade *I*.

Heuristically, grade *I* is a measure of the number of independent 'indeterminates' that may be found in *I*.

Corollary A.78. *Let I be an ideal contained in the Jacobson radical of the Noetherian ring R, and let*

$$0 \to E \longrightarrow F \longrightarrow G \to 0$$

be an exact sequence of finitely generated R–modules. Then

> *If I–depth* $F < I$*–depth* G*, then I–depth* $E = I$*–depth* F;
>
> *If I–depth* $F > I$*–depth* G*, then I–depth* $E = I$*–depth* $G + 1$;
>
> *If I–depth* $F = I$*–depth* G*, then I–depth* $E \geq I$*–depth* G.

Proof. Let $\mathbb{K}(\mathbf{x})$ be the Koszul complex on a set $\mathbf{x}$ of generators of I. Tensoring the exact sequence of modules with $\mathbb{K}(\mathbf{x})$ gives the exact sequence of chain complexes,

$$0 \to \mathbb{K}(\mathbf{x};E) \longrightarrow \mathbb{K}(\mathbf{x};F) \longrightarrow \mathbb{K}(\mathbf{x};G) \to 0.$$

The assertions will follow from a scan of the long homology exact sequence and the interpretation of depth given in the previous proposition. □

The next result is the basis for several inductive arguments with ordinary Koszul complexes.

Proposition A.79. *Let* $\mathbb{C}$ *be a chain complex and let* $\mathbb{F} = \{F_1, F_0\}$ *be a chain complex of free modules concentrated in degrees 1 and 0. Then for each integer* $q \geq 0$ *there is an exact sequence*

$$0 \to H_0(H_q(\mathbb{C}) \otimes \mathbb{F}) \longrightarrow H_q(\mathbb{C} \otimes \mathbb{F}) \longrightarrow H_1(H_{q-1}(\mathbb{C}) \otimes \mathbb{F}) \to 0.$$

Proof. Construct the exact sequence of chain complexes

$$0 \to \widehat{\mathbb{F}_0} \xrightarrow{f} \mathbb{F} \xrightarrow{g} \widehat{\mathbb{F}_1} \to 0,$$

$$(\widehat{\mathbb{F}_0})_0 = F_0$$
$$(\widehat{\mathbb{F}_0})_1 = 0$$
$$(\widehat{\mathbb{F}_1})_0 = 0$$
$$(\widehat{\mathbb{F}_1})_1 = F_1,$$

and f and g are the obvious injection and surjection mappings. Tensoring with $\mathbb{C}$ and writing the homology exact sequence, we get

$$H_{q+1}(\mathbb{C} \otimes \widehat{\mathbb{F}_1}) \xrightarrow{\partial} H_q(\mathbb{C} \otimes \widehat{\mathbb{F}_0}) \rightarrow H_q(\mathbb{C} \otimes \mathbb{F}) \rightarrow H_q(\mathbb{C} \otimes \widehat{\mathbb{F}_1}) \xrightarrow{\partial} H_{q-1}(\mathbb{C} \otimes \widehat{\mathbb{F}_0}),$$

where the connecting homomorphism ∂ is up to a sign the differentiation of $\mathbb{F}$ tensored with $H_q(\mathbb{C})$. Noting that $H_{q+1}(\mathbb{C} \otimes \widehat{\mathbb{F}_1}) = H_q(\mathbb{C}) \otimes F_1$, and $H_q(\mathbb{C} \otimes \widehat{\mathbb{F}_0}) = H_q(\mathbb{C}) \otimes F_0$, we obtain the desired exact sequence. $\square$

Corollary A.80 (Rigidity of the Koszul Complex). *Let* $\mathbf{x} = \{x_1, \ldots, x_n\}$ *be a sequence of elements contained in the Jacobson radical of* R, *and let* M *be a finitely generated module. If* $H_q(\mathbb{K}(\mathbf{x}; M)) = 0$, *then* $H_i(\mathbb{K}(\mathbf{x}; M)) = 0$ *for* $i \geq q$.

Proof. Denote $\mathbf{y} = \{x_1, \ldots, x_{n-1}\}$, and $a = x_n$. In Proposition A.79, set $\mathbb{C} = \mathbb{K}(\mathbf{y}; M)$, $\mathbb{F} = \mathbb{K}(a)$, so that $\mathbb{C} \otimes \mathbb{F} = \mathbb{K}(\mathbf{x}; M)$. For each $i \geq 0$, we have the exact sequence

$$0 \rightarrow H_0(H_i(\mathbb{K}(\mathbf{y}; M)) \otimes \mathbb{K}(a)) \rightarrow H_i(\mathbb{K}(\mathbf{x}; M)) \rightarrow H_1(H_{i-1}(\mathbb{K}(\mathbf{y}; M)) \otimes \mathbb{K}(a)) \rightarrow 0.$$

If $H_q(\mathbb{K}(\mathbf{x}; M)) = 0$, then

$$H_0(H_q(\mathbb{K}(\mathbf{y}; M)) \otimes \mathbb{K}(a)) = H_q(\mathbb{K}(\mathbf{y}; M))/aH_q(\mathbb{K}(\mathbf{y}; M)) = 0,$$

which by Nakayama lemma implies $H_q(\mathbb{K}(\mathbf{y}; M)) = 0$. Inducting on n, we get

$$H_i(\mathbb{K}(\mathbf{y}; M)) = 0 \quad \text{for } i \geq q.$$

Taking this into the exact sequence gives that $H_i(\mathbb{K}(\mathbf{x}; M)) = 0$ for $i \geq q$. $\square$

Taylor Resolutions

Let $I = \subset k[x_1, \ldots, x_n]$ be the ideal generated by the monomials $(\mathbf{f}) = \{f_1, \ldots, f_s\}$. We describe a canonical free resolution of I associated to $\mathbf{f}$ ([Tay66]), that resembles the Koszul complex (for more details, see [Ei95, p. 439]).

Let

$$\mathbb{T}(\mathbf{f}) = \{\wedge R^s, \partial\},$$

where the differential is defined as follows. A basis element e_M of $\wedge^k R^s$ is indexed by a subset M of $[1, \ldots, s]$, of cardinality k; let $f_M = \mathrm{lcm}(f_i \in M)$. For each subset N of cardinality $k-1$, set

$$c_{M,N} = \begin{cases} 0 & \text{if } N \not\subset M, \\ (-1)^\ell f_M/f_N & \text{if } M = N \cup \{\text{the } \ell^{\text{th}} \text{ element of } M\}. \end{cases}$$

Now define

$$\partial(e_M) = \sum_{|N|=k-1} c_{M,N} e_N.$$

Theorem A.81. *Let* $\mathbf{f}$ *be a set of monomials. Then* $\mathbb{T}(\mathbf{f})$ *is a free resolution of* $R/(\mathbf{f})$.

Finite Free Resolutions

Some rings, or modules, allow the construction of projective resolutions with natural choices. Suppose $(R, \mathfrak{m})$ is a Noetherian local ring and M is a finitely generated module. If the $R/\mathfrak{m}$–vector space $M/\mathfrak{m}M$ has dimension n, by Nakayama lemma there exists an exact sequence

$$0 \to K \longrightarrow F_0 = R^n \longrightarrow M \to 0,$$

where $K \subset \mathfrak{m}F_0$. Iterating, this means that there exists a projective resolution of M,

$$\mathbb{F}_\bullet: \qquad \cdots \longrightarrow F_n \xrightarrow{\varphi_n} F_{n-1} \longrightarrow \cdots \longrightarrow F_1 \xrightarrow{\varphi_1} F_0 \longrightarrow M \to 0,$$

in which the F_i are free modules, and the matrix representation of φ_i has all entries in $\mathfrak{m}$. For obvious reasons, these are called *minimal free resolutions*.

When $R = R_0 \oplus R_1 \oplus \cdots$ is a graded ring, with R_0 a local ring, and M is a graded R–module, a free graded resolution of M with the attributes above will also exist.

The following result describes how a finite free resolution is anchored on its left end.

Theorem A.82 (McCoy Theorem). *Let R be a commutative ring and $\varphi: R^m \to R^n$ be a homomorphism of free R–modules. Denote by I the ideal generated by the $m \times m$ minors of a matrix representation of φ. Then φ is injective if and only if $0:I = 0$. In particular, if $(R, \mathfrak{m})$ is a local ring, $0:\mathfrak{m} \neq 0$, and all entries of φ lie in $\mathfrak{m}$, then φ is not injective.*

Proof. If $v = (a_1, \ldots, a_m)$ is a nonzero vector in the kernel of φ, by Cramer rule it follows that I is annihilated by a_i for each i.

For the converse, denote by $I_t(\varphi)$ the ideal generated by the $t \times t$ minors of φ. We may assume that for some $t \leq m$, $0:I_{t-1}(\varphi) = 0$ and $0:I_t(\varphi) \neq 0$. If $t = 1$, for any annihilator r of $I_1(\varphi)$, we have $\varphi(rR^m) = 0$, so we may take $t \geq 2$. Consider the system of linear equations

$$a_{11}x_1 + a_{12}x_2 + \cdots + a_{1m}x_m = 0$$
$$\vdots$$
$$a_{n1}x_1 + a_{n2}x_2 + \cdots + a_{nm}x_m = 0.$$

Let $0 \neq r \in 0:I_t(\varphi)$; we may assume that r does not annihilate one minor of size $t-1$, say the upper-left minor of size $t-1$. A nonzero solution can be now obtained: set $x_{t+1} = \cdots = x_m = 0$, and let x_i, for $i \leq t$, be the minor defined by the ith column of the upper-left $(t-1) \times t$ submatrix. Then $r \cdot (x_1, \ldots, x_m)$ solves the first $t-1$ equations by Cramer rule, and the remaining equations because $r \cdot I_t(\varphi) = 0$. $\qquad\square$

The Rank of a Module

Given a module M over a commutative ring R there are several possible ways to define the *rank* of M. If R is an integral domain with field of fractions K, one usually sets

$$\text{rank}(M) = \dim_K(M \otimes_R K).$$

More generally, one case define the *rank function* of M as the numerical function

$$r_M : \text{Spec}(R) \mapsto \mathbb{Z}, \quad r_M(\mathfrak{p}) = \text{rank}(M/\mathfrak{p}M).$$

Another approach is influenced by the theorem above: Given a finitely generated module M one looks for the largest integer r for which there is an embedding

$$R^r \hookrightarrow M.$$

McCoy's theorem guarantees that a free module has the expected rank. We leave to the reader the proof that r is always bounded by the number of generators of M.

Auslander–Buchsbaum Equality

The following is an explanation of the difference between the (finitistic) global dimension of a ring and the projective dimension of one of its modules ([Mat80]).

Theorem A.83. *Let $(R, \mathfrak{m})$ be a Noetherian local ring and let M be a finitely generated R–module. If $\text{proj dim}_R M < \infty$ then*

$$\boxed{\text{proj dim}_R M + \text{depth } M = \text{depth } R.} \qquad (A.23)$$

Proof. We induct on $r = \text{codim } R$ and $p = \text{proj dim}_R M$. We may assume that $p > 0$. Let

$$0 \to F_p \xrightarrow{\psi} F_{p-1} \longrightarrow \cdots \longrightarrow F_1 \xrightarrow{\varphi} F_0 \longrightarrow M \to 0 \qquad (A.24)$$

be a minimal free resolution of M.

We compare the value of the formula (A.23) for M and for its first module of syzygies $K = \varphi(F_1)$. If $r = 0$, the maximal ideal $\mathfrak{m}$ has a non trivial annihilator and therefore ψ, with its entries all in $\mathfrak{m}$, cannot be injective.

Suppose first that $\mathfrak{m}$ is not an associated prime of M. Since $r > 0$, there exists $x \in \mathfrak{m}$ which is regular on M and R, and therefore on all the modules in (A.24). Tensoring the resolution with $R/(x)$ gives a minimal resolution of the same length for M/xM as an $R/(x)$–module. We now use the formula for $R/(x)$.

For the second case, suppose $\mathfrak{m}$ is an associated prime of M, and let $x \in \mathfrak{m}$ be a regular element of R. Tensoring the exact sequence

$$0 \to K \longrightarrow F_0 \longrightarrow M \to 0,$$

by $R/(x)$, gives the exact sequence (using, say, the snake lemma, or the reader's favorite tool from homological algebra)

$$0 \to {}_x M \longrightarrow K/xK \longrightarrow F_0/xF_0 \longrightarrow M/xM \to 0,$$

where ${}_x M$ is the set of elements of M annihilated by x. Note that m is an associated prime of this submodule. We thus have that m is an associated prime of K/xK. On the other hand, a minimal resolution of K/xK over $R/(x)$ is obtained by tensoring that part of (A.24) that resolves K. From the previous case, proj $\dim_{R/(x)} K/xK = $ depth $R/(x) = r - 1$, and therefore proj $\dim_R M = r$, as desired. $\square$

Depth and Ext

The depth of a module can also be read off the vanishing properties of derived functors, particularly Ext and cohomology with support. The former provides several useful formulas. We point out one of these whose proof is left to the reader.

Proposition A.84. *Let R be a Noetherian ring, I an ideal, M a finitely generated R–module such that $M \neq I \cdot M$. Suppose that $\mathbf{x} = x_1, \dots, x_q$ is a regular M–sequence contained in I. Then for any integer n*

$$\mathrm{Ext}_R^n(R/I, M) \simeq \mathrm{Ext}_R^{n-q}(R/I, M/\mathbf{x}M).$$

Moreover, if $\mathbf{x}$ is a maximal regular M–sequence, then

$$\mathrm{Ext}_R^n(R/I, M) = 0 \text{ for } n < q,$$
$$\mathrm{Ext}_R^q(R/I, M) = (\mathbf{x}M : I)/\mathbf{x}M \neq 0.$$

Acyclicity of Free Complexes

We next give the most widely used tool to check the acyclicity of free complexes. A fuller discussion may be found in [BH93] and we content ourselves with the skeleton of a proof.

Let $\varphi : R^m \to R^n$ be a homomorphism between free modules. Picking bases it can be represented by an $n \times m$ matrix $[a_{ij}]$, and rank φ is the largest integer r such that $[a_{ij}]$ has a nonzero minor of order r. It is easily seen to be the least integer r such that r^{th} exterior power $\wedge^r \varphi : \wedge^r(R^m) \to \wedge^r(R^n)$ is nonzero. This identification ensures that r is well defined. We denote by $I_r(\varphi)$ the ideal generated by all the minors of order r.

Theorem A.85 (Buchsbaum–Eisenbud). *Let R be a Noetherian ring and let*

$$\mathbb{F}_\bullet : \qquad 0 \to F_n \xrightarrow{\varphi_n} F_{n-1} \longrightarrow \cdots \longrightarrow F_1 \xrightarrow{\varphi_1} F_0 \to 0$$

be a complex of finitely generated free R–modules. For each $i = 1, \dots, n$, denote by $r_i = r_i(\mathbb{F}) = \sum_{j=i}^n (-1)^{j-i} \mathrm{rank}\, F_j$. The following conditions are equivalent:

(a) $\mathbb{F}$ *is acyclic.*
(b) grade $I_{r_i}(\varphi_i) \geq i$, *for* $i = 1, \ldots, n$.

Proof. (a) $\Rightarrow$ (b): Let $\mathfrak{p}$ be any associated prime ideal of (0) in the ring R. Localizing $\mathbb{F}$ at $\mathfrak{p}$ gives a complex $\mathbb{F} \otimes R_\mathfrak{p}$ of free modules which is a free resolution of $L = \operatorname{coker} \varphi_1$. By Theorem A.83, L must be a free $R_\mathfrak{p}$–module. Thus the complex $\mathbb{F}_\mathfrak{p}$ splits completely which means that $r_i = \operatorname{rank} \varphi_i$ and all ideals $I_{r_i}(\varphi_i)$ localize to $R_\mathfrak{p}$ and consequently must contain regular elements. If x is a regular element of R in $\bigcap_i I_{r_i}(\varphi_i)$, tensoring $\mathbb{F}$ by $R/(x)$ gives the exact sequence

$$0 \to F_n \otimes R/(x) \longrightarrow F_{n-1} \otimes R/(x) \longrightarrow \cdots \longrightarrow F_1 \otimes R/(x),$$

on which we use the induction hypothesis since $r_i(\mathbb{F}) = r_i(\mathbb{F} \otimes R/(x))$.

(b) $\Rightarrow$ (a): By induction on all such shorter complexes, we assume that the sub-complex

$$0 \to F_n \xrightarrow{\varphi_n} F_{n-1} \longrightarrow \cdots \longrightarrow F_r \xrightarrow{\varphi_r} F_{r-1},$$

is acyclic except possibly at F_r. By Theorem A.82 and the hypothesis on $I_{r_n}(\varphi_n)$, φ_n is an injective homomorphism, so that $1 \leq r < n$. Set $B = \operatorname{image} \varphi_{r+1}$, $Z = \ker \varphi_r$ and consider the natural exact sequence

$$0 \to B \longrightarrow Z \longrightarrow H = H_r(\mathbb{F}) \to 0.$$

To show that $H = 0$, suppose otherwise and let $\mathfrak{p}$ be a minimal associated prime ideal of H. Localizing at $\mathfrak{p}$ we may assume that $\mathfrak{p}$ is the unique maximal ideal of R, and that depth $R \geq n - r$. Since $n \geq 2$, and using the depth lemma repeatedly we obtain that depth $B \geq 2$, depth $Z \geq 2$ and therefore depth $H > 0$, which contradicts the choice of $\mathfrak{p}$. $\square$

Regular Local Rings

Two early triumphs of the application of homological algebra to commutative algebra were:

Theorem A.86 (Serre). *Let* $(R, \mathfrak{m})$ *be a Noetherian local ring. The following conditions are equivalent:*

(a) R *is a regular ring.*
(b) $v(\mathfrak{m}) = \dim R$.
(c) $\mathfrak{m}$ *is generated by a regular sequence.*

Corollary A.87. *Let* R *be a regular local ring and let* $\mathfrak{p}$ *be a prime ideal. Then* $R_\mathfrak{p}$ *is a regular local ring.*

Proof. Given an $R_\mathfrak{p}$–module E, it has a natural R–module structure such that $E \otimes_R R_\mathfrak{p} = E$. It follows that tensoring a finite R–projective resolution of E with $R_\mathfrak{p}$ yields a finite $R_\mathfrak{p}$–projective resolution of E. $\square$

Theorem A.88 (Auslander-Buchsbaum). *Any regular local ring R is a factorial domain.*

Exercise A.89. Let I be an ideal of the Noetherian ring R and let M be a finitely generated module with $M/IM \neq 0$. If x is an element in the Jacobson radical of R, prove that (I,x)–depth $M \leq 1 + I$–depth M.

Exercise A.90. Let R be a commutative, Noetherian ring and E a finitely generated R–module. Suppose that the symmetric algebra $A = S_R(E)$ is regular (i.e. every localization of A is a regular local ring). Prove that R is regular and E is a projective R–module.

Exercise A.91. Let $(R, \mathfrak{m})$ be a Noetherian local ring and let E be a finitely generated R–module with the condition (S_r) of Serre. If $\mathfrak{m} = (a_1, \ldots, a_n)$ and $t_1, \ldots, t_n$ is a set of indeterminates over R, set $A = R[t_1, \ldots, t_n]$, $S = A_{\mathfrak{m}A}$ and $E' = S \otimes_R E$. If $x = \sum a_i t_i$, show that E'/xE' has the condition (S_r) on the punctured spectrum of S.

Exercise A.92. Let R be a Noetherian integral domain and let $I = (a_0, \ldots, a_n)$ be an ideal of grade at least 2. Prove that the polynomial $a_n x^n + \cdots + a_0$ is a prime element of $R[x]$ if it is irreducible over the field of fractions of R.

A.7 Cohen–Macaulay Rings and Modules

Cohen–Macaulay rings[4] are arguably the most important class of Noetherian rings. It includes the class of all regular rings, such as rings of polynomials over a field or the integers, rings of formal powers series over fields and convergent power series. They are a meeting ground for algebraic, analytic and geometric techniques. Most Cohen–Macaulay rings however are "singular", which rules out many geometric-analytic approaches but not wild enough to forbid them all. It may be said that their singularities are regular. Their significance arose also from the fact that they turn out in the solution of many important problems, such as the classical rings of invariants of reductive groups ([HR74]).

But what is a Cohen–Macaulay ring? The spirit of the answer to this question, paraphrasing Mel Hochster, is that they provide a setting for proving interesting and difficult results. Before we give the technical definition, let us give an indication, in the setting of affine domains, of what those rings are like. Let A be an affine domain over a field k, and let $R = k[z_1, \ldots, z_d]$ be one of its Noether normalizations. Then A is Cohen–Macaulay if and only if A is a free module over the polynomial ring R. Note how this permits, in analogy with the study of rings of integers of number fields, the introduction of many constructions–ramification locus, differents—that reflect how (or how many) of the primes of A (in other words, the points of the associated variety) lie over the primes of R (that is, of the points of affine space).

- Cohen–Macaulay rings and modules
- Gorenstein rings
- Canonical modules

---

[4] The terminology honors I. S. Cohen (1917-1955) and F. S. Macaulay (1862-1937).

Cohen–Macaulay Rings

Definition A.93. *A Noetherian ring R is* Cohen–Macaulay *if* height $I = $ grade I *for each ideal.* [5]

Proposition A.94. *Let R be a Noetherian ring. If* height $\mathfrak{m} = $ grade $\mathfrak{m}$ *for each maximal ideal, then R is Cohen–Macaulay. In particular, if R is a local ring, it suffices to test this equation for the maximal ideal.*

Proof. If $\mathfrak{p}$ is maximal among the prime ideals with height $\mathfrak{p} > $ grade $\mathfrak{p}$, let $\mathfrak{p} \subset \mathfrak{m} = $ maximal ideal; we may assume that R is local (why?). Let $x \in \mathfrak{m} \setminus \mathfrak{p}$; then grade$(\mathfrak{p}, x) \leq 1 + $ grade $\mathfrak{p}$ by Exercise A.89, while height$(\mathfrak{p}, x) \geq 1 + $ height $\mathfrak{p}$, thus contradicting height$(\mathfrak{p}, x) = $ grade$(\mathfrak{p}, x)$.   □

Corollary A.95. *Let $\mathfrak{p} \subset \mathfrak{q}$ be immediate primes (i.e. no other prime in–between) in a Cohen–Macaulay ring R. Then* height $\mathfrak{q} = 1 + $ height $\mathfrak{p}$. *In particular, all saturated chains of prime ideals between two fixed primes have the same length.*

Remark A.96. Let R be a Cohen–Macaulay ring. If S is a multiplicative set (resp. x is a regular element) of R, then $S^{-1}R$ (resp. $R/(x)$) is also Cohen–Macaulay. The power series ring $R[[t]]$ is Cohen–Macaulay as t lies in the Jacobson radical.

As for the ring of polynomials $R[t]$, the situation is more interesting. Let $\mathfrak{m}$ be a maximal ideal of $R[t]$ and localize at $\mathfrak{p} = \mathfrak{m} \cap R$, in other words, we may assume that R is local and $\mathfrak{m}$ contracts to the maximal ideal of R. Then $\mathfrak{m} = (\mathfrak{p}, f)$, where f may be taken to be a nonzero monic polynomial and height $\mathfrak{m} = 1 + $ height $\mathfrak{p}$. Finally, note that $\mathfrak{p}$–depth $R = \mathfrak{p}$–depth $R[t]/(f)$, since $R[t]/(f)$ is a free R–module.

Corollary A.97 (Macaulay Theorem). *Let k be a field and let $I = (f_1, \ldots, f_m) \subset k[x_1, \ldots, x_n]$ be an ideal of codimension m. Then every primary component of I has codimension m.*

It follows from the properties of the Koszul complex that the grade of a maximal ideal $\mathfrak{m}$ (but not for an arbitrary prime ideal) is the depth of the local ring $R_\mathfrak{m}$. Thus the notion of a Cohen–Macaulay ring is a local property. We use this fact to justify the definition of a Cohen–Macaulay module M: For each maximal ideal (resp. for each prime ideal) $\mathfrak{m}$, $\mathfrak{m}$–depth $M_\mathfrak{m} = \dim M_\mathfrak{m}$.

This permits giving a description of an affine Cohen–Macaulay ring A in module-theoretic terms.

Theorem A.98. *Let A be an affine algebra and let $k[\mathbf{z}] = k[z_1, \ldots, z_d] \hookrightarrow A$ be a Noether normalization. If A is equidimensional then A is a Cohen–Macaulay ring if and only if A is a free $k[\mathbf{z}]$–module.*

---

[5] Perhaps the fastest definition of Cohen–Macaulay local ring is through a theorem of Paul Roberts [PRo87]: a Noetherian local ring R is Cohen–Macaulay if and only if it admits a nonzero finitely generated module E of finite injective dimension.

Proof. For any maximal ideal $\mathfrak{p}$ of $k[\mathbf{z}]$, $\dim A_\mathfrak{p} = d$. For A to be Cohen–Macaulay as a ring amounts to say that it is Cohen–Macaulay as a $k[\mathbf{z}]$–module. But $k[\mathbf{z}]$ is a ring of global dimension d, and we can apply Theorem A.83 to have that $A_\mathfrak{p}$ is a free $k[\mathbf{z}]_\mathfrak{p}$–module and thus A is a projective $k[\mathbf{z}]$–module (necessarily free by the theorem of Quillen–Suslin [Kun85]).   □

In general, a Cohen–Macaulay affine ring A will break up into a direct product of affine rings, $A = A_1 \times \cdots \times A_r$, each of which is equi–dimensional.

Example A.99. The finiteness of the projective dimension of a ring over another may come in very restrictive form. Here is one instance (whose converse will not always hold). Let A be an affine algebra over a field k and consider the sequence

$$0 \to \mathbb{D} \longrightarrow A^e = A \otimes_k A \longrightarrow A \to 0, \ x \otimes y \mapsto xy.$$

Then if the projective dimension of A over A^e is finite then A is Cohen–Macaulay. We can work with a localization of A; assume $\dim A = d$ and depth $A = c$. Then depth $A^e = 2c$. If proj $\dim_{A^e}(A)$ is finite it is at least d since $\mathbb{D}$ has height d. From the Auslander–Buchsbaum equality we have $d + c \leq 2c$, and therefore $d = c$.

Hilbert–Burch Theorem

Let M be a finitely generated module over a Noetherian ring R. Suppose M has a projective resolution

$$0 \to R^m \xrightarrow{\varphi} R^n \longrightarrow M \to 0.$$

If $m = n$ or M is a torsionfree R–module, the maximal minors of φ play decisive roles in the structure of M. We consider the elementary but important general description of ideals of projective dimension one.

Theorem A.100 (Hilbert–Burch). *Let $I = (a_1, \ldots, a_n)$ be an ideal of a Noetherian ring R with a free resolution of length one,*

$$0 \to R^{n-1} \xrightarrow{\varphi} R^n \longrightarrow I \to 0.$$

There exists a regular element d of R such that $I = d \cdot \Delta$, where Δ is the ideal generated by the minors of order $n - 1$ of any matrix representation of φ.

Proof. A sketch of the proof goes as follows. First, that I contains regular elements and Δ has grade two follow easily from formula (A.23). Assume the ith basis element of R^n maps to a_i. Let

$$\varphi = \begin{bmatrix} a_{1,1} & \cdots & a_{1,n-1} \\ \vdots & \ddots & \vdots \\ a_{n,1} & \cdots & a_{n,n-1} \end{bmatrix}$$

be a matrix representation of φ, and denote by $\Delta_i = (-1)^i \det \varphi_i$ the signed determinant of the submatrix φ_i obtained by deleting the ith row of φ. Since

$$(a_1, \cdots, a_n) \cdot \varphi = 0, \quad \text{for } 1 \leq i < j \leq n,$$

it follows that

$$a_i \cdot \Delta_j = a_j \cdot \Delta_i.$$

These equations give an isomorphism between $(a_1, \ldots, a_n)$ and $(\Delta_1, \ldots, \Delta_n)$. □

Cohen–Macaulay Modules

There is also a notion of Cohen–Macaulay module which in fairness must be independent of the base ring.

Definition A.101. *A finitely generated R–module M is* Cohen–Macaulay *if*

$$I\text{--depth } M = \text{height } (I/J)$$

for every ideal $I \supset J = $ *annihilator* M.

For the purpose of this definition, we may as well assume that M is a faithful R–module. This definition is easier to manage when R is a local ring, or R is a graded ring and M is a graded module. (We only state the local version.)

Definition A.102. *Let M be a finitely generated module over a local ring* $(R, \mathfrak{m})$. *A system of parameters is a set* $\mathbf{x} = \{x_1, \ldots, x_d\}$ *of elements in* $\mathfrak{m}$, $d = \dim M$, *such that* $\ell(M/(\mathbf{x})M) < \infty$.

These sets are often obtained by taking general linear combinations of the generators of $\mathfrak{m}$. It is useful in the following well-used test of the Cohen–Macaulay property (see [BH93, Corollary 4.6.11]).

Theorem A.103. *Let* $(R, \mathfrak{m})$ *be a Noetherian local ring, M a finitely generated R–module of positive rank, and* $(\mathbf{x})$ *an ideal generated by a system of parameters of R. Then M is Cohen–Macaulay if and only if* $\ell(M/(\mathbf{x})M) = e(\mathbf{x}, R) \cdot \text{rank } M$.

Here $e(\mathbf{x}, R)$ is the multiplicity of the ideal $(\mathbf{x})$. If R is a Cohen–Macaulay ring this is simply $\ell(R/(\mathbf{x}))$. The other notion, of the rank of a module, is defined in terms of the module $M \otimes_R K$, where K is the total ring of fractions of R. Thus, M has rank n if $M \otimes_R K \simeq K^n$; in particular, if R is a domain then every module has a rank.

Depth and Hyperplane Sections

We describe a useful aspect of the relation between the depth of a module E and the depth of the modules E/xE for families of elements x.

Proposition A.104. *Let* $(R, \mathfrak{m})$ *be a Noetherian local ring with infinite residue field, of Krull dimension d, let E be a finitely generated R–module and let s be an integer* $s < d$. *Suppose that for each subset* $x_1, \ldots, x_s$ *of a system of parameters of R contained in* $\mathfrak{m} \setminus \mathfrak{m}^2$,

$$\text{depth } E/(x_1, \ldots, x_s)E \geq 1.$$

Then depth $E \geq s + 1$. *In particular if* $s = d - 1$ *then E is Cohen–Macaulay.*

Proof. We look at the case $s = 1$. If depth $E > 0$, we choose $x_1 \in \mathfrak{m} \setminus \mathfrak{m}^2$ which is regular on E. The assertion is then clear. (The assumption that $R/\mathfrak{m}$ is infinite allows for the choice of x_1.)

Suppose then that $\mathfrak{m}$ is an associated prime of E. Denote by E_0 the submodule of E with support in $\mathfrak{m}$, that is, E_0 is the subset of elements of E annihilated by some power of $\mathfrak{m}$. This implies that in the exact sequence

$$0 \to E_0 \longrightarrow E \longrightarrow F \to 0,$$

the module F has positive depth. Let x_1 be a minimal parameter that is regular on F. Tensoring the exact sequence by $R/(x_1)$ (or using the kernel-cokernel sequence induced by multiplication by x_1) we obtain the exact sequence

$$0 \to E_0/x_1 E_0 \longrightarrow E/x_1 E \longrightarrow F/x_1 F \to 0.$$

However, by assumption the module $E/x_1 E$ has positive depth so cannot contain a nonzero module, to wit $E_0/x_1 E_0$, supported on $\mathfrak{m}$, so that by Nakayama lemma $E_0 = 0$. The general case follows by using descending induction on the module $E/(x_1, \ldots, x_{s-1})E$. $\qquad\square$

Perfect Ideals

The following isolates an important class of Cohen–Macaulay rings which are quotients of a given Cohen–Macaulay ring.

Definition A.105. *Let R be a Cohen–Macaulay local ring and let I be an ideal with finite projective dimension. By (A.23),*

$$\dim R/I \geq \operatorname{depth} R/I = \dim R - \operatorname{proj} \dim_R R/I.$$

In the case of equality, R/I is a Cohen–Macaulay ring. The ideal I is then said to be perfect.

This means that there is a free resolution

$$0 \to F_r \longrightarrow F_{r-1} \longrightarrow \cdots \longrightarrow F_1 \longrightarrow F_0 = R \longrightarrow R/I \to 0, \qquad (A.25)$$

where $r = \operatorname{codim} I$. An example ($r = 2$) is given by Theorem A.100, in the case of an $n \times (n-1)$ matrix whose maximal minors generate an ideal of height two.

It is a property of a perfect ideal I that if the resolution A.25 is dualized with respect to $\operatorname{Hom}_R(\cdot, R)$, the new complex is a resolution of the r-dimensional cohomology

$$0 \to F_0^* \longrightarrow F_1^* \longrightarrow \cdots \longrightarrow F_{r-1}^* \longrightarrow F_r^* \longrightarrow \operatorname{Ext}_R^r(R/I, R) \to 0. \qquad (A.26)$$

Conversely, the ring R/I can be recovered from $\operatorname{Ext}_R^r(R/I, R)$:

$$R/I = \operatorname{Ext}_R^r(\operatorname{Ext}_R^r(R/I, R), R).$$

Gorenstein Rings

If $(A, \mathfrak{m})$ is an Artinian local ring which is a finite algebra over a field k, the dual vector space $A^* = \mathrm{Hom}_k(A, k)$ has a natural structure of an A–module: If $f \in A^*, r \in A$, then $(r \cdot f)(x) = f(rx)$. A^* is an injective A–module since

$$\mathrm{Hom}_A(E, A^*) = \mathrm{Hom}_k(E, k).$$

It is easy to see that A^* is the injective envelope of $A/\mathfrak{m}$.

The condition that A be Gorenstein is then: $A^* \simeq A$, that is, A^* is generated by a single element. It is easy to describe the generator of A^*, being enough to pick $f \in A^*$ which does not vanish on the one-dimensional subspace $0 :_A \mathfrak{m}$ (the socle of A).

Definition A.106. *Let R be a regular local ring or a polynomial ring over a field k. A perfect ideal I of codimension r is called a* Gorenstein ideal *if*

$$R/I \simeq \mathrm{Ext}^r_R(R/I, R).$$

The significance of this definition in terms of the resolution (A.25) is the following. If R is a local ring or R is a graded ring and I is a homogeneous ideal, the modules F_i in a minimal resolution of R/I are essentially unique. Thus for a Gorenstein ideal I the modules F_i must satisfy

$$F_i \simeq F_{r-i}, \; \forall i.$$

(In the graded case this is only true after a uniform shift in the grading of the F_i's.)

Example A.107. The premier example of a Gorenstein ring is a complete intersection: If R is a regular local ring and $\mathbf{f} = f_1, \ldots, f_m$ is a regular sequence, then $A = R/(\mathbf{f})$ is a Gorenstein ring.

On the other hand, $I = (x^3, y^3, z^4, xy^2 - xz^2, x^2z^2 - y^2z^2)$ is a Gorenstein ideal but not a complete intersection.

There are many other ways in which Gorenstein rings arise without being complete intersections. For example, let A be a finite dimensional k–algebra and denote $A^* = \mathrm{Hom}_k(A, k)$. Then the idealization of A^*,

$$B = A \oplus A^*,$$

is a Gorenstein ring. We leave to the reader to prove this assertion and to establish when these algebras are complete intersections.

There is a general and intrinsic definition of Gorenstein ring or ideal.

Definition A.108. *A Noetherian local ring R is a* Gorenstein ring *if it has finite injective dimension as a module over itself.*

The following lists some fundamental properties of these rings (see [BH93]):

Theorem A.109. *Let* $(R, \mathfrak{m})$ *be a Noetherian local ring of dimension d. Then*

(a) *R is a Gorenstein ring if and only if R is Cohen–Macaulay and for one (any) system of parameters* $\mathbf{x} = x_1, \ldots, x_d$,

$$((\mathbf{x}) : \mathfrak{m}) / (\mathbf{x}) \simeq R/\mathfrak{m}.$$

(b) *If R is a Gorenstein local ring and I is a Cohen-Macaulay ideal of codimension r, not necessarily perfect, then* R/I *is a Gorenstein ring if and only if*

$$\mathrm{Ext}_R^r(R/I, R) \simeq R/I.$$

(c) *If R is a Gorenstein local ring the mapping*

$$M \mapsto \mathrm{Ext}_R^r(M, R)$$

is a self-dual functor on the category of finitely generated Cohen-Macaulay R-modules of codimension r. In particular, if $(R, \mathfrak{m})$ *is a Cohen–Macaulay local ring of dimension 1, then it is Gorenstein if and only if the functor* $\mathrm{Hom}_R(\cdot, R)$ *is self–dual on the category of torsionfree R-modules.*

The condition (c) is interesting because it tells how the ring R interacts with the category of modules, in words, it permits us to construct interesting functors. These properties are characteristic of Gorenstein rings. It was Grothendieck who realized that suitably modified, (c) will still hold for many Cohen-Macaulay local rings.

Gorenstein Ideals of Codimension 3

In addition to perfect, Cohen–Macaulay ideals of codimension two, which are completely described by Theorem A.100, another family of perfect ideals has a beautiful determinantal description.

Let R be a Gorenstein local ring and let I be a perfect, Gorenstein ideal of codimension 3. This implies that the ideal $I = (a_1, \ldots, a_n)$ has a minimal free resolution,

$$0 \to R \xrightarrow{\psi} R^n \xrightarrow{\varphi} R^n \longrightarrow I \to 0, \tag{A.27}$$

in which the matrix representation of ψ is $[a_1, \ldots, a_n]$. In fact, the following holds (see [BH93]):

Theorem A.110 (Buchsbaum–Eisenbud). *If I is as above, there exists a minimal resolution such that:*

(a) *The mapping* φ *is skew–symmetric (i.e. has a matrix representation with this property), and*

(b) *I is the ideal* $P_{n-1}(\varphi)$ *generated by the Pfaffians of the submatrices obtained by deleting the ith row and column of* φ, *for* $i = 1 \ldots n$.

Conversely, if φ *is skew–symmetric and* $P_{n-1}(\varphi)$ *has codimension 3 (its maximum value), then* $P_{n-1}(\varphi)$ *is a Gorenstein ideal. (In particular, n must be odd.)*

The Canonical Module of a Local Ring or Graded Ring

Definition A.111. *Let R be a Gorenstein local ring and let I be an ideal of codimension r, defining the ring $A = R/I$. The module*

$$\omega_A = \mathrm{Ext}_R^r(A, R).$$
(A.28)

is the canonical module *of A.*

In the case of a finite k–algebra A the canonical module is $\omega_A = A^* = \mathrm{Hom}_k(A, k)$. As the notation indicates, ω_A depends only on A, not especially on R and I. Its basic property is the following extension of Theorem A.109(c).

Theorem A.112. *If A is a Cohen-Macaulay local ring with a canonical module ω_A, then the mapping*
$$M \mapsto \mathrm{Ext}_A^r(M, \omega_A)$$
is a self-dual functor on the category of finitely generated Cohen-Macaulay A-modules of codimension r.

Remark A.113. If $r = \mathrm{codim}\, I$, and $\mathbf{x} = x_1, \dots, x_r$ is a regular sequence contained in I, then
$$\omega_A \simeq ((\mathbf{x}):I)/(\mathbf{x}),$$
from the way these Ext's are calculated.

If A is Cohen-Macaulay, ω_A is a Cohen–Macaulay module of the same dimension as A. In general the canonical module of a ring A retains many of the most interesting properties of A, and quite often improves on them.

The case of graded rings is rich in numerical information. If $R = k[z_1, \dots, z_d]$ is a ring of polynomials over a field k, graded in the usual manner, the canonical module is $R[-d]$, not R itself, to ensure naturality in the category of graded modules. (Sometimes the shift is ignored harmlessly.)

If A is a finitely generated algebra over a field k and $R = k[z_1, \dots, z_d]$ is a Noether normalization, then

$$\omega_A = \mathrm{Hom}_R(A, R[-d])$$
(A.29)

which extends the formula in the case of fields (see [HK71] for details).

Suppose further that A is a graded ring, $A = A_0 + A_1 + \cdots$, and R is such that $z_i \in A_1$. If A is Cohen–Macaulay, A is a free R–module,

$$A \simeq \bigoplus_i R[-d_i],$$

so that

$$\omega_A = \mathrm{Hom}_R(A, R[-d]) \simeq \bigoplus_i R[d_i - d].$$

If

$$H_A(\mathbf{t}) = \frac{h_0 + h_1 \mathbf{t} + \cdots + h_r \mathbf{t}^r}{(1-\mathbf{t})^d}, \ h_r \neq 0,$$

is the Hilbert–Poincaré function of A, this representation of ω_A gives also

$$H_{\omega_A}(\mathbf{t}) = \frac{\sum_i \mathbf{t}^{d-d_i}}{(1-\mathbf{t})^d} = (1/\mathbf{t})^s \frac{h_r + h_{r-1}\mathbf{t} + \cdots + h_0 \mathbf{t}^r}{(1-\mathbf{t})^d}, \ s = \sup\{d_i\} - d.$$

Reading Depth

If A is a Cohen–Macaulay local ring with a canonical module ω_A, depths of modules can be expressed as follows.

Proposition A.114. *For any finitely generated A–module M,*

$$\mathrm{depth}\, M = \dim A - \sup\{ \, r \mid \mathrm{Ext}_A^r(M, \omega_A) \neq 0 \, \}.$$

Proof. The proof is immediate (but a pleasant calculation). A fuller explanation is given in Theorem A.132.   □

Exercise A.115 (Peskine). Let R be a local Cohen–Macaulay ring of Krull dimension $d > 0$, with a canonical module ω_R. Suppose that ω_R is isomorphic to an ideal of R (equivalently, suppose the total ring of fractions of R is a Gorenstein ring). Prove that $S = R/\omega_R$ is a Gorenstein ring of Krull dimension $d - 1$.

Exercise A.116. Let R be a local Cohen–Macaulay ring and let $x_1, \ldots, x_n$ be a regular sequence. Prove that the ideal I generated by all products $x_{i_1} \cdots x_{i_k}$ of k distinct x_i is perfect.

Exercise A.117. Let R be a ring of polynomials and let I be a monomial ideal. Prove that if I is a Cohen–Macaulay ideal then its radical $\sqrt{I}$ is also Cohen–Macaulay (*Hint*: use polarization).

A.8 Local Cohomology

Let R be a commutative ring and $I \subset R$ an ideal. For any R–module M, the assignment

$$\Gamma_I : M \rightsquigarrow \{ \, x \in M \mid I^{n(x)} x = 0 \, \}$$

is one of the most multifaceted functors of homological algebra and is widely used in commutative algebra. We briefly sketch some of its basic properties (for complete details, see [BH93], [G67], [Har77], or [HK71]).

The module $\Gamma_I(M)$ is called the module of *sections of M with support in the closed set* $V(I)$. It is clear that $\Gamma_I(\cdot)$ is a left exact functor on the category R–mod, taking values in the subcategory of R–modules with support in $V(I)$.

The right derived functors of $\Gamma_I(\cdot)$ are defined in the usual manner: For a module M, let

$$\mathbb{E}_\bullet : \qquad 0 \to E_0 \longrightarrow E_1 \longrightarrow E_2 \longrightarrow \cdots$$

be an injective resolution of M. Then

$$R\Gamma_I^*(M) = H^*(\Gamma_I(\mathbb{E})),$$

which are normally written $H_I^i(M)$.

The most important case is when $(R, \mathfrak{m})$ is a local ring and $I = \mathfrak{m}$. The modules $H_{\mathfrak{m}}^i(M)$ are then called the *local cohomology modules* of M.

Elementary Properties

There is another formulation of the local cohomology modules that is useful to derive several of its properties.

Proposition A.118. (a) *If R is a Noetherian ring, then* $H_I^i(M) = H_{\sqrt{I}}^i(M)$.

(b) *If W is a multiplicative set of R, then* $(H_I^i(M))_W = H_{I_W}^i(M_W)$.

(c) *If* $\varphi : R \mapsto S$ *is a homomorphism of rings and M is an S–module, then for an R–ideal I,* $H_I^i(M) = H_{IS}^i(M)$.

(d) $\Gamma_I(M) = \lim_{n \to \infty} \operatorname{Hom}_R(R/I^n, M)$.

Each of these properties attests to the suppleness of $\Gamma_I(\cdot)$. For example, we could use (d) to show that

$$H_I^i(M) = \lim_{n \to \infty} \operatorname{Ext}_R^i(R/I^n, M).$$

The next result follows from [G67, Lemma 1.8], but the special case here and its proof is [Bro83, Lemma 3.9].

Proposition A.119. *Let R be a commutative Noetherian ring, I an ideal,* $x \in R$ *and M an R–module. There exists a natural exact sequence*

$$0 \to H_{(I,x)}^0(M) \to H_I^0(M) \to H_{I_x}^0(M_x) \to H_{(I,x)}^1(M) \to \cdots$$
$$\cdots \to H_{(I,x)}^i(M) \to H_I^i(M) \to H_{I_x}^i(M_x) \to H_{(I,x)}^{i+1}(M) \to \cdots.$$

Proof. Let $\mathbb{E}_\bullet$ be an injective resolution of M,

$$0 \to M \longrightarrow E_0 \longrightarrow E_1 \longrightarrow \cdots.$$

Each E_i is a direct sum of injective modules of the form $E(R/\mathfrak{q})$, the injective envelope of $R/\mathfrak{q}$ for some prime $\mathfrak{q}$. As a consequence, there is direct sum decomposition

$$0 \to \Gamma_{xR}(E_i) \longrightarrow E_i \longrightarrow (E_i)_x \to 0.$$

Thus for each $i \geq 0$, we have the exact sequence

$$0 \to \Gamma_I(\Gamma_{xR}(E_i)) \longrightarrow \Gamma_I(E_i) \longrightarrow \Gamma_I((E_i)_x) \to 0.$$

Since $\Gamma_I \circ \Gamma_{xR} = \Gamma_{(I,x)}$, these exact sequences give rise to the short exact sequence of complexes

$$0 \to \Gamma_{(I,x)}(\mathbb{E}_\bullet) \longrightarrow \Gamma_I(\mathbb{E}_\bullet) \longrightarrow \Gamma_I((\mathbb{E}_\bullet)_x) \to 0.$$

The assertion of the proposition is precisely the long exact cohomology sequence that results. $\qquad\square$

This proposition can be used to study inductively the depth and the dimension of a module.

Depth and Vanishing of Cohomology

A property of local cohomology modules is that they read depth (and dimension, later) of modules.

Theorem A.120. *Let R be a Noetherian ring, I an ideal and M a finitely generated R–module, and suppose $M \neq IM$. Then*

$$I\text{–depth}(M) = \inf\{\, i \mid H_I^i(M) \neq 0 \,\}.$$

Proof. Let us argue by induction on $s = I$–depth(M). If $s = 0$, it is clear that $\Gamma_I(M) \neq 0$.

Suppose then that $x \in I$ is a regular element on M,

$$0 \to M \xrightarrow{\;x\;} M \longrightarrow M/xM = \overline{M} \to 0.$$

The exact sequence of cohomology

$$H_I^{s-2}(\overline{M}) \longrightarrow H_I^{s-1}(M) \xrightarrow{\;x\;} H_I^{s-1}(M) \longrightarrow H_I^{s-1}(\overline{M}) \longrightarrow H_I^s(M)$$

gives that $H_I^{s-2}(\overline{M}) = 0$ by the induction hypothesis (note that $\overline{M} \neq I\overline{M}$), and since the modules $H_I^i(M)$ have support in $V(I)$, multiplication by x can only be injective on $H_I^{s-1}(M)$ if this module vanishes. Finally, the rest of the sequence says that $H_I^s(M) \neq 0$ since $H_I^{s-1}(\overline{M})$ embeds in it. $\qquad\square$

Mayer–Vietoris Sequence

This is a marvelous tool for studying connectedness and in induction arguments.

Proposition A.121. *Let R be a Noetherian ring, I and J ideals of R and M an R–module. There is an exact sequence*

$$H_{I \cap J}^{i-1}(M) \to H_{I+J}^i(M) \to H_I^i(M) \oplus H_J^i(M) \to H_{I \cap J}^i(M) \to H_{I+J}^{i+1}(M).$$

Proof. Consider commutative diagrams of inverse systems

$$0 \to R/I^{n+1} \cap J^{n+1} \to R/I^{n+1} \oplus R/J^{n+1} \to R/I^{n+1} + J^{n+1} \to 0$$
$$\downarrow \qquad\qquad \downarrow \qquad\qquad \downarrow$$
$$0 \to \quad R/I^n \cap J^n \quad \to \quad R/I^n \oplus R/J^n \quad \to \quad R/I^n + J^n \quad \to 0.$$

Applying $\mathrm{Hom}_R(\cdot, M)$, we get long exact sequences of direct systems of modules such as $\mathrm{Ext}^i_R(R/I^n, M)$, $\mathrm{Ext}^i_R(R/(I^n \cap J^n), M)$ and $\mathrm{Ext}^i_R(R/(I^n + J^n), M)$. But the filtration $\{I^n \cap J^n,\ n \geq 0\}$ is cofinal for $\{(IJ)^n,\ n \geq 0\}$, while $\{I^n + J^n,\ n \geq 0\}$ is cofinal for $\{(I+J)^n,\ n \geq 0\}$, the latter by the Artin–Rees lemma. Taking the limit we get the claimed exact sequence of cohomology. $\qquad\square$

Comparison of a-invariants

To show the power of these notions we give an application to the estimation of a–invariants.

Theorem A.122. *Let $R \subset S$ be finitely generated graded algebras over a field k. Suppose the embedding $R \subset S$ is pure, R and S are Cohen–Macaulay and $\dim S = \dim R$. Then R is Cohen–Macaulay and $a(R) \leq a(S)$.*

Proof. The assumption that S is *pure* over R means that for any R–module E, the natural mapping $E \mapsto S \otimes_R E$ is an embedding. It is satisfied if R is a direct summand of the R–module S.

Our aim is to discuss one main idea of [Ke91], together with the results above, in order to compare $a(R)$ to $a(S)$.

Let J be the irrelevant maximal ideal of R and denote $\dim R = \dim S = d$. Then $a(R)$ (see Remark B.25) is the degree of the highest nonzero component of

$$H^d_J(R) = \lim_{n \to \infty} \mathrm{Ext}^d_R(R/J^n, R).$$

On the other hand, from the purity of the embedding $R \hookrightarrow S$ it follows easily that we also have an embedding (in all dimensions in fact)

$$\mathrm{Ext}^d_R(R/J^n, R) \hookrightarrow \mathrm{Ext}^d_R(R/J^n, S),$$

and therefore we have the inclusion

$$H^d_J(R) \hookrightarrow H^d_{JS}(S).$$

To prove the claim we show that $H^d_{JS}(S)$ is a homomorphic image of $H^d_{\mathfrak{m}}(S)$, where $\mathfrak{m}$ is the irrelevant maximal ideal of S. For that it will suffice to show that $H^d_{JS}(S)$ is a homomorphic image $H^d_{(J,x)S}(S)$ for any homogeneous element $x \in \mathfrak{m}$.

From Proposition A.119, we have the exact sequence

$$H^d_{(J,x)S}(S) \longrightarrow H^d_{JS}(S) \longrightarrow H^d_{JS_x}(S_x) \longrightarrow H^{d+1}_{(J,x)S}(S) = 0.$$

To see that the homomorphism

$$H^d_{(J,x)S}(S) \longrightarrow H^d_{JS}(S)$$

of graded S–modules is a surjection, we may localize S at $\mathfrak{m}$. But its cokernel will be $H^d_{JS_x}((S_{\mathfrak{m}})_x)$, which vanishes since $\dim(S_{\mathfrak{m}})_x < d$.   □

Čech Complex

The most useful and revealing aspects of local cohomology are those obtained from the construction of the Čech complexes. This is very straightforward: If $x \in R$,

$$C(x): \qquad 0 \to R \longrightarrow R_x \to 0, \qquad (A.30)$$

where R_x is the localization of R at the powers of x. $C(x)$ is graded so that R is the component of degree 0 and R_x the component of degree 1. For a sequence $\mathbf{x} = (x_1, \ldots, x_n)$, the Čech complex is defined by

$$C(\mathbf{x}) = \bigotimes_{i=1}^n C(x_i), \qquad (A.31)$$

more concretely,

$$0 \to R \to \bigoplus_{1 \le i \le n} R_{x_i} \to \bigoplus_{1 \le i < j \le n} R_{x_i x_j} \to \cdots \to R_{x_1 \cdots x_n} \to 0 \qquad (A.32)$$

Theorem A.123. *Let R be a Noetherian ring and I and ideal generated by a sequence $\mathbf{x} = (x_1, \ldots, x_n)$. Then for any R–module M,*

$$H^i_I(M) \simeq H^i(C(\mathbf{x}) \otimes_R M), \ \forall i \ge 0. \qquad (A.33)$$

Before we establish this formula and derive several applications, let us examine the complex $C(\mathbf{x})$ in the case of a regular sequence $\mathbf{x}$. For a single element, $C(x)$ is a direct limit of complexes

$$
\begin{array}{ccc}
0 \to R & \xrightarrow{x^{r+1}} & R \to 0 \\
\| & & \uparrow x \\
0 \to R & \xrightarrow{x^r} & R \to 0
\end{array}
$$

for $r \ge 0$. For a sequence, $\mathbf{x} = x_1, \ldots, x_n$, $C(\mathbf{x})$ is the direct limit of Koszul complexes

$$C(\mathbf{x}) = \lim_{r \to \infty} \mathbb{K}(x_1^r, \ldots, x_n^r; R).$$

Remark A.124. The following are two important properties of this construction:

(a) The cohomology modules of $C(\mathbf{x})$ are supported in the closed set $V(\mathbf{x})$.
(b) If the sequence is regular, each of these Koszul complexes is acyclic.

Proposition A.125. *If* x *is generated by a regular sequence, then* $C(\mathbf{x})$ *is a flat resolution of* $H^0(C(\mathbf{x}))$.

Proof. By the preceding observation, $C(\mathbf{x})$ is an acyclic complex whose components are direct sums of localizations of R.   □

Proposition A.126. *Let* R *be a Noetherian ring,* $\mathbf{x} = x_1,\ldots,x_n$ *a sequence in* R, *and let* M *be an* R–*module.*

(a) *If* $M_{x_i} = 0$ *for all* $x_i \in \mathbf{x}$, *then* $H^i(C(\mathbf{x}) \otimes M) = 0, \ i > 0$.
(b) *If* $M_{x_i} = M$ *for some* x_i, *then* $H^i(C(\mathbf{x}) \otimes M) = 0, \ i \geq 0$.

Proof. We first consider the case of a regular sequence **x**.

(a) Let $K = H^n(C(\mathbf{x}))$. By the preceding result and structure of the Čech complex, all the components of $C(\mathbf{x}) \otimes_R M$, except in dimension 0, vanish.

(b) From the definition of $C(\mathbf{x})$, K is a direct limit of $K_r = R/(x_1^r,\ldots,x_n^r), r \geq 1$. It follows that if, say, multiplication by x_1 is an isomorphism on M,

$$x_1 : M \longrightarrow M,$$

the mapping induced on each

$$x_1 : \mathrm{Tor}_i(K_r, M) \longrightarrow \mathrm{Tor}_i(K_r, M)$$

is an isomorphism but also a nilpotent endomorphism. Therefore $\mathrm{Tor}_i(K_r, M)$ vanishes, and hence their limit $\mathrm{Tor}_i(K, M) = H^{n-i}(C(\mathbf{x}) \otimes_R M)$ will vanish as well.

We now remove the assumption on **x** with a change of rings. Consider the ring of polynomials $B = R[T_1,\ldots,T_n]$ and define the action of the T_i's on M, by $T_i m = x_i m$ for $m \in M$. Let $C_B(\mathbf{T})$ be the Čech complex built on the sequence $\mathbf{T} = T_1,\ldots,T_n$ over the ring B. Now it suffices to note that

$$C_R(\mathbf{x}) \otimes_R M \simeq C_B(\mathbf{T}) \otimes_B M,$$

to establish the claim.   □

Corollary A.127. *Let* R *be a Noetherian ring and* x *a sequence in* R. *For any injective module* M, $H^i(C(\mathbf{x}) \otimes_R M) = 0$ *for* $i > 0$.

Proof. Any injective module over R is a direct sum of indecomposable injective modules, each of which is the injective envelope $E(R/\mathfrak{p})$ of a module $R/\mathfrak{p}$, where $\mathfrak{p}$ is a prime ideal. But for $x \in R$, multiplication by x on a given $E = E(R/\mathfrak{p})$ is an isomorphism if $x \notin \mathfrak{p}$ or $E_x = 0$ since E is supported on $\{\mathfrak{p}\}$. The assertion follows since homology commutes with taking direct limits.   □

Proof of Theorem A.123. Let $I = (\mathbf{x})$, where **x** is the sequence $x_1,\ldots,x_n$. For any R–module,

$$\Gamma_I(M) = H^0(C(\mathbf{x}) \otimes M),$$

by the definition of the Čech complex. On the other hand, it follows by Corollary A.127 that

$$H^i(C(\mathbf{x}) \otimes M) = 0$$

if M is an injective module for $i > 0$. The assertion follows now from the usual yoga of derived functors.   □

Local Rings and Graded Rings

We point out the nature of the local cohomology modules in two cases of interest. We begin with an important to understand aspests of local cohomology.

Matlis Duality

Let $(R, \mathfrak{m})$ be a Noetherian complete local ring and let $E = E(R/\mathfrak{m})$ be the injective envelope of the residue field of R. There is a remarkable equivalence between two categories (see [BH93, Chapter 3]):

$$A = \text{the category of Noetherian } R\text{–modules}$$
$$B = \text{the category of Artinian } R\text{–modules}$$

A is given as all the homomorphic images of free R–modules R^n, $n \in \mathbb{N}$, while B consists of the submodules of E^n, $n \in \mathbb{N}$.

Theorem A.128 (Matlis Duality). *The mapping*

$$\text{Hom}_A(\cdot, E): \quad A \mapsto B$$

defines a self-dual equivalence of categories.

Local Rings

First suppose that $(R, \mathfrak{m})$ is a Noetherian local ring and M is a finitely generated R–module. The module M admits a (minimal) resolution by injective modules,

$$\mathbb{E}_\bullet: \qquad 0 \to E_0 \longrightarrow E_1 \longrightarrow \cdots,$$

with

$$E_i = \bigoplus_{\mathfrak{p}} E(R/\mathfrak{p})^{\beta_i(\mathfrak{p}, M)}.$$

Here, the $\beta_i(\mathfrak{p}, M)$ (the Bass numbers of M) are finite and given by

$$\beta_i(\mathfrak{p}, M) = \dim_{k(\mathfrak{p})} \text{Ext}_R^i(R/\mathfrak{p}, M)_\mathfrak{p}$$

where $k(\mathfrak{p})$ is the residue field of $R/\mathfrak{p}$.

By previous observations, it follows that

$$\Gamma_\mathfrak{m}(E_i) = E(R/\mathfrak{m})^{\beta_i(\mathfrak{m}, M)}.$$

Corollary A.129. *Under these conditions, the modules $H^i_m(M)$ are Artinian. Furthermore, if R is a Gorenstein ring of dimension d, then $H^d_m(R) = E(R/m)$.*

Proof. Note that $\Gamma_m(\mathbb{E}_\bullet)$ is a complex of finite direct sums of $E(R/m)$, each of which is an Artinian module. For the second assertion, we observe that the complex $\Gamma_m(\mathbb{E}_\bullet)$ has only one component, $E(R/m)$, concentrated in dimension d. $\square$

Graded Rings

Another case of significance is that of a graded ring (say $\mathbb{N}$–graded by simplicity but more refined gradings are allowed) and graded modules.

If the sequence $x_1, \ldots, x_n \in R$ consists of homogeneous elements, the Čech complex $C(\mathbf{x})$ has a natural structure of $\mathbb{Z}$–graded modules.

Corollary A.130. *Suppose R is a standard graded algebra, m is its irrelevant maximal ideal and M is a finitely generated, graded R–module. Then the modules $H^i_m(M)$ are graded and*

$$\ell(H^i_m(M)_j) < \infty, \ \forall j,$$
$$H^i_m(M)_j = 0, \ j \gg 0.$$

Proof. Both assertions follow from Corollary A.129 and the underlying graded structure.

Dimension

Local cohomology is sensitive enough to capture the gross size of modules.

Theorem A.131. *Let (R, m) be a Noetherian local and let M be a nonzero, finitely generated R–module of Krull dimension d. Then*

$$H^i_m(M) = 0, \ i > d,$$
$$H^d_m(M) \neq 0.$$

Proof. The first assertion is clear since $H^i_m(M)$ can be defined over the ring $R/ann(M)$, which has dimension $\dim M$, and the Čech complex built on a sequence of $\dim M$ parameters for $m/ann(M)$.

For the other assertion, we have already discussed one case where this holds, when $M = R$ is a Gorenstein ring. The general argument is more technical (but not harder) and goes as follows.

We may assume that M is a faithful R–module. First, note that for the top dimension, $H^d_m(M) \simeq H^d_m(R) \otimes_R M$, from the definition of $C(\mathbf{x})$.

Next, pass to the ring $\widehat{R}$, the m–adic completion of R, replace M by $M \otimes_R \widehat{R}$ and use Cohen's structure theorem to map a Gorenstein complete local ring S of the same Krull dimension onto $\widehat{R}$.

Finally, with $E = H^d_{\mathfrak{m}_S}(S)$, we must prove that $E \otimes_S \widehat{M} \neq 0$, to which end it suffices to verify

$$\text{Hom}_S(E \otimes_S \widehat{M}, E) = \text{Hom}_S(\widehat{M}, \text{Hom}_S(E, E)) = \text{Hom}_S(\widehat{M}, S) \neq 0.$$

The equalities follow from adjointness and $\text{Hom}_S(E, E) = S$ from Matlis duality. Since $\dim \widehat{M} = \dim S$, we have necessarily $\text{Hom}_S(\widehat{M}, S) \neq 0$. $\qquad\qquad\square$

Local Duality

For a local ring $(R, \mathfrak{m})$ and a finitely generated R–module M, we have the following picture of the vanishing of its local cohomology modules:

$$H^i_{\mathfrak{m}}(M) = 0, \text{ depth } M < i, \quad \text{or} \quad i > \dim M,$$
$$H^i_{\mathfrak{m}}(M) \neq 0, \ i = \text{depth } M, \quad \text{or} \quad \dim M.$$

There remains to understand it in the range $\text{depth } M < i < \dim M$. Some of the information will be provided by local duality theory, which we briefly describe (see [BH93] for fuller details).

Theorem A.132 (Grothendieck). *Let $(R, \mathfrak{m})$ be a local complete Gorenstein ring of dimension d, and let E denote the injective envelope of $R/\mathfrak{m}$. For a finitely generated R–module M and all integers i there exist natural isomorphisms*

$$H^i_{\mathfrak{m}}(M) \simeq \text{Hom}_R(\text{Ext}^{d-i}_R(M, R), E). \tag{A.34}$$

There are several extensions of this result to more general Cohen–Macaulay rings and to graded rings. It is this last that is truly rich (see also Chapter 8).

One of its applications is the following criterion for a module to be Cohen–Macaulay, whose proof is left as an exercise for the reader.

Corollary A.133 (Hartshorne–Ogus). *With R as above, let M be a finitely generated reflexive R–module. Then M is Cohen–Macaulay if and only if for each prime ideal $\mathfrak{p}$*

$$\text{depth } M_{\mathfrak{p}} + \text{depth } \text{Hom}_R(M, R)_{\mathfrak{p}} \leq \dim R_{\mathfrak{p}} + 2.$$

Matlis duality throws some light into the nature of the local cohomology modules.

Corollary A.134. *Let $(R, \mathfrak{m})$ be a local ring and let M be a finitely generated R–module. Then the modules $H^i_{\mathfrak{m}}(M)$ are Artinian S–modules, where S is the $\mathfrak{m}$–adic completion of R.*

Exercise A.135. Let $(R, \mathfrak{m})$ be a Noetherian local ring and let M be a finitely generated R–module of dimension d. If M satisfies the condition S_r of Serre and $r < d$, prove that $H^r_{\mathfrak{m}}(M)$ is a module of finite length.

A.9 Linkage Theory

One of our basic operations, used throughout the book, has been that of the quotient of two ideals, $I:J$. The part of commutative algebra that deals with predicting properties of some structured quotients is called *linkage theory*. In geometric terms, it is about the following problem. Let X be an algebraic variety and Y a closed subvariety of X contained in the closed subvariety Z. What are the properties, with respect to Z and Y, of the closed subvarieties W of Z such that $Z = Y \cup W$?

General Properties

We assume throughout that R is a Cohen–Macaulay Noetherian ring.

Definition A.136. *Let I and J be two ideals of R. I and J are said to be* directly linked *if there exists a regular sequence $\mathbf{z} = z_1, \dots, z_n \subset I \cap J$ such that $I = (\mathbf{z}):J$ and $J = (\mathbf{z}):I$. Furthermore, they are said to be* geometrically linked ideals *if they are unmixed ideals of the same grade n, without common components and $I \cap J = (\mathbf{z})$ for some regular sequence of grade n.*

In both cases, the schemes $V(I)$ and $V(J)$ together define a complete intersection: $V(I) \cup V(J) = V(\mathbf{z})$. Another notion of linkage, *residual intersection*, replaces the regular sequence $\mathbf{z}$ by more general ideals. (There is a copious literature on linkage theory that the reader may consult; hopefully [Ulr92] will be published.)

Example A.137. A simple instance is

$$(x^a, y^b) : (x, y) = (x^a, y^b, x^{a-1} y^{b-1}).$$

Theorem A.138. *Let R be a local Gorenstein ring and let I be an unmixed ideal of grade g and let $\mathbf{z} = z_1, \dots, z_g$ be a regular sequence. If $J = (\mathbf{z}):I$ then $I = (\mathbf{z}):J$. Furthermore, if I is a Cohen–Macaulay ideal then J is a Cohen–Macaulay ideal.*

Proof. Since I and $(\mathbf{z}):J$ are unmixed ideals of the same codimension as $(\mathbf{z})$, it suffices to prove equality of their primary components. We may indeed assume that $(R, \mathfrak{m})$ is a local ring and $\mathbf{z}$ is a system of parameters. We may pass to $\bar{R} = R/(\mathbf{z})$, so that $\bar{J} = 0 : \bar{I}$. But the functor $\operatorname{Hom}_{\bar{R}}(\cdot, \bar{R})$ is self–dual, so that in particular $\operatorname{Hom}_{\bar{R}}(\bar{I}, \bar{R}) \simeq \bar{R}/\bar{J}$ and therefore

$$0 : \bar{J} \simeq \operatorname{Hom}_{\bar{R}}(\bar{R}/\bar{J}, \bar{R}) \simeq \operatorname{Hom}_{\bar{R}}(\operatorname{Hom}_{\bar{R}}(\bar{I}, \bar{R}) . \bar{R}) \simeq \bar{I}.$$

The proof of the second assertion is similar, using a few additional elements of the theory of the canonical module of a ring (see [HK71]). First, the reduction to the case $g = 0$ can be carried out as above; we use the notation $\bar{R}$ for R, etc. The canonical module of R/I

$$\omega_{R/I} = \operatorname{Hom}_R(R/I, R) = J$$

is a Cohen–Macaulay module of dimension $\dim R$. We must show that R/J is a Cohen–Macaulay module (of dimension $\dim R$). We use that $\operatorname{Hom}_R(\cdot,R)$ is a self–dualizing functor on the category of finitely generated Cohen–Macaulay modules of maximal dimension. For that consider the exact sequence

$$0 \to J \longrightarrow R \longrightarrow R/J \to 0,$$

to which we apply $\operatorname{Hom}_R(\cdot,R)$ and obtain the exact sequence

$$0 \to \operatorname{Hom}_R(R/J,R) = I \longrightarrow R \longrightarrow R/I \longrightarrow \operatorname{Ext}^1_R(R/J,R) \to 0,$$

since $\operatorname{Hom}_R(\operatorname{Hom}_R(R/I,R),R) = R/I$. It follows that $\operatorname{Ext}^1_R(R/J,R) = 0$, since all the maps are the natural ones. $\qquad\square$

Linkage and Resolutions

Parts of the following result can be traced to Dubreil; in this format and proof they are due to Ferrand and Peskine–Szpiro (see [PSz74] for more details and attributions). It exhibits a projective resolution of $(\mathbf{z}):I$ whenever a resolution of I is known. These resolutions provide means to better control the linkage process, in particular the numerical data of the modules of syzygies.

Theorem A.139. *Let R be a Gorenstein local ring, let I be a perfect ideal of height g and let $\mathbb{F}$ be a minimal free resolution of R/I. Let $\mathbf{z} = z_1,\dots,z_g \subset I$ be a regular sequence, let $\mathbb{K} = \mathbb{K}(\mathbf{z};R)$, and let $u\colon \mathbb{K} \to \mathbb{F}$ be the comparison mapping induced by the inclusion $(\mathbf{z}) \subset I$. Then the dual $\mathbb{C}(u^*)[-g]$ of the mapping cone of u, modulo the subcomplex $u_0\colon R \to R$, is a free resolution of length g of $R/(\mathbf{z}):I$.*

Proof. The mapping cone $\mathbb{C}(f)$ of a chain mapping

$$f\colon (\mathbb{M},\partial) \longrightarrow (\mathbb{N},\partial')$$

is the complex

$$C(f)_{k+1} = M_k \oplus N_{k+1} \xrightarrow{\varphi_{k+1}} M_{k-1} \oplus N_k = C(f)_k,$$

where

$$\varphi_{k+1} = \begin{bmatrix} -\partial_k & 0 \\ f_k & \partial'_{k+1} \end{bmatrix}.$$

This construction yields the exact sequence of complexes

$$0 \to \mathbb{N} \longrightarrow \mathbb{C}(f) \longrightarrow \mathbb{M}[-1] \to 0.$$

We use it on the dual of the comparison mapping $u\colon \mathbb{K} \to \mathbb{F}$ induced by the natural surjection $R/(\mathbf{z}) \to R/I$. We have the exact sequence of complexes

$$0 \to \mathbb{K}^*[-g] \longrightarrow \mathbb{C}(u^*)[-g] \longrightarrow \mathbb{F}^*[-g-1] \to 0.$$

Since $\mathbb{F}^*[-g]$ and $\mathbb{K}^*[-g]$ are acyclic, $H_i(\mathbb{C}(u^*)[-g]) = 0$ for $i \geq 2$, while in lower dimension we have the exact sequence

$$0 \to H_1(\mathbb{C}(u^*)[-g]) \to H_0(\mathbb{F}^*[-g-1]) \xrightarrow{\psi} H_0(\mathbb{K}^*[-g]) \to H_0(\mathbb{C}(u^*)[-g]) \to 0.$$

But the mapping ψ is the inclusion

$$\psi \colon ((\mathbf{z}):I)/(\mathbf{z}) \hookrightarrow R/(\mathbf{z}),$$

showing that $\mathbb{C}(u^*)[-g]$ is a free resolution of $H_0(\mathbb{C}(u^*)[-g]) = R/(\mathbf{z}):I.$ □

The following consequence is a fuller explanation of Theorem 4.5.

Corollary A.140. *Let* $\mathbf{u}$ *and* $\mathbf{v}$ *be regular sequences of length n and let* φ *be an* $n \times n$ *matrix such that* $[\mathbf{v}]^t = \varphi[\mathbf{u}]^t$. *Then* $(\mathbf{v}, \det \varphi)$ *is a perfect ideal of grade n, and* $(\mathbf{v}):(\mathbf{u}) = (\mathbf{v}, \det \varphi)$.

Invariants of Linkage

There are many connections between directly linked ideals, $I \sim_{(\mathbf{z})} J$. A refinement is provided by the more general notion of linkage: I is linked to J if there exists a sequence of direct links

$$I = I_0 \sim I_1 \sim \cdots \sim I_n = J.$$

The literature is very rich in studies of which properties of an ideal are passed along a chain of direct links. It is often significant that the number of steps in the chain (e.g. odd/even) plays a role in connecting the properties of I to those of J.

The following (and there is a graded version as well) is part of one of the earliest discovered invariants ([Ra079]).

Proposition A.141. *Let* S *be a Gorenstein local ring of dimension* $d \geq 2$ *and let* I *be a height unmixed ideal of codimension* $d - 2$. *Set* $A = S/I$ *and denote by* $\widetilde{A}$ *its* S_2*-ification. Then for any ideal* J *linked to* I, *defining similarly* $B = S/J$ *and* $\widetilde{B}$, *then* $\widetilde{A}/A \simeq \widetilde{B}/B$ *if* I *is linked to* J *by an even number of direct links, or* $\widetilde{B}/B \simeq \operatorname{Ext}_S^d(\widetilde{A}/A, S)$ *if they are linked by an odd number of direct links. In particular the two modules have the same length.*

Proof. It is easy to reduce to the case $d = 2$. We consider the case of a direct link first. The hypothesis is that $J = \operatorname{Hom}_S(S/I, S)$. Note that this means that J is the canonical module of $A = S/I$.

Apply the functor $\operatorname{Hom}_S(\cdot, S)$ to the short exact sequence

$$0 \to J \longrightarrow S \longrightarrow B \to 0.$$

One gets the exact sequence

$$0 \to \operatorname{Hom}_S(S/J, S) \longrightarrow \operatorname{Hom}_S(S, S) \longrightarrow \operatorname{Hom}_S(J, S) \longrightarrow \operatorname{Ext}_S^1(S/J, S) \to 0.$$

Note that each $\varphi \in \mathrm{Hom}_S(J,S)$ maps J into the ideal of S that is annihilated by I, in other words, we have $\mathrm{Hom}_S(J,S) = \mathrm{Hom}_S(J,J)$ and therefore obtain the exact sequence

$$0 \to A = S/I \longrightarrow \mathrm{Hom}_S(\omega_A, \omega_A) = \widetilde{A} \longrightarrow \mathrm{Ext}_S^1(B,S) = \mathrm{Ext}_S^2(\widetilde{B}/B, S) \to 0.$$

Both assertions follow immediately from this calculation. $\square$

B

Hilbert Functions

By **Jürgen Herzog**

The following notes are intended to give an introduction to graded rings and modules, and their Hilbert functions. In the first sections we consider G-graduations and G-filtrations where G is an arbitrary abelian group. This allows us to treat filtrations arising from term orders, and to explain why many properties behave well when passing from the ideal of initial forms to the ideal itself. When dealing with Hilbert functions we concentrate on $\mathbb{Z}$-graduations. We give complete proofs of most of the results, except for Macaulay's and Green's theorem whose detailed proofs can be found in [BH93, Section 4.2]. However these theorems will be used in Green's proof of Gotzmann's regularity and persistence theorems which will be presented in the last section.

B.1 G-Graded Rings and G-Filtrations

Term orders give rise to graduations and filtrations which are more general than the usual $\mathbb{Z}$-graduations and $\mathbb{Z}$-filtrations. In order to cover these cases we define

Definition B.1. (a) *Let G be an abelian group. A G-graded ring is a commutative ring together with a family of subgroups $\{R_g\}_{g \in G}$ such that*
   (i) $R = \bigoplus_{g \in G} R_g$;
   (ii) $R_g R_h \subset R_{g+h}$ *for all $g, h \in G$.*
(b) *Let R be a G-graded ring. A G-graded module is an R-module M together with a family of subgroups $\{M_g\}_{g \in G}$ such that*
   (i) $M = \bigoplus_{g \in G} M_g$;
   (ii) $R_g M_h \subset M_{g+h}$ *for all $g, h \in G$.*

An element $x \in M_g$ is called *homogeneous of degree g*; the degree of x is denoted by $\deg x$. An arbitrary element $x \in M$ can be uniquely written as $x = \sum_{g \in G} x_g$ with $x_g \in M_g$ for all $g \in G$. One call x_g the *g-th homogeneous component of x*.

A $\mathbb{Z}$-graded ring R is called *positively graded* if R is generated over R_0 by homogeneous elements of positive degree, and is called *homogeneous*, or *standard*, if the generators all have degree 1.

The polynomial ring $R = k[x_1, \ldots, x_n]$, k an arbitrary commutative ring, may be given various structures of a positively graded ring: one sets $R_0 = k$, and assigns to each variable x_i an integer a_i as its degree.

The polynomial ring admits a natural $\mathbb{Z}^n$-graduation as well. The multi-exponent of a monomial $u \in R$ is an element in $\mathbb{N}^n$. We set

$$R_a = \begin{cases} 0 & \text{if } a \notin \mathbb{N}^n, \\ kx^a & \text{if } a \in \mathbb{N}^n. \end{cases}$$

This graduation is also called the *fine graduation*.

Let N and M be G-graded R-modules. A *homogeneous homomorphism from N to M* is an R-module homomorphism $\phi: N \to M$ such that $\phi(N_g) \subset M_g$ for all $g \in G$. The G-graded R-modules together with the homogeneous homomorphisms form an abelian category.

A submodule N of M is called a *G-graded submodule of M*, if N is graded and the inclusion map $N \subset M$ is homogeneous, or equivalently, if $N_g \subset M_g$ for all $g \in G$. One easily checks that this is the case if and only if the following condition holds: Whenever $x \in M$ belongs to N, then all homogeneous components x_g of x belong to N as well. Still another way to express that N is a G-graded submodule of M is given by the condition that $M_g \cap N = N_g$ for all $g \in G$. A graded submodule I of R is called a *graded* or *homogeneous ideal* of R.

Given a G-graded submodule N of M, the factor module M/N admits a natural graduation: one sets $(M/N)_g = (M_g + N)/N$ for all $g \in G$; observe that $(M/N)_g \simeq M_g/N_g$.

For a graded ideal I of R, the factor ring R/I is not just a G-graded R-module but again a G-graded ring, since $(R/I)_g(R/I)_h \subset (R/I)_{g+h}$ for all g, $h \in G$. The most important examples of graded rings arise in this way – they are of the form $k[x_1, \ldots, x_n]/I$ where k is a field and I is an ideal generated by homogeneous polynomials. If I is generated by monomials, then $k[x_1, \ldots, x_n]/I$ is actually a $\mathbb{Z}^n$-graded ring.

G-filtrations

We now introduce filtered rings and modules. Their Rees rings and associated graded rings yield important examples of graded rings.

Let $(G, <)$ be a totally ordered group. In other words, G is a group whose underlying set is endowed with a total order $<$ compatible with the group operation: $g_1 < g_2$ implies $g_1 + h < g_2 + h$. Basically we have in mind the following examples: (i) The integers with their natural order, and (ii) $\mathbb{Z}^n$ together with a total order.

Definition B.2. (a) *Let R be a k-algebra where k is an arbitrary (commutative) ring, and let M be an R-module. An (ascending) G-filtration F of M is a family $F = \{F_g M\}_{g \in G}$ of k-submodules of M satisfying:*

(i) $F_g M \subset F_h M$ for all $g, h \in G$ with $g < h$;

(ii) $\bigcup_{g \in G} F_g M = M$.

(b) *Let F be G-filtration of R. Then (R, F) is called a G-filtered k-algebra if*

(iii) $F_g R F_h R \subset F_{g+h} R$ for all $g, h \in G$;

(iv) $1 \in F_0 R$.

(c) *Let (R, F) be a G-filtered k-algebra, and let M an R-module with a G-filtration (which we again denote by F). Then (M, F) is called a G-filtered R-module if*

(v) $F_g R F_h M \subset F_{g+h} M$ for all $g, h \in G$.

Let M and N be R-modules, F a G-filtration on M, F' a G-filtration on N. An R-module homomorphism $\phi: M \to N$ is called compatible with the corresponding G-filtrations, if $\phi(F_g M) \subset F'_g N$ for all $g \in G$. A *homomorphism of G-filtered k-algebras (modules)* i s a k-algebra (module) homomorphism compatible with the corresponding G-filtrations.

Let F be a G-filtration on an R-module M. Then F is called *orderly* if for all non-zero elements $x \in M$ there is a smallest element $g \in G$ such that $x \in F_g M$. This element g is called the *order of x* and is denoted by $\mathrm{ord} x$. One sets $\mathrm{ord} 0 = -\infty$ and understands that $-\infty < g$ for all $g \in G$.

If $\bigcap_{g \in G} F_g M = 0$, then F is called *separated*. It is clear that orderly filtrations are separated. On the other hand, for $\mathbb{Z}$-filtrations these two notions coincide.

Attached to a G-filtered k-algebra R with filtration F there are two G-graded rings, the *(extended) Rees ring of R with respect F*

$$R(F, R) = \bigoplus_{g \in G} F_g R,$$

and the *associated graded ring of R with respect to F*

$$\mathrm{gr}_F(R) = \bigoplus_{g \in G} F_g R / F_{<g} R,$$

where $F_{<g} R = \bigcup_{h < g} F_h R$.

Note that both rings have a natural k-algebra structure. Similarly one defines the G-graded $R(F, R)$-module $R(F, M) = \bigoplus_{g \in G} F_g M$, and the G-graded $\mathrm{gr}_F(R)$-module

$$\mathrm{gr}_F(M) = \bigoplus_{g \in G} F_g M / F_{<g} M.$$

It is obvious that any k-algebra homomorphism of filtered k-algebras induces a homogeneous homomorphism of the corresponding Rees rings or associated graded rings. A similar statement holds for homomorphisms of filtered modules.

Let us consider some examples. The most common $\mathbb{Z}$-filtration is the I-adic filtration: $F_i M = I^{-i} M$, $i \in \mathbb{Z}$, where M is an R-module and $I \subset R$ an ideal. Here we use the convention that $I^j = R$ if $j < 0$. We denote the corresponding Rees ring by $R(I)$, and write $R(I, M)$ for the $R(I)$-module $R(F, M)$.

It is clear that $R(I)$ may be identified with the subring $R[It, t^{-1}]$ of $R[t, t^{-1}]$ where t is an indeterminate over R. The associated graded ring, which in this case is denoted by $\mathrm{gr}_I(R)$, is equal to $\bigoplus_{i \geq 0} I^i / I^{i+1}$.

Another important class of filtrations are the so-called *degree filtrations*. Given a G-graded ring R, and a G-graded R-module M, where G is a totally ordered group, we define a G-filtration, the *degree filtration*, by setting $F_g R = \bigoplus_{h \leq g} R_h$, and similarly $F_g M = \bigoplus_{h \leq g} M_h$. With this filtration, R is a G-filtered R_0-algebra.

A remarkable fact of this filtration is that $R \simeq \operatorname{gr}_F R$ as a G-graded ring, and $M \simeq \operatorname{gr}_F M$ as a G-graded R-module.

Let us consider two more specific examples: Let $R = k[x_1, \ldots, x_n]$ be the polynomial ring with the standard $\mathbb{Z}$-grading, F be the corresponding degree filtration. Then $R(F)$ is isomorphic to the subring $k[x_1 t, \ldots, x_n t, t]$ of $k[x_1, \ldots, x_n, t]$ where t is an additional indeterminate. Thus we see that $R(F)$ may be identified with a polynomial ring.

The second example of a degree filtration comes from Gröbner basis theory, and will be further studied in the next section. Let $<$ be a term order for the polynomial ring $R = k[x_1, \ldots, x_n]$. Such an order induces a total order on the semigroup $\mathbb{N}^n$ of multi-exponents of the monomials in $x_1, \ldots, x_n$, and it has a unique extension to $\mathbb{Z}^n$. For the corresponding degree filtration F one has $F_g R = \bigoplus_{h \leq g} k x^h$ where, by definition, $x^h = 0$ if $h \notin \mathbb{N}^n$. We say that F *is defined by a term order*.

Unfortunately, the associated Rees ring need not to be Noetherian. Indeed, $R(F, R)$ may be identified with the subring $\bigoplus_{g \in \mathbb{Z}^n} F_g R T^g$ of $R[t_1, t_1^{-1}, \ldots, t_n, t_n^{-1}]$, and as such is a semigroup ring. Now it is easily seen that for the standard term orders the underlying semigroup is not finitely generated, so that $R(F, R)$ is not Noetherian in these cases.

When passing to factor modules one considers the induced filtration: Let M be an R-module, F a G-filtration on M, and N a submodule. There are induced G-filtrations on N and M/N:

$$F_g N = F_g M \cap N, \qquad F_g(M/N) = (F_g M + N)/N, \qquad g \in G.$$

It is then clear that the inclusion map $N \to M$ and the canonical epimorphism $M \to M/N$ are compatible with these filtrations.

Next suppose that M is a G-filtered R-module, N a submodule, and that the quotient module M/N is equipped with the induced G-filtration. Then M/N is a G-filtered module, and that the canonical epimorphism $\varepsilon : M \to M/N$ induces a homomorphism of G-graded modules

$$in(\varepsilon) : \operatorname{gr}_F(M) \to \operatorname{gr}_F(M/N).$$

We want to describe the kernel of $in(\varepsilon)$. Suppose F is orderly, and $x \in M$ is a non-zero element of order g. The element $in(x) \in \operatorname{gr}_F(M)$ with homogeneous components

$$in(x)_h = \begin{cases} 0 & \text{if } h \neq g, \\ x + F_{<g} M & \text{if } h = g, \end{cases}$$

is called the *initial form of x with respect to F*. Note that $in(x)$ is homogeneous of degree g.

We leave the easy proof of the following statement to the reader.

Proposition B.3. *With the notation introduced, suppose that F is orderly. Then $in(\varepsilon)$ is an epimorphism, and $\ker(in(\varepsilon))$ is the submodule $in(N)$ of $\mathrm{gr}_F(M)$ generated by all elements $in(x)$, $x \in N$; in particular,*

$$\mathrm{gr}_F(M/N) \simeq \mathrm{gr}_F(M)/in(N).$$

Sometimes we write $in_F(N)$ to indicate that initial forms are taken with respect to F.

Let M be an R-module with G-filtration F. We let

$$O_M(G) = \{g \in G : \mathrm{gr}_F(M)_g \neq 0\} \cup \{-\infty\}.$$

The subset $O_M(G)$ of G is the set of orders of M. It inherits a total order from G.

Proposition B.4. *Let M be a G-filtered R-module whose filtration is orderly and such that any strictly descending sequence $g_1 > g_2 > \cdots$ of elements in $O_M(G)$ terminates. Let N be a submodule of M, and $x_1, \ldots, x_n$ elements of N such that their initial forms $in(x_1), \ldots, in(x_n)$ generate $in(N)$. Then the elements $x_1, \ldots, x_n$ generate N.*

Proof. Suppose $N \neq Rx_1 + \cdots + Rx_n$, and consider the subset

$$U = \{\mathrm{ord}\, x : x \in N \setminus (Rx_1 + \cdots + Rx_n)\}$$

of $O_M(G)$. Then $U \neq \emptyset$, and hence the hypothesis on $O_M(G)$ implies that U contains a minimal element, say g_0. Let $x \in N \setminus (Rx_1 + \cdots + Rx_n)$ with $\mathrm{ord}\, x = g_0$. There exist elements $a_i \in R$ such that $in(x) = in(a_1)in(x_1) + \cdots + in(a_n)in(x_n)$ with $\mathrm{ord}\, a_i + \mathrm{ord}\, x_i = g_0$ for $i = 1, \ldots, n$. We set $x' = x - \sum_{i=1}^n a_i x_i$. Then $x' \in N \setminus (Rx_1 + \cdots + Rx_n)$, but $\mathrm{ord}\, x' < \mathrm{ord}\, x$, which is a contradiction. $\qquad\square$

B.2 The Study of R via $\mathrm{gr}_F(R)$

Let R be the polynomial ring, and F the $\mathbb{Z}^n$-filtration given by a term order on R. Then any strictly decreasing sequence $g_1 > g_2 > \cdots$ of elements in $O_R(\mathbb{Z}^n)$ terminates. Indeed, $(x^{g_1}) \subset (x^{g_1}, x^{g_2}) \subset \cdots$ is a strictly ascending chain of ideals which must terminate, because for this argument we may assume that the base ring k is Noetherian which implies that R is Noetherian. Thus B.3 says that $\mathrm{gr}_F(R/I) \simeq R/in(I)$ where $in(I)$ is the ideal generated by the initial monomials of the polynomials in I, and B.4 implies that I is generated by a Gröbner basis.

In Gröbner bases theory knowledge about $in(I)$ gives information about I. The reason is that many properties of the associated graded ring $\mathrm{gr}_F(R)$ are shared by the k-algebra R itself, as we shall see now.

A first simple example of how properties of $\mathrm{gr}_F(R)$ are inherited by R is given by

Proposition B.5. *Let R be a G-filtered k-algebra, M a G-filtered R-module, and suppose that the corresponding filtrations on R and M are orderly.*

(a) *If a is an element in R such that $in(a)$ is a non-zerodivisor on $\mathrm{gr}_F(M)$, then a is a non-zerodivisor on M.*

(b) *If $\mathrm{gr}_F(R)$ is a domain or reduced, then so is R.*

Proof. (a) Let $x \in M$, $x \neq 0$. Then $in(x) \neq 0$, and so $in(a)in(x) \neq 0$, from which one concludes that $in(a)in(x) = in(ax)$. Therefore $in(ax) \neq 0$, and so $ax \neq 0$.

(b) By (a), if $\mathrm{gr}_F(R)$ is a domain, then so is R. Now suppose $\mathrm{gr}_F(R)$ is reduced, and suppose there exists a non-zero element a in R such that $a^n = 0$. Then $in(a) \neq 0$, but $in(a)^n = 0$, a contradiction. $\qquad\square$

The next proposition makes clear how M and $\mathrm{gr}_F(M)$ are related algebraically. For that we identify $R(F,R)$ with the subring $\bigoplus_{g \in G} F_g R t^g$ of the group ring $R[G] = \bigoplus_{g \in G} R t^g$. By Definition B.2, (iv) one has $1 \in F_g R$ for all positive $g \in G$. Therefore the set S of all t^g, $g > 0$, is multiplicatively closed.

Proposition B.6. *Let R be a G-filtered k-algebra, and M a G-filtered R-module.*

(a) *$S^{-1}R(F,R)$ is isomorphic to the group ring $R[G]$ of G over R, and*

$$S^{-1}R(F,M) \simeq M \otimes_R R[G].$$

(b) *Assume there is a (unique) smallest positive element $g_0 \in G$. (Such an element exists for any order on $\mathbb{Z}^n$ which is induced by a term order). Then t^{g_0} is a non-zerodivisor on $R(F,M)$ and*

$$\mathrm{gr}_F(M) \simeq R(F,M)/t^{g_0}R(F,M).$$

Proof. (a) Let $a \in R$ and $g \in G$. There exists an element $h \in G$ such that $a \in F_h R$. If $h \leq g$, then $at^g \in F_g R t^g$, and if $h \geq g$, then $at^g \in F_h R t^g = (t^{h-g})^{-1} F_h R t^h$. Thus in any case, $at^g \in S^{-1}R(F,R)$, as desired. The statement for $R(F,M)$ is proved in the same way.

(b) The element t^{g_0} is a non-zerodivisor on $M \otimes_R R[G]$. By (a), $R(F,M)$ is an $R(F,R)$-submodule of $M \otimes_R R[G]$, and so t^{g_0} is a non-zerodivisor on $R(F,M)$ as well.

The existence of a smallest positive element $g_0 \in G$ is equivalent to the condition that for all $g \in G$, there is a largest element $g' < g$. Indeed, given g_0 one has $g' = g - g_0$ for all $g \in G$.

Hence

$$\mathrm{gr}_F(M) = \bigoplus_{g \in G} F_g M / F_{g'} M,$$

and the asserted isomorphism follows. $\qquad\square$

Many properties of rings behave well under localization or reduction modulo a non-zerodivisor. Therefore B.6 is the appropriate tool to compare $\mathrm{gr}_F(R)$ and R. However there is one serious obstruction. As we observed already in the previous section, $R(F,R)$ need not to be Noetherian, and this even happens for filtrations defined by monomial orders, a case which we are especially interested in. We shall see below that this difficulty can be avoided by passing from the $\mathbb{Z}^n$-filtration to a suitable $\mathbb{Z}$-filtration. Thus in the next result we concentrate on $\mathbb{Z}$-filtrations.

If R is a Cohen-Macaulay ring, not necessarily local, we define the *type* $r(R)$ of R to be the supremum of the local types $r(R_\mathfrak{p})$, $\mathfrak{p}$ a maximal prime ideal of R. Then $r(S^{-1}R) \leq r(R)$ for any multiplicatively closed set $S \subset R$; see [BH93, Corollary 3.12].

Proposition B.7. *Suppose R is a $\mathbb{Z}$-filtered k-algebra with $F_0 R = k$, and such that $R(F,R)$ is Noetherian. Suppose further that $\mathrm{gr}_F(R)$ is Cohen-Macaulay of type r. Then $R_\mathfrak{p}$ is Cohen-Macaulay of type $\leq r$ for all $\mathfrak{p} \in \mathrm{Spec}(k)$ with*

$$\mathfrak{p} \supseteq \sum_{i>0} F_{-i} R F_i R.$$

In particular, if $\mathrm{gr}_F(R)$ is Gorenstein, then so is $R_\mathfrak{p}$.

Example B.8. (a) For the I-adic filtration, B.7 says that if the algebra $\mathrm{gr}_I(R)$ is Cohen-Macaulay of type r, then $R_\mathfrak{p}$ is Cohen-Macaulay of type $\leq r$ for all $\mathfrak{p} \supseteq I$.

(b) Suppose k is a field, then necessarily $\mathfrak{p} = 0$, and the condition on $\mathfrak{p}$ formulated in B.7 implies that $F_i R = 0$ for all $i < 0$. In particular, suppose $R = k[x_1, \ldots, x_n]/I$ is an affine k-algebra, $k[x_1, \ldots, x_n]$ is positively graded, and the filtration F on R is induced by the degree filtration on $k[x_1, \ldots, x_n]$. Applying B.7 we obtain: If $k[x_1, \ldots, x_n]/in(I)$ is Cohen-Macaulay of type r, then $k[x_1, \ldots, x_n]/I$ is Cohen-Macaulay of type $\leq r$.

Proof. The $k_\mathfrak{p}$-algebra $R_\mathfrak{p}$ is again $\mathbb{Z}$-filtered with filtration $F_i' R_\mathfrak{p} = (F_i R)_\mathfrak{p}$. Thus we may as well assume that k is local and $\mathfrak{p}$ is the maximal ideal $\mathfrak{m}$ of k. Then our hypothesis on $\mathfrak{p}$ implies that the graded ideal $\widetilde{\mathfrak{m}}$ with homogeneous components

$$\widetilde{\mathfrak{m}} = \begin{cases} \mathfrak{m} & \text{if } i = 0, \\ F_i R & \text{if } i \neq 0, \end{cases}$$

is the unique graded maximal ideal of $R(F,R)$.

Using the notation in B.6 we have $g_0 = 1$, and $\mathrm{gr}_F(R) = R(F,R)/tR(F,R)$, where $t \in \widetilde{\mathfrak{m}}$ is a non-zerodivisor of $R(F,R)$. Thus we see that $R(F,R)$ is Cohen-Macaulay of type r. By B.6, $R(F,R) \simeq R[t,t^{-1}]$, so that $R[t,t^{-1}]$ is Cohen-Macaulay of type $\leq r$. Finally, since the extension $R[t,t^{-1}]$ is faithfully flat, R itself is Cohen-Macaulay of type $\leq r$; see [BH93, Proposition 1.2.16]. $\qquad\square$

Proposition B.9. *Let k be a field, $R = k[x_1, \ldots, x_n]$ the polynomial ring over k, $I \subset R$ an ideal, and F the $\mathbb{Z}^n$-filtration defined by a monomial order $<$ on R. Then R admits a positive $\mathbb{Z}$-graduation such that*

$$in_{F'}(I) = in_F(I)$$

where F' denotes the degree filtration belonging to this $\mathbb{Z}$-graduation. In other words, there is an isomorphism of k-algebras

$$\mathrm{gr}_F(R/I) \simeq \mathrm{gr}_{F'}(R/I).$$

Combining this result with B.8(b) we obtain

Corollary B.10. *With the notation introduced, let* $in(I)$ *be the ideal of initial terms with respect to a given monomial order. If* $R/in(I)$ *is Cohen-Macaulay of type* r, *then* R/I *is Cohen-Macaulay of type* $\leq r$.

For the proof of B.9 we shall need the following

Lemma B.11. *Let* $<$ *be the total order on* $\mathbb{Z}^n$ *which is induced by a term order on the ring* $k[x_1,\ldots,x_n]$. *Given elements* $g_1,\ldots,g_m \in \mathbb{Z}^n$ *with* $g_i > 0$ *for* $i = 1,\ldots,m$, *there exists an element* $g = (a_1,\ldots,a_n) \in \mathbb{Z}^n$ *with* $a_i > 0$ *for* $i = 1,\ldots,n$, *such that*

$$\langle g, g_i \rangle > 0 \quad for \quad i = 0,\ldots,m.$$

Here $\langle , \rangle$ *denotes the usual scalar product in* $\mathbb{R}^n$.

Proof. Let $C = \sum_{i=1}^m \mathbb{R}_+ g_i$ be the cone spanned by the g_i, and D the cone spanned by the set

$$\mathbb{Z}^n_- = \{h : h = (h_1,\ldots,h_n),\ h_i \leq 0 \quad \text{for all } i\}.$$

Since the order $<$ on $\mathbb{Z}^n$ is induced by a term order, we have that all elements in $\mathbb{N}^n \setminus \{0\}$ are positive. Hence all elements in $\mathbb{Z}^n_- \setminus \{0\}$ are negative.

A cone is called *positive* if it does not contain any non-trivial linear subspace. Note that C and D are both positive. This is clear for D, and follows for C since otherwise $\mathbb{R}g_i$ would be a subset of C for some i, and this would imply that $-g_i \in C$, contradicting the fact that the elements in $\mathbb{Z}^n \cap C$ are positive.

We claim that $C \cap D = \{0\}$. Indeed, both C and D are *rational* cones, i.e. intersections of finitely many rational half-spaces, $H^+ = \{v \in \mathbb{R}^n : \langle v, a \rangle \geq 0\}$, $a \in \mathbb{Z}^n$. So $C \cap D$ is a rational cone, and if it is not trivial it contains an element $c \in \mathbb{Z}^n$, $c \neq 0$. But then $c > 0$ since $c \in C$, and $c < 0$ since $c \in D$, a contradiction.

As the positive finitely generated cones C and D intersect only trivially, there exists a rational hyperplane $H = \{v \in \mathbb{R}^n : \langle v, g \rangle = 0\}$, $g \in \mathbb{Z}^n$, separating these cones. That is, we have

$$D \subseteq \{v \in \mathbb{R}^n : \langle v, g \rangle < 0\},$$

and

$$C \subseteq \{v \in \mathbb{R}^n : \langle v, g \rangle > 0\}.$$

Let $e_1,\ldots,e_n$ be the canonical basis of $\mathbb{R}^n$. Since all $-e_i$ belong to $\mathbb{Z}^n_-$, the first inclusion implies that $\langle -e_i, g \rangle < 0$, that is, $\langle e_i, g \rangle > 0$ for all i. In other words, all components of g are positive. The second inclusion implies that $\langle g_i, g \rangle > 0$ for all i, as desired. $\square$

Proof. [Proof of B.9] According to [Ei95, Proposition 15.26] there exists a finite set

$$\{(u_1, v_1),\ldots,(u_s, v_s)\}$$

of pairs of monomials with $u_i > v_i$ for all i satisfying the following property: If F' is the degree filtration of a $\mathbb{Z}$-graduation of R for which $\deg u_i > \deg v_i$ for $i = 1,\ldots,s$, then $in_{F'}(I) = in_F(I)$.

We apply the previous lemma to find such a $\mathbb{Z}$-graduation which in addition makes R a positively graded k-algebra: Let b_i be the multi-exponent of u_i, and c_i the multi-exponent of v_i. Then $g_i = b_i - c_i > 0$ for $i = 1, \ldots, s$, and by B.11 there exists an element $g = (a_1, \ldots, a_n) \in \mathbb{Z}^n$ with $a_i > 0$ for $i = 1, \ldots, n$, and $\langle g, g_i \rangle > 0$ for all i. Now we define the positive graduation on R by setting $\deg x_i = a_i$ for $i = 1, \ldots, n$. Then for $i = 1, \ldots, s$ we get $\deg u_i - \deg v_i = \langle g, g_i \rangle > 0$, as desired. $\qquad\square$

The result of Proposition B.9 can also be phrased as follows: Let I be an ideal in the polynomial ring $R = k[x_1, \ldots, x_m]$, and let $in(I)$ be the ideal of initial forms of I with respect to some term order. Then there is a flat one-parameter family with special fibre $R/in(I)$ and general fibre R/I.

Indeed referring to the notation of (B.6) and (B.9) we see that there is an element $t \in S$, $S = R(F', R/I)$, such that S is a flat $k[t]$-algebra (since t is a non-zerodivisor of S), and such that

(i) $S/(t) \simeq R/in(I)$, and
(ii) $S/(t-c) \simeq R/I$ for all $c \in k$, $c \neq 0$.

This observation has an interesting consequence: Let M be a finitely generated R-module. We set $\mathfrak{m} = (x_1, \ldots, x_n)$, and let

$$\beta_i(M) = \dim_k \mathrm{Tor}_i^R(R/\mathfrak{m}, M)$$

for all $i \geq 0$. One calls $\beta_i(M)$ the i-th Betti-number of M (with respect to $\mathfrak{m}$).

Corollary B.12. *Let $I \subset R$ be an ideal, and $<$ a term order on R. Then*

$$\beta_i(R/I) \leq \beta_i(R/in_<(I))$$

for all $i \geq 0$.

Proof. By Proposition B.9 there exists a positive $\mathbb{Z}$-graduation ($\deg x_i = a_i > 0$ for $i = 1, \ldots, n$) such that $in_{F'}(I) = in_F(I)$, where F' denotes de degree filtration of this graduation, and F the filtration defined by the monomial order. Note that $R(F', R)$, as a graded k-algebra, is isomorphic to the positively graded polynomial ring $k[t, x_1 t^{a_1}, \ldots, x_n t^{a_n}]$, where $\deg t = 1$ and $\deg x_i t^{a_i} = a_i$ for $i = 1, \ldots, n$. Furthermore, the natural epimorphism $R(F', R) \to R(F', R/I)$ is homogeneous.

Let F be the minimal graded free $R(F', R)$-resolution of $R(F', R/I)$. Since t is a homogeneous non-zerodivisor of both algebras we see that F/tF is a graded minimal free resolution of $R/in_<(I) = \mathrm{gr}_{F'}(R/I)$ over $R = \mathrm{gr}_{F'}(R)$. Hence

$$\beta_i(R/in_<(I)) = \mathrm{rank}_R F_i/tF_i = \mathrm{rank}_{R(F', R)} F_i.$$

On the other hand $t - 1$ is a (non-homogeneous) non-zerodivisor of $R(F', R/I)$ and $R(F', R)$, too. Thus $F/(t-1)F$ is a free resolution of

$$R/I = R(F', R/I)/(t-1)R(F', R/I)$$

over $R = R(F', R)/(t-1)R(F', R)$. It follows that $\beta_i(R/I) \leq \mathrm{rank}_{R(F', R)} F_i$, as desired. $\qquad\square$

Note that B.12 yields another proof of B.10.

B.3 The Hilbert–Samuel Function

Let M be a G-graded module over the G-graded ring R. Each homogeneous component M_g of M is an R_0-module. We denote by $\ell(M_g)$ the R_0-length of M_g. If $\ell(M_g) < \infty$ for all $g \in G$, we may define the $\mathbb{Z}$-valued function

$$H(M, \cdot): G \to \mathbb{Z}, \qquad g \mapsto H(M, g) = \ell(M_g),$$

which is called the *Hilbert function* of M.

Now assume that R is a Noetherian $\mathbb{Z}^n$-graded ring. Then R_0 is Noetherian and R is finitely generated as R_0-algebra. Assume R is generated over R_0 by homogeneous elements $x_1, \ldots, x_m$ with $\deg x_i \in \mathbb{N}^n \setminus \{0\}$. Then for any finitely generated $\mathbb{Z}^n$-graded R-module M, all its homogeneous components M_g are finitely generated R_0-modules. Indeed, if $m_1, \ldots, m_k$ are homogeneous generators of M, then the elements $u m_i$ generate M_g where u is a monomial in the x_j of degree $g - \deg m_i$, and there are only finitely many such monomials. Hence if we assume in addition that R_0 is of finite length, then all M_g have finite length, and so the Hilbert function of M is defined.

For the rest of this section these will be our standard assumptions. For simplicity we will also assume that the degrees of all generators of M belong to $\mathbb{N}^n$. Then the generating function of $H(M, g)$,

$$H_M(t_1, \ldots, t_n) = \sum_{g \in \mathbb{Z}^n} H(M, g) \mathbf{t}^g$$

is a formal power series in the variables $t_1, \ldots, t_n$ with coefficients in $\mathbb{Z}$, and is called the *Hilbert-Poincaré series* of M, or simply the *Hilbert series* of M.

The finiteness assumptions on R and M are reflected by

Proposition B.13. $H_M(\mathbf{t})$ *is a rational function of the form*

$$\frac{Q_M(\mathbf{t})}{\prod_{i=1}^m (1 - \mathbf{t}^{g_i})},$$

where $g_i = \deg x_i$ *for* $i = 1, \ldots, m$.

Proof. We prove the assertion by induction on m. If $m = 0$, then M is a finitely generated R_0-module, and hence has finite length, so that $M_g \neq 0$ for at most finitely many $g \in \mathbb{Z}^n$. This implies that $H_M(\mathbf{t})$ is a polynomial.

Now suppose that $m > 0$. Multiplication with x_m gives rise to an exact sequence of $\mathbb{Z}^n$-graded modules

$$0 \to K \longrightarrow M(-g_m) \xrightarrow{x_n} M \longrightarrow C \to 0.$$

Here $M(-g_m)$ is the twisted $\mathbb{Z}^n$-graded R-module with homogeneous components

$$M(-g_m)_h = M_{h-g_m}, \ h \in \mathbb{Z}^n.$$

Since K and C are finitely generated $R/(x_m)$-modules, and since $R/(x_m)$ is generated over R_0 by the images of the x_i, $i = 1, \ldots, m-1$, we may apply our induction hypothesis and deduce from the exact sequence that $(1 - t^{g_m})H_M(t)$ is a rational function with denominator $\prod_{i=1}^{m-1}(1 - t^{g_i})$. Hence the assertion follows. $\qquad\square$

Let us now assume that R is $\mathbb{Z}$-graded. Then B.13 has an immediate consequence for the Hilbert function.

Corollary B.14. *If R is $\mathbb{Z}$-graded, then for large n, $H(M,n)$ is a quasi-polynomial, i.e., there exist a positive integer g and polynomials P_j, $j = 0, \ldots, g-1$, such that for all $n \in \mathbb{Z}$ one has $H(M,n) = P_j(n)$ when $n \equiv j \mod g$. Indeed, we may choose g as the least common multiple of $g_1, \ldots, g_m$ where $g_i = \deg x_i$ for $i = 1, \ldots, m$.*

This corollary says in particular, that if R is homogeneous, then for large n, $H(M,n)$ is a polynomial. We denote this polynomial by $P_M(n)$; it is called the *Hilbert polynomial of M*.

Proof. Let $\xi \in \mathbb{C}$ be a primitive g-th root of unity. Then

$$\prod_{i=1}^{m}(1 - t^{g_i}) = \prod_{j=0}^{g-1}(1 - \xi^j t)^{h_j}$$

with $\sum_{j=0}^{g-1} h_j = \sum_{i=1}^{m} g_i$.

By partial fraction decomposition, there exist polynomials $Q_j(t) = \sum_{k=0}^{n_j} a_{jk} t^k$ such that

$$\frac{Q_M(t)}{\prod_{i=1}^{m}(1 - t^{g_i})} = \sum_{j=0}^{g-1} \frac{Q_j(t)}{(1 - \xi^j t)^{h_j}}.$$

Expanding $Q_j(t)/(1 - \xi^j t)^{h_j}$, we get a power series $\sum_{n\geq 0} b_{jn} \xi^{jn} t^n$ where for $n \geq n_j$, b_{jn} equals the polynomial

$$S_j(n) = \sum_{k=0}^{n_k} a_{jk}\binom{h_j + n - k - 1}{h_j - 1}.$$

Hence we conclude that

$$H(M,n) = \sum_{j=0}^{g-1} S_j(n)\xi^{jn}$$

for large n. This easily implies all assertions. $\qquad\square$

The Hilbert series of a module M encodes an important invariant of M, its dimension.

Proposition B.15. *The dimension of M is the pole order of $H_M(t)$ at 1.*

Proof. It is possible (see [BH93, Lemma 1.5.10]) to find a homogeneous system of parameters $y_1,\ldots,y_d$ of M with, say $\deg y_i = g_i > 0$ for $i = 1,\ldots,d$. Just as in B.13 one now proves (by induction) on d that

$$H_M(t) = \frac{Q_M(t)}{\prod_{i=1}^{d}(1 - t^{g_i})} \qquad (B.1)$$

where $Q_M(t)$ is a polynomial with $Q_M(1) \neq 0$.

So the pole order of $H_M(t)$ is the maximal number s such that $(1 - t)^s$ divides $\prod_{i=0}^{d}(1 - t^{g_i})$, and this number is obviously d. $\qquad\square$

Corollary B.16. *Suppose R is homogeneous, and $\dim M = d$. Then*

(a) $H_M(t) = Q_M(t)/(1 - t)^d$;
(b) *the Hilbert polynomial P_M of M is of degree $d - 1$ with leading coefficient*

$$Q_M(1)/(d - 1)!.$$

Proof. (a) We may assume that the group of units of R_0 is infinite. Otherwise we adjoin a variable Y of degree 0 to R, and consider the ring $R(Y)$ which is the localization of $R[Y]$ with respect to all polynomials whose coefficients contain a unit. Replacing M by $M(Y) = M \otimes R(Y)$ does neither change the dimension of M nor its Hilbert function, but now $R_0(Y)$ (which is the group of units of $R(Y)$) is certainly infinite. In this situation (cf. [BH93, Proposition 1.5.12]) M has a system of parameters consisting of elements of degree 1, so that $H_M(t) = Q(t)/(1 - t)^d$.

(b) If $Q_M(t) = \sum_{k=0}^{s} h_k t^k$, then for $n \geq s$ we have

$$H(M,n) = \sum_{k=0}^{s} h_k \binom{d+n-k-1}{d-1}.$$

Thus we see that for $n \geq s$, $H(M,n)$ is a polynomial in n of degree $d - 1$ with leading coefficient $(\sum_{k=0}^{s} h_k)/(d - 1)! = Q_M(1)/(d - 1)!$. $\qquad\square$

We set $e(M) = Q(1)$, and call it the *multiplicity* of M. Thus B.16(b) says that $e(M)/(d - 1)!$ is the leading coefficient of the Hilbert polynomial of M.

The proof of B.15 shows us somewhat more: In case M is Cohen-Macaulay, the system of parameters $y_1,\ldots,y_d$ is M-regular, and this implies that $Q_M(t)$ is the Hilbert series of $M/(y_1,\ldots,y_d)M$. It is clear that any Hilbert series has non-negative coefficients, and that the Hilbert series of a homogeneous algebra actually has positive coefficients. Thus we obtain

Proposition B.17. *Suppose M is a Cohen-Macaulay module. Then $Q_M(t)$ is a polynomial with non-negative coefficients. Moreover, if R is homogeneous and Cohen-Macaulay of dimension d, then*

$$H_R(t) = \frac{Q_R(t)}{(1 - t)^d},$$

and all coefficients of $Q_R(t)$ are positive.

The Hilbert-Samuel Function

Let R be a Noetherian ring, I an ideal of R such that $\ell(R/I) < \infty$, and let M be a finitely generated R-module. Then the graded components $I^n M/I^{n+1}M$ of the associated graded module $\mathrm{gr}_I(M)$ all have finite length. Therefore, B.16 implies that for large n, the function $\lambda(n) = \ell(I^n M/I^{n+1}M)$ is a polynomial of degree $\dim \mathrm{gr}_I(M) - 1$.

Definition B.18. *The function* $\chi_M^I(n) = \ell(M/I^{n+1}M)$ *is called the* Hilbert-Samuel *function of M (with respect to I).*

We call $e(I, M) = e(\mathrm{gr}_I(M))$ the *multiplicity of M (with respect to I).*

Since $\chi_M^I(n) = \sum_{r=0}^n \lambda(r)$, the following result is an immediate consequence of B.16(b), and of the fact that $\dim M = \dim \mathrm{gr}_I(M)$ if $(R, \mathfrak{m})$ is local; see [BH93, Theorem 4.4.6].

Proposition B.19. *Let $(R, \mathfrak{m})$ be a Noetherian local ring, and $M \neq 0$ a finitely generated R-module of dimension d. Then*

(a) *for large n, the Hilbert-Samuel function $\chi_M^I(n)$ is a polynomial of degree d;*
(b) $e(I, M) = \lim_{n \to \infty}(d\,!/n^d)\ell(M/I^{n+1}M)$.

Suppose we are given a field k, and an ideal $I \subset (x_1, \ldots, x_m)$ in the polynomial ring $R = k[x_1, \ldots, x_m]$, and we want to compute the Hilbert-Samuel function of R/I with respect to the (maximal) ideal $\mathfrak{m}$ which is generated by the residues of the x_i in R/I. For an ideal generated by monomials such a computation is much easier than for a general ideal. We may achieve this more pleasant situation as follows: Given a term order $<$ on R, we define a filtration F' associated with the order $<$, but opposite to the degree filtration, by setting $F_g'R = \bigoplus_{h \geq -g} kx^h$. Then the initial form of a polynomial $f \in R$ with respect to this filtration is the *smallest* (with respect to $<$) monomial occurring in f. To distinguish it from the usual initial form we denote it by $in'(f)$. Furthermore, for any ideal I we let $in'(I)$ be the ideal in R generated by all elements $in'(f)$.

Proposition B.20. *Suppose the term order $<$ on R is degree compatible, that is, for all monomials u and v in R one has $u < v$ if $\deg u < \deg v$. Let $\mathfrak{n}$ be the maximal ideal of $R/in'(I)$ which is generated by the images of the x_i in $R/in'(I)$. Then*

$$\chi_{R/I}^{\mathfrak{m}}(n) = \chi_{R/in'(I)}^{\mathfrak{n}}(n) \quad \text{for all} \quad n.$$

In particular,

$$\dim(R/I)_{\mathfrak{m}} = \dim(R/in'(I))_{\mathfrak{n}} = \dim(R/in'(I)),$$
$$e(\mathfrak{m}, R/I) = e(\mathfrak{n}, R/in'(I)).$$

The reader may wonder why we consider $in'(I)$ instead of $in(I)$ in B.20. The simplest possible example shows that this is unavoidable: Let $R = k[x_1]$ and $I = (x_1 + x_1^2)$; then $in(I) = (x_1^2)$, so that $1 = e(\mathfrak{m}, R/I) < e(\mathfrak{n}, R/in(I)) = 2$.

Also our hypothesis on the term order is needed. Because if we choose for example the lexicographical term order on $R = k[x_1, x_2]$, and let $I = (x_1^2 + x_2^3)$, then $in'(I) = (x_2^3)$ and $2 = e(\mathfrak{m}, R/I) < e(\mathfrak{n}, R/in(I)) = 3$.

In the corollary below we will see however that for graded ideals we may take the usual initial forms and that there will be no restrictions on the term order.

Proof. Let x^{g_n} be the smallest monomial of degree at least $n+1$. Then, by our assumption on the order $<$,

$$(x_1, \ldots, x_m)^{n+1} = F'_{-g_n} R.$$

Denoting the induced filtration on $\bar{R} = R/I$ again by F', we see that $\mathfrak{m}^{n+1} = F'_{-g_n} \bar{R}$. By B.3 we have $R/in'(I) \simeq \mathrm{gr}_{F'}(\bar{R})$. Hence

$$\chi^{\mathfrak{n}}_{\bar{R}/in'(I)}(n) = \mathrm{length}(\bar{R}/F'_{-g_n}\bar{R}) = \mathrm{length}(\bar{R}/\mathfrak{m}^{n+1}) = \chi^{\mathfrak{m}}_{R/I}(n).$$

The equality

$$\dim(R/in'(I))_{\mathfrak{n}} = \dim R/in'(I)$$

follows from [BH93, Exercise 1.5.25]. $\square$

Corollary B.21. *Let I be a graded ideal of $R = k[x_1, \ldots, x_m]$, and let $in(I)$ denote the ideal of initial forms with respect to an arbitrary term order $<$. Then*

$$H(R/I, n) = H(R/in(I), n) \quad \text{for all} \quad n \geq 0,$$

and in particular

$$\dim R/I = \dim R/in(I) \quad \text{and} \quad e(R/I) = e(R/in(I)).$$

Proof. We define a new term order $\prec$ by the following rule: $u \prec v$ if either $\deg u < \deg v$, or $\deg u = \deg v$ and $u > v$. Noting that the ideal of initial forms of I is generated by the initial forms of the homogeneous elements of I, we see that $in'_\prec(I) = in_<(I)$. Thus we are in the situation of B.20, and the proof follows from the same arguments as in that case. $\square$

B.4 Hilbert Functions, Resolutions and Local Cohomology

Let k be a field, $S = k[x_1, \ldots, x_m]$ a polynomial ring over k, $R = S/I$ a homogeneous k-algebra, and M a finitely generated graded R-module. Then M, as an S-module, admits a finite graded free resolution

$$0 \to \bigoplus_j S(-j)^{b_{pj}} \longrightarrow \cdots \longrightarrow \bigoplus_j S(-j)^{b_{0j}} \longrightarrow M \to 0.$$

Since the Hilbert series is additive on short exact sequences, and since

$$H_{S(-j)}(t) = \frac{t^j}{(1-t)^m},$$

we deduce from this resolution that

$$H_M(t) = \frac{S_M(t)}{(1-t)^m} \quad \text{with} \quad S_M(t) = \sum_{i=0}^{p}(-1)^i \sum_j b_{ij} t^j.$$

This argument provides a new proof of the fact that the Hilbert series is a rational function, and would equally well work for any positively graded k-algebra.

For simplicity let us assume that all generators of M have non-negative degrees. Then $S_M(t)$ is a polynomial which is determined by the shifts in the resolution. But since alternating sums are taken, terms in $S_M(t)$ which correspond to different shifts may cancel. Thus the numerical data of the graded S-resolution determine the Hilbert function of M, but not conversely.

Definition B.22. *The degree of the rational function $H_M(t)$ is the a-invariant of M and is denoted $a(M)$: Thus*

$$a(M) = \deg S_M(t) - m = \deg Q_M(t) - d$$

where $d = \dim M$.

The significance of the a-invariant becomes apparent in

Proposition B.23. *One has $H(M, a(M)) \neq P_M(a(M))$, and $H(M,n) = P_M(n)$ for all $n > a(M)$.*

Proof. By B.16 we have $H_M(t) = (\sum_{i=0}^{s} h_i t^i)/(1-t)^d$, $d = \dim M$, and $a(M) = s - d$. Therefore,

$$H(M,n) = \sum_{i=1}^{s} \binom{n-i+d-1}{d-1} h_i \quad \text{for} \quad n \geq s-d+1.$$

Hence for these n, $H(M,n)$ is a polynomial function, and thus coincides with $P_M(n)$, while

$$P_M(s-d) - H(M,s-d) = (-1)^{d-1} h_s \neq 0.$$

$\square$

The difference $H(M,n) - P_M(n)$ can be measured in terms of local cohomology. We denote by $\mathfrak{m}$ the graded maximal ideal of R. Note that the local cohomology modules $H_{\mathfrak{m}}^i(M)$ are naturally graded. We quote the following result from [BH93, Theorem 4.3.5(b)]:

Proposition B.24. $H(M,n) - P_M(n) = \sum_{i \geq 0}(-1)^i \dim_k H_{\mathfrak{m}}^i(M)_n$ *for all $n \in \mathbb{Z}$.*

Observe that the alternating sum in B.24 is finite since $H_{\mathfrak{m}}^i(M) = 0$ for $i > \dim M$. Furthermore, since the local cohomology modules are Artinian there exists an integer c such that $H_{\mathfrak{m}}^i(M)_n = 0$ for all i and all $n \geq c$. It is clear that $c > a(M)$.

Remark B.25. Note that if M is Cohen–Macaulay of dimension d, then

$$a(M) = \sup\{n \mid H_{\mathfrak{m}}^d(M)_n \neq 0.\}$$

A major measure for the complexity of the resolution of M is the following.

Definition B.26. *The* Castelnuovo-Mumford regularity *of M is the integer*

$$\operatorname{reg} M = \max\{j - i \colon b_{ij} \neq 0\}.$$

In other words, $\operatorname{reg} M = \max\{s(\operatorname{Tor}_i^S(M,k)) - i \colon i \in \mathbb{Z}\}$ where for a graded module N with $N_j = 0$ for large j, we set $s(N) = \max\{j \colon N_j \neq 0\}$. Let q be an integer. The module M is called *q-regular* if $q \geq \operatorname{reg}(M)$, equivalently, if $\operatorname{Tor}_i^S(M,k)_{j+i} = 0$ for all i and all $j > q$.

Eisenbud and Goto [EiG84] gave an interesting interpretation of regularity. Denoting by $M_{\geq q}$ the truncated graded R-module $\bigoplus_{j \geq q} M_j$, one has:

Theorem B.27. *The following conditions are equivalent:*

(a) *M is q-regular;*
(b) *$H_{\mathfrak{m}}^i(M)_{j-i} = 0$ for all i and all $j > q$;*
(c) *$M_{\geq q}$ admits a linear S-resolution, i.e., a graded resolution of the form*

$$0 \to S(-q-l)^{c_l} \longrightarrow \cdots \longrightarrow S(-q-1)^{c_1} \longrightarrow S(-q)^{c_0} \longrightarrow M_{\geq q} \to 0.$$

Proof. (a) $\Leftrightarrow$ (c): By definition, the module $M_{\geq q}$ has a linear resolution if and only if for all i

$$\operatorname{Tor}_i^S(M_{\geq q}, k)_r = H_i(\mathbf{x}; M_{\geq q})_r = 0$$

for $r \neq i + q$. Here $H(\mathbf{x}; M)$ denotes the Koszul homology of M with respect to the sequence $\mathbf{x} = x_1, \ldots, x_m$.

Since $(M_{\geq q})_j = 0$ for $j < q$, we always have $H_i(\mathbf{x}; M_{\geq q})_r = 0$ for $r < i + q$, while for $r > i + q$

$$H_i(\mathbf{x}; M_{\geq q})_r = H_i(\mathbf{x}; M)_r = \operatorname{Tor}_i^S(M, k)_r.$$

Thus the desired result follows.

(b) $\Rightarrow$ (c): We may assume $q = 0$, and $M = M_{\geq 0}$. Then it is immediate that $H_{\mathfrak{m}}^0(M)$ is concentrated in degree 0. This implies that $M = H_{\mathfrak{m}}^0(M) \oplus M/H_{\mathfrak{m}}^0(M)$. The first summand is a direct summand of copies of k. Hence M is 0-regular if and only if $M/H_{\mathfrak{m}}^0(M)$ is 0-regular. In other words we may assume that depth $M > 0$. Without loss of generality we may further assume that k is infinite. Then there exists an element $y \in S$ of degree 1 which is M-regular. From the cohomology exact sequence associated with

$$0 \to M(-1) \xrightarrow{\ y\ } M \longrightarrow M/yM \to 0$$

we see that M/yM is 0-regular. By induction on the dimension on M, we may suppose that M/yM has linear S/yS-resolution. But if $\mathbb{F}$ is a minimal graded free S-resolution, then $\mathbb{F}/y\mathbb{F}$ is a minimal graded S/yS-resolution of M/yM. This implies that $\mathbb{F}$ is a linear S-resolution of M.

(c) $\Rightarrow$ (b): Again we may assume $q = 0$, and $M = M_{\geq 0}$. Then M has a linear resolution

$$\cdots \longrightarrow S(-2)^{c_2} \longrightarrow S(-1)^{c_1} \longrightarrow S^{c_0} \longrightarrow M \to 0.$$

Computing $\mathrm{Ext}^i_S(M,S)$ with this resolution we see at once that $\mathrm{Ext}^i_S(M,S)_j = 0$ for $j < -i$. By duality (see [BH93, Section 3.6]) there exists an isomorphism of graded R-modules

$$H^i_{\mathfrak{m}}(M) \simeq \mathrm{Hom}_k(\mathrm{Ext}^{m-i}_S(M,S(-m)),k).$$

Therefore, $H^i_{\mathfrak{m}}(M)_{j-i} = 0$ for all $j > 0$, as desired. $\qquad\square$

Corollary B.28. $a(M) \leq \mathrm{reg}\, M - \mathrm{depth}\, M$, and equality holds if M is Cohen-Macaulay.

Proof. Let $q = \mathrm{reg}\, M$, $t = \mathrm{depth}\, M$, and $n > q - t$; then $n > q - i$ for all $i \geq t$. So that, by B.27(b), $H^i_{\mathfrak{m}}(M)_n = 0$ for all $i \geq t$. Since $H^i_{\mathfrak{m}}(M) = 0$ for $i < t$ anyhow, we deduce from B.24 that $H(M,n) = P_M(n)$. Thus B.23 concludes the proof of the inequality.

Suppose now that M is Cohen-Macaulay. We claim that $s(\mathrm{Tor}^S_i(M,k)) - i$ is a non-decreasing function of i. But then, if $p = \mathrm{proj\, dim}\, M$, we get

$$\mathrm{reg}\, M = s(\mathrm{Tor}^S_p(M,k)) - p = \deg S_M(\mathbf{t}) - (m - t) = a(M) + t$$

since, by Auslander-Buchsbaum's equality (Theorem A.83), $p = m - t$.

Suppose the claim is false, and let F be the graded minimal free resolution of M. Then there exists an integer i, and a homogeneous generator of F_i whose degree is greater than or equal to the degrees of all generators of F_{i+1}. This implies that the matrix describing the map $F_{i+1} \to F_i$ has a zero column. Thus in the S-dual F^* of F there appears a matrix with zero row. However since M is Cohen-Macaulay, F^* is a minimal free resolution of $\mathrm{Ext}^p_S(M,S)$, a contradiction. $\qquad\square$

B.5 Lexsegment Ideals and Macaulay Theorem

In this section we discuss Macaulay's theorem which characterizes the numerical functions which are the Hilbert function of a homogeneous k-algebra R. A detailed proof can be found in [BH93, Section 4.2].

Let R, S and I be as in the previous section. We fix the following term order $<$ on S: $x^a < x^b$ if the last non-zero component of $(b_1 - a_1, \ldots, b_m - a_m, \Sigma b_i - \Sigma a_i)$ is positive. This order is the deglex order with $x_m > x_{m-1} > \cdots > x_1$.

In Section B.3 we have seen that $H(R,n) = H(S/\mathrm{in}(I),n)$ for all n, so that in particular any possible Hilbert function is realized as the Hilbert function of a ring defined by monomials. Actually the monomials can be chosen to generate a lexsegment ideal; see the proof of B.31. A homogeneous ideal $I \subset S$ is called a *lexsegment ideal* if for all homogeneous components $I_n \neq 0$ of I there exists a monomial u_n such that I_n is the k-vector space spanned by the monomials $v \geq u_n$.

We shall have to consider the following numerical function $\mathbb{N} \to \mathbb{N}$, $a \mapsto a^{\langle n \rangle}$, which for a given $n \in \mathbb{N}$ is defined as follows: First we expand a into its *n-th Macaulay representation*:

$$a = \binom{k(n)}{n} + \binom{k(n-1)}{n-1} + \cdots + \binom{k(1)}{1} \tag{B.2}$$

where $k(n) > k(n-1) > \cdots > k(1) \geq 0$. Such an expansion is unique, and the coefficients $k(i)$, $i = 1, \ldots, n$, are called the *n-th Macaulay coefficients* of a.

Now we define

$$a^{\langle n \rangle} = \binom{k(n)+1}{n+1} + \binom{k(n-1)+1}{n} + \cdots + \binom{k(1)+1}{2}. \tag{B.3}$$

The Macaulay expansion of an integer a corresponds to the natural decomposition of the set of the first a monomials of degree n. Indeed, this set can be written as the union

$$L_u = \bigcup_{i=1}^{n} [x_1, \ldots, x_{j(i)-1}]_i x_{j(i+1)} \cdots x_{j(n)},$$

where $u = x_{j(1)} x_{j(2)} \cdots x_{j(n)}$, $j(1) \leq j(2) \leq \ldots \leq j(n)$, is the $(a+1)$-th monomial of degree n in $k[x_1, x_2, \ldots]$. Here we denote by $[x_1, \ldots, x_l]_i$ the set of all monomials of degree i in the variables $x_1, \ldots, x_l$. Thus each set

$$[x_1, \ldots, x_{j(i)-1}]_i x_{j(i+1)} \cdots x_{j(n)}$$

in the union of L_u consists of $\binom{k(i)}{i}$ elements where $k(i) = j(i) + i - 2$. Summing up the single terms we get the Macaulay expansion of a.

Let us denote by R_u the set of monomials of degree n with $v \geq u$. Then if the ideal I is spanned in degree n by the elements of R_u, it follows that $H(R, n) = |L_u|$. One easily checks that $\{x_1, \ldots, x_m\} R_u = R_{x_1 u}$, and that $|L_{x_1 u}| = |L_u|^{\langle n \rangle}$.

Thus as a consequence of this discussion we obtain:

Proposition B.29. *Let I be a lexsegment ideal, then for $R = S/I$ we have:*

(a) $H(R, n+1) \leq H(R, n)^{\langle n \rangle}$ *for all n;*
(b) $H(R, n+1) = H(R, n)^{\langle n \rangle}$ *for a given n if and only if $I_{n+1} = (x_1, \ldots, x_m) I_n$.*

Surprisingly the estimate B.29(a) for the growth of the Hilbert function is not only valid when R is defined by a lexsegment ideal but quite general, and actually characterizes the numerical functions which are Hilbert functions.

Theorem B.30 (Macaulay). *Let $h: \mathbb{N} \to \mathbb{N}$ be a numerical function. The following conditions are equivalent:*

(a) *there exists a homogeneous k-algebra R with $H(R, n) = h(n)$ for all $n \geq 0$;*
(b) $h(0) = 1$, *and $h(n+1) \leq h(n)^{\langle n \rangle}$ for all $n \geq 1$.*

Corollary B.31. *Let $P_R(t)$ be the Hilbert polynomial of the homogeneous k-algebra $R = S/I$. Then there exist uniquely determined integers $a_1 \geq a_2 \geq \cdots \geq a_s \geq 0$ such that*

$$P_R(n) = \binom{n+a_1}{a_1} + \binom{n+a_2-1}{a_2} + \cdots + \binom{n+a_s-(s-1)}{a_s} \quad \text{for all} \quad n \in \mathbb{Z}.$$

In the next section we will give an interpretation of the number s appearing in the formula for $P_R(t)$ in this formula.

Proof. For each n with $I_n \neq 0$ there exists a monomial u_n with $|R_{u_n}| = \dim_k I_n$. It follows from the discussions preceding B.29 and from Macaulay's theorem that $\{x_1, \ldots, x_m\} R_{u_n} \subset R_{u_{n+1}}$ for all n in question. Therefore if we let J be the lexsegment ideal spanned by the R_{u_n} we get $\dim_k J_n = \dim_k I_n$, and hence $H(R,n) = H(S/J,n)$ for all n. Thus we may assume that I itself is a lexsegment ideal.

Since I is finitely generated there exists an integer n_0 such that

$$I_{n+1} = (x_1, \ldots, x_m) I_n \quad \text{for all} \quad n \geq n_0.$$

Applying B.29 we see that

$$H(R, n+1) = H(R,n)^{\langle n \rangle} \quad \text{for all} \quad n \geq n_0. \tag{B.4}$$

Let

$$\binom{k(n_0)}{n_0} + \binom{k(n_0-1)}{n_0-1} + \cdots + \binom{k(j)}{j}$$

the n_0-th Macaulay expansion of $H(R,n_0)$ where we skipped the terms which are zero (so that $k(n_0) > k(n_0-1) > \cdots > k(j) \geq j$). Then it follows from (B.4) that

$$H(R,n) = \binom{n-n_0+k(n_0)}{n} + \binom{n-n_0+k(n_0-1)}{n-1} + \cdots + \binom{n-n_0+k(j)}{n-n_0+j}$$

for all $n \geq n_0$. Hence if we set $a_i = k(n_0 - i + 1) - n_0 + i - 1$ for $i = 1, \ldots, s$ with $s = n_0 - j + 1$, we obtain the desired presentation of $P_R(n)$. The uniqueness of such a presentation follows from the fact that it can be converted in a Macaulay presentation and vice versa. $\square$

Remark B.32. Given s as in B.31, then $P_R(n)^{\langle n \rangle} = P_R(n+1)$ for all $n \geq s$.

Proof. For $n \geq s$ the n-th Macaulay expansion of $P_R(n)$ is given by

$$\binom{n+a_1}{n} + \binom{n+a_2-1}{n-1} + \cdots + \binom{n+a_s-(s-1)}{n-(s-1)}.$$

Therefore

$$\begin{aligned}
P_R(n)^{\langle n \rangle} &= \binom{n+1+a_1}{n+1} + \binom{n+a_2}{n} + \cdots + \binom{n+1+a_s-(s-1)}{n+1-(s-1)} \\
&= \binom{n+1+a_1}{a_1} + \binom{n+a_2}{a_2} + \cdots + \binom{n+1+a_s-(s-1)}{a_s} \\
&= P_R(n+1),
\end{aligned}$$

to prove the assertion. $\square$

B.6 The Theorems of Green and Gotzmann

Green's theorem describes how the Hilbert function behaves when one passes from R to R/hR where h is a general linear form. This result will be used in the proofs of Gotzmann's [Go78] regularity and persistence theorem which give deep insights in the nature of the Hilbert polynomial. The proofs which we will present are due to Green [Gre89].

As before let S be the polynomial ring in m variables defined over a field k, $I \subset S$ a graded ideal and $R = S/I$. Gotzmann's regularity theorem is a statement about regularity of the ideal sheaf I associated with I. Different ideals may yield the same ideal sheaf: the ideal $\bar{I} = \ker(S \to R/H_{\mathrm{m}}^0(R))$ is called the *saturation of I*. The ideal sheafs associated with two ideals coincide if and only if their saturations coincide.

Theorem B.33 (Gotzmann Regularity Theorem). *Write the Hilbert polynomial* $P_R(\mathbf{t})$ *of* $R = S/I$ *in the unique form*

$$P_R(\mathbf{t}) = \binom{t+a_1}{a_1} + \binom{t+a_2-1}{a_2} + \cdots + \binom{t+a_s-(s-1)}{a_s}$$

with $a_1 \geq a_2 \geq \cdots \geq a_s \geq 0$, *as described in B.31. Then the saturation* $\bar{I}$ *of* I *is s-regular.*

In the course of the proof we will use the following theorem of Green whose proof can be found in [Gre89] or [BH93, Section 4.2].

Theorem B.34 (Green). *Let* R *be a homogeneous k-algebra, k an infinite field, and* $h \in R$ *a general linear form. Then*

$$H(R/hR, n) \leq H(R, n)_{\langle n \rangle}$$

for $n \geq 1$.

We need to explain the function $a \mapsto a_{\langle n \rangle}$ which appears in B.34, and which is defined for any positive integer a. If a has n-th Macaulay coefficients $k(n), \ldots, k(1)$, then we set

$$a_{\langle n \rangle} = \binom{k(n)-1}{n} + \binom{k(n-1)-1}{n-1} + \cdots + \binom{k(1)-1}{1}.$$

Proof. [Proof of B.33] We prove the theorem by induction on the dimension of S. The arguments will provide a new proof of B.31. For $m = 1$ the assertion is trivial. Now let $m > 1$, and let h be a general linear form. Since $P_R(\mathbf{t}) = P_{S/\bar{I}}(\mathbf{t})$ we may assume that $I = \bar{I}$. We may further assume that $\bar{I} \neq S$. Then depth $R > 0$, and h is R-regular. Hence we get an exact sequence

$$0 \to R(-1) \xrightarrow{h} R \longrightarrow R/hR \to 0$$

yielding the equation

$$P_{R/hR}(n) = P_R(n) - P_R(n-1). \tag{B.5}$$

Set $\tilde{R} = R/hR$, $\tilde{S} = S/hS$; then $\tilde{R} = \tilde{S}/\tilde{I}$ for some ideal $\tilde{I} \subset \tilde{S}$. Let J be the saturation of $\tilde{I}$. Then $P_{\tilde{R}}(n) = P_{\tilde{S}/J}(n)$, and by our induction hypothesis,

$$P_{\tilde{S}/J}(n) = \binom{n+b_1}{b_1} + \binom{n+b_2-1}{b_2} + \cdots + \binom{n+b_r-(r-1)}{b_r}, \tag{B.6}$$

and J is r-regular.

(B.5) and (B.6) imply that

$$P_R(n) = \binom{n+a_1}{a_1} + \binom{n+a_2-1}{a_2} + \cdots + \binom{n+a_r-(r-1)}{a_r} + c,$$

where c is a constant, and where $a_i = b_i + 1$ for all i.

We claim that $c \geq 0$, and that $\tilde{I}$ is s-regular for $s = r + c$. These two claims complete the proof. Indeed, we may set $a_{r+1} = \cdots = a_{r+c} = 0$.

In order to prove the first claim, suppose that $c < 0$. Then for $n \gg 0$ we have

$$H(R,n) < \binom{n+a_1}{a_1} + \binom{n+a_2-1}{a_2} + \cdots + \binom{n+a_r-(r-1)}{a_r}. \tag{B.7}$$

Set b for the right hand side of this inequality. Then

$$b = \binom{n+a_1}{n} + \binom{n+a_2-1}{n-1} + \cdots + \binom{n+a_r-(r-1)}{n-(r-1)},$$

so that

$$\begin{aligned} b_{\langle n \rangle} &= \binom{n+b_1}{n} + \binom{n+b_2-1}{n-1} + \cdots + \binom{n+b_r-(r-1)}{n-(r-1)} \\ &= \binom{n+b_1}{b_1} + \binom{n+b_2-1}{b_2} + \cdots + \binom{n+b_r-(r-1)}{b_r}. \end{aligned}$$

Observing that $n + a_r - (r-1) > n - (r-1)$ one deduces from (B.7) (see [BH93, 4.2.1]) that $H(R,n)_{\langle n \rangle} < b_{\langle n \rangle}$. Therefore by Green's theorem,

$$H(\tilde{R},n) < \binom{n+b_1}{b_1} + \binom{n+b_2-1}{b_2} + \cdots + \binom{n+b_r-(r-1)}{b_r}.$$

This contradicts (B.6).

For the proof of the second claim note first that

$$H_{\mathfrak{m}}^i(J) = H_{\mathfrak{m}}^i(\tilde{I}) \text{ for } i > 1.$$

Therefore, since J is r-regular we deduce from the local cohomology sequence associated with

$$0 \to I(-1) \xrightarrow{h} I \longrightarrow \tilde{I} \to 0$$

that $H^i_{\mathfrak{m}}(I)_{j-i} = 0$ for all $i > 2$ and $j > r$ (and thus for $j > s$).

It remains to be shown that $H^2_{\mathfrak{m}}(I)_{j-2} = 0$ for $j > s$. Suppose this is not the case, and let j be the largest such number. Then by B.24,

$$H(R, j-2) - P_R(j-2) = -H^1_{\mathfrak{m}}(R)_{j-2} < 0$$

since $H^0_{\mathfrak{m}}(R) = 0$ and $H^{i-1}_{\mathfrak{m}}(R)_{j-2} = H^i_{\mathfrak{m}}(I)_{j-2} = 0$ for $i > 2$, as we have already seen. By our choice of j we have $H^1_{\mathfrak{m}}(R)_{j-1} = 0$, so that

$$H(R, j-2) < P_R(j-2) \quad \text{but} \quad H(R, j-1) = P_R(j-1).$$

If $j = s+1$, then $j-2 = s-1$, and

$$P_R(s-1) = \binom{s-1+a_1}{s-1} + \cdots + \binom{1+a_{s-1}}{1} + 1,$$

so that

$$H(R, j-2) = H(R, s-1) \leq \binom{s-1+a_1}{s-1} + \cdots + \binom{1+a_{s-1}}{1}.$$

Hence Macaulay's theorem implies that

$$H(R, j-1) \leq H(R, j-2)^{\langle j-2 \rangle} \leq \binom{s+a_1}{s} + \cdots + \binom{2+a_{s-1}}{2}$$
$$= P_R(s) - (a_s + 1) < P_R(s) = P_R(j-1)$$

which is a contradiction.

If $j > s+1$, then $P_R(j-2)^{\langle j-2 \rangle} = P_R(j-1)$ (see B.32). We apply again Macaulay's theorem, and get

$$H(R, j-1) \leq H(R, j-2)^{\langle j-2 \rangle} < P_R(j-1)$$

which leads to the same contradiction. $\square$

In the proof of B.31 we have seen that $H(R, n+1) = H(R, n)^{\langle n \rangle}$ for all large n. But may it happen that $H(R, n+1) = H(R, n)^{\langle n \rangle}$, and $H(R, r+1) < H(R, r)^{\langle r \rangle}$ for some r and n with $r > n$? The following result gives an answer to this question.

Theorem B.35 (Gotzmann Persistence Theorem). *Suppose that for some n*

$$H(R, n+1) = H(R, n)^{\langle n \rangle},$$

and I is generated by elements of degree $\leq n$. Then $H(R, r+1) = H(R, r)^{\langle r \rangle}$ for all $r \geq n$.

Proof. We prove the theorem by induction on $m = \dim S$. If $m = 1$, I is principal, and hence the assertion is trivial. Now let us assume that $m > 1$. Let

$$H(R,n) = \binom{k(n)}{n} + \cdots + \binom{k(1)}{1}$$

be the n-th Macaulay expansion of $H(R,n)$. Macaulay's theorem implies that

$$H(R,r) \le \binom{r-n+k(n)}{r} + \cdots + \binom{r-n+k(1)}{r-n+1} \tag{B.8}$$

for all $r \ge n$, and it amounts to show that equality holds.

Let h be a general linear form. Then we get

$$(H(R,n)_{\langle n\rangle})^{\langle n\rangle} \ge H(R/hR,n)^{\langle n\rangle} \ge H(R/hR,n+1) \tag{B.9}$$
$$\ge H(R,n+1) - H(R,n) = ((H(R,n)_{\langle n\rangle})^{\langle n\rangle}.$$

The first inequality is Green's theorem, the second is Macaulay's, the third follows from the exact sequence

$$R(-1) \xrightarrow{h} R \longrightarrow R/hR \to 0,$$

and the last equality results from our assumption that $H(R,n+1) = H(R,n)^{\langle n\rangle}$.

Since the first and last term in this chain of inequalities coincide we must have equality everywhere. In particular, we have $H(R/hR,n+1) = H(R/hR,n)^{\langle n\rangle}$. Since the defining ideal of R/hR is again generated by elements of degree $\le n$, we may apply our induction hypothesis, and hence get that $H(R/hR,r+1) = H(R/hR,r)^{\langle r\rangle}$ for all $r \ge n$.

We also deduce from (B.9) that

$$H(R/hR,n+1) = ((H(R,n)_{\langle n\rangle})^{\langle n\rangle} \tag{B.10}$$
$$= \binom{k(n)}{n+1} + \binom{k(n-1)}{n-1} + \cdots + \binom{k(1)}{2}.$$

Therefore

$$P_{R/hR}(r) = \binom{r+(k(n)-n-1)}{k(n)-n-1} + \cdots + \binom{r+(k(1)-2)-(n-1)}{k(1)-2}$$

for all r. Hence if J is the saturation of the defining ideal of R/hR, then J is n-regular by Gotzmann's regularity theorem. We denote by $\bar{I}$ the saturation of I and set $\bar{R} = S/\bar{I}$. It follows just as in the proof of the regularity theorem that

$$P_{\bar{R}}(r) = \binom{r+(k(n)-n)}{k(n)-n} + \cdots + \binom{r+(k(1)-1)-(n-1)}{k(1)-1} + c \tag{B.11}$$

with $c \ge 0$.

Suppose $c > 0$; then, since $P_R(n) = P_{\bar{R}}(n)$, (B.8) implies

$$P_R(r) = \binom{r-n+k(n)}{r} + \cdots + \binom{r-n+k(1)}{r-n+1} + c$$
$$\geq H(R,r) + c > H(R,r)$$

for all $r \geq n$. This is a contradiction.

Now (B.11) and Gotzmann's regularity theorem imply that $\bar{I}$ is n-regular, so that $H(\overline{R},r) = P_{\overline{R}}(r)$ for all $r \geq n$.

Thus for all $r \geq n$ we obtain the following string of inequalities

$$H(\overline{R},r) \leq H(R,r) \leq P_R(r) = P_{\overline{R}}(r) = H(\overline{R},r).$$

Hence we must have equality everywhere, and this proves the theorem. $\square$

C

Using *Macaulay 2*

By **David Eisenbud, Daniel R. Grayson, and
Michael E. Stillman**

Macaulay 2 is a computer software package devoted to supporting research in algebraic geometry and commutative algebra. It was written in the years 1993-1997 by Daniel R. Grayson and Michael E. Stillman with funding from the National Science Foundation and includes subroutine packages for factoring contributed by Gert-Martin Greuel, Ruediger Stobbe and Michael Messollen. In this appendix we will illustrate some of the concepts presented in this book through computations accomplished with *Macaulay 2*.

Copies of *Macaulay 2* for use in educational institutions may be obtained from the web site at http://www.math.uiuc.edu/Macaulay2. The source code is available there, as well as versions already compiled for various types of computers and operating systems, including: Linux, Windows 95, or NeXTstep on a PC; Hewlett Packard workstations with HP-UX; Sun workstations with Solaris or SunOS; IRIS workstations with IRIX; and Silicon Graphics workstations. The version for the Macintosh PowerPC should be available soon.

The current version of *Macaulay 2* at the time of writing is 0.8.17, indicating that the design of the language is not in its final state; up-to-date versions of this appendix will be included in distributions of *Macaulay 2*.

1. Elementary uses of *Macaulay 2*
    a. First steps
    b. A monomial curve example
    c. Random regular sequences
    d. Division with remainder
    e. Elimination theory
    f. Quotients and saturation
2. Local cohomology of graded modules

    a. Mathematical background
    b. *Macaulay 2* routines
    c. Example: the rational quartic curve in $\mathbb{P}^3$
  3. Zariski cohomology of coherent sheaves
    a. Mathematical background
    b. *Macaulay 2* routines
    c. Example: the Hodge diamond of a K3 surface

C.1 Elementary Uses of *Macaulay 2*

In this section we introduce a number of basic operations using Gröbner bases, and at the same time become familiar with a range of useful *Macaulay 2* constructs.

First Steps

Start *Macaulay 2* with the command M2, and you will be presented with "i1 :" as input prompt. An expression entered at the keyboard will be evaluated – no punctuation is required at the end of the line.

```
i1 : 2+2

o1 = 4
```

The answer, 4, is displayed after the output label "o1 =". Multiplication is indicated with the traditional *.

```
i2 : 1*2*3*4

o2 = 24
```

Powers are obtained as follows.

```
i3 : 2^100

o3 = 1267650600228229401496703205376
```

Because some answers can be very long, it is a good idea to run the program in a window which does not wrap output lines, and allows the user to scroll horizontally to see the rest of the output.

```
i4 : 100!

o4 = 93326215443944152681699238856266700490715968264381621468592963 ...
```

Here is some arithmetic with fractions.

```
i5 : 3/5 + 7/11

     68
o5 = --
     55
```

```
o5 : QQ
```

Notice how the system displays the type of thing the output value is on an additional output line: QQ. (We reserve single letter symbols such as Q for use as variables in rings. Hence we must use QQ to stand for the field of rational numbers; it may remind you of the "blackboard bold" font of AMSTeX.

It is possible to obtain integer quotients and remainders with the following operators.

```
i6 : 1234//100

o6 = 12

i7 : 1234%100

o7 = 34
```

The operator for equality testing is a double equal sign.

```
i8 : 2+2==4

o8 = true
```

Multiple expressions may be separated by semicolons.

```
i9 : 1;2;3*4

o11 = 12
```

A semicolon at the end of the line suppresses the printing of the value.

```
i12 : 4*5;
```

The output from the previous line can be obtained with oo, even if a semicolon prevented it from being printed.

```
i13 : oo

o13 = 20
```

Lines before that can be obtained with ooo and oooo. Alternatively, the symbol labeling an output line can be used to retrieve the value.

```
i14 : o11 + 1

o14 = 13
```

A list of expressions can be formed with braces.

```
i15 : {1, 2, s}

o15 = {1,2,s}

o15 : List
```

A function can be created with the arrow operator.

```
i16 : cube = i -> i^3

o16 = cube

o16 : Function
```

To evaluate a function, place its argument to the right of the function.

```
i17 : cube 5

o17 = 125
```

Functions of more than one variable take a parenthesized sequence of arguments.

```
i18 : mult = (x,y) -> x * y
```

```
o18 = mult

o18 : Function

i19 : mult(6,9)

o19 = 54
```

The function `apply` can be used to apply a function to each element of a list.

```
i20 : apply({1,2,3,4}, cube)

o20 = {1,8,27,64}

o20 : List
```

The function `scan` is analogous to `apply` except that no value is returned. It may be used to implement loops in programs.

```
i21 : scan({1,2,3,4}, i -> print (i, i^3))
1,1
2,8
3,27
4,64
```

Most computations with polynomials take place in rings that may be specified in usual mathematical notation.

```
i22 : R = ZZ/5[x,y,z]

o22 = R

o22 : PolynomialRing
```

Here ZZ refers to the ring of integers.

```
i23 : (x+y)^5

        5   5
o23 = x  + y

o23 : R
```

A free module can be created as follows.

```
i24 : F = R^3

        3
o24 = R

o24 : R - module, free
```

A list of indices can be used to produce the homomorphism corresponding to the corresponding basis vectors.

```
i25 : F_{1,2}

o25 = | 0 0 |
      | 1 0 |
      | 0 1 |

                 3       2
o25 : Matrix R  <--- R
```

We can create a homomorphism between free modules with `matrix` by providing a list of the rows of the matrix, each of which is in turn a list of ring elements.

```
i26 : ff = matrix {{x,y,z}}

o26 = | x y z |
```

```
o26 : Matrix R  <--- R
```
Use image to get the image.
```
i27 : image ff

o27 = image | x y z |
```
```
o27 : R - module, submodule of R
```
The corresponding ideal can be obtained with `ideal`.
```
i28 : ideal (x,y,z)

o28 = ideal | x y z |

o28 : Ideal of R
```
We may use `kernel` to compute the kernel of `ff`.
```
i29 : kernel ff

o29 = image | 0  -y -z |
            | -z  x  0 |
            | y   0  x |
```
```
                                     3
o29 : R - module, submodule of R
```
The answer comes out as a module which is expressed as the image of a homomorphism whose matrix is displayed. In case the matrix itself is desired, it can be obtained with `generators`.
```
i30 : generators oo

o30 = | 0  -y -z |
      | -z  x  0 |
      | y   0  x |
```
```
              3        3
o30 : Matrix R  <--- R
```
We may use `rank` to compute the rank.
```
i31 : rank kernel ff

o31 = 2
```
A presentation for the kernel can be obtained with `presentation`.
```
i32 : presentation kernel ff

o32 = | x  |
      | z  |
      | -y |
```
```
              3        1
o32 : Matrix R  <--- R
```
We can produce the cokernel with `cokernel`.
```
i33 : cokernel ff

o33 = cokernel | x y z |
```
```
                                    1
o33 : R - module, quotient of R
```
The direct sum is formed as follows.
```
i34 : N = kernel ff ++ cokernel ff
```

```
o34 = subquotient(| 0  -y -z 0 |,| 0 0 0 |)
                  | -z  x  0 0 |  | 0 0 0 |
                  | y   0  x 0 |  | 0 0 0 |
                  | 0   0  0 1 |  | x y z |
```

```
                                 4
o34 : R - module, subquotient of R
```

The answer is expressed in terms of the subquotient function, which produces subquotient modules. Each subquotient module is accompanied by its matrix of generators and its matrix of relations. These matrices can be recovered with the functions generators and relations.

```
i35 : generators N
```

```
o35 = | 0  -y -z 0 |
      | -z  x  0 0 |
      | y   0  x 0 |
      | 0   0  0 1 |
```

```
            4        4
o35 : Matrix R  <--- R
```

```
i36 : relations N
```

```
o36 = | 0 0 0 |
      | 0 0 0 |
      | 0 0 0 |
      | x y z |
```

```
            4        3
o36 : Matrix R  <--- R
```

The function prune can be used to convert a subquotient module to a quotient module.

```
i37 : prune N
```

```
o37 = cokernel | 0 0 0 -y |
               | 0 0 0 z  |
               | 0 0 0 x  |
               | z y x 0  |
```

```
                              4
o37 : R - module, quotient of R
```

To compute the Gröbner basis of the ideal $(x^2, xy + y^2)$ we proceed as follows. We form the ideal:

```
i38 : I = ideal(x^2,x*y+y^2)
```

```
o38 = ideal | x2 xy+y2 |
```

```
o38 : Ideal of R
```

```
i39 : p = generators gb I
```

```
o39 = | xy+y2 x2 y3 |
```

```
            1        3
o39 : Matrix R  <--- R
```

From this result we can for example compute the codimension, dimension, degree, and the whole Hilbert function and polynomial.

As a parenthetical remark, we note that *Macaulay 2* tends to display a matrix in a compact horizontal format in which the exponents are written to the right of the corresponding variables. Hence 3xy2+y3z would represent $3xy^2 + y^3z$. Here is a way to see the matrix above in a more legible two-dimensional form.

```
i40 : expression p
```

$$o40 = \begin{vmatrix} & 2 & 2 & 3 \\ x\ y + y & x & y \end{vmatrix}$$

```
o40 : MatrixExpression
```

A Monomial Curve Example

This will be more fun if we work with an example having some meaning. We choose to work with the ideal defining the rational quartic curve in $\mathbb{P}^3$ given parametrically in an affine representation by

$$t \mapsto (t, t^3, t^4).$$

(The reader who doesn't understand this terminology may ignore it for the moment, and treat the ideal given below as a gift from the gods.) First we introduce our favorite field.

```
i41 : KK = ZZ/31991;
```

We obtain the ideal by first making the polynomial ring in 4 variables (the homogeneous coordinate ring of $\mathbb{P}^3$).

```
i42 : R = KK[a..d];
```

Then we use the *Macaulay 2* function monomialCurve, which we shall treat for now as a black box.

```
i43 : I = monomialCurve(R,{1,3,4})

o43 = ideal | bc-ad c3-bd2 ac2-b2d b3-a2c |

o43 : Ideal of R
```

The codimension of the ideal I is 2, and its dimension is 2.

```
i44 : codim I

o44 = 2

i45 : dim I

o45 = 2
```

If we wanted to regard I as a module instead of an ideal, and take the codimension of the support of that module, we would write

```
i46 : codim module I

o46 = 0
```

The dimension of I as an ideal is by definition the same as the dimension of the module R/I. If we simply write R/I, *Macaulay 2* will interpret it as a ring, so we write R^1 for the free module of rank 1 over R, and the dimension may be computed again as

```
i47 : dim (R^1/I)
```

```
o47 = 2
```

We could also make the generators of *I* into a matrix and take the cokernel to get the same thing:

```
i48 : M = coker generators I

o48 = cokernel | bc-ad c3-bd2 ac2-b2d b3-a2c |
                                              1
o48 : R - module, quotient of R

i49 : codim M

o49 = 2

i50 : dim M

o50 = 2
```

And similarly for the degree:

```
i51 : degree I

o51 = 4

i52 : degree M

o52 = 4
```

As one might expect, the degree of the quartic is 4! The Hilbert polynomial may be obtained as

```
i53 : hilbertPolynomial( M, Projective=>false)

o53 = 4 $i + 1

o53 : QQ[$i]
```

Here `Projective=>false` is an "option" that tells *Macaulay 2* to display the Hilbert polynomial as an ordinary polynomial. We see from the polynomial that we are dealing with a curve of degree 4 and genus 0. The default display would show the polynomial as a linear combination of the Hilbert polynomials of projective spaces of different dimensions:

```
i54 : hilbertPolynomial M

o54 = - 3 P  + 4 P
           0      1

o54 : ProjectiveHilbertPolynomial
```

Similarly, we can display the Hilbert series as a rational function:

```
i55 : hilbertSeries M

            3       2
        - $T  + 2 $T  + 2 $T + 1
o55 = ---------------------------
                          2
               (- $T + 1)

o55 : Divide
```

Another way to get information about the module M is to see its free resolution

```
i56 : Mres = res M
```

```
           1      4      4      1
o56 = R  <-- R  <-- R  <-- R

           0      1      2      3

o56 : ChainComplex
```

To get more precise information about `Mres`, we could do

```
i57 : betti Mres
      total: 1 4 4 1
          0: 1 . . .
          1: . 1 . .
          2: . 3 4 1
```

The display is chosen for compactness: the first line gives the total Betti numbers, the same information given when we type the resolution. The remaining lines express the degrees of each of the generators of the free modules in the resolution. The j-th column after the colons (counting from 0) gives the degrees of generators of the j-th module; an n in the j-th column in the row headed by d means that the j-th free module has n generators of degree $d + j$. Thus for example in our case, the generator of the third free module in the resolution has degree $3 + 2 = 5$.

Random Regular Sequences

An interesting and illustrative open problem is to understand the initial ideal (and the Gröbner basis) of a "generic" regular sequence. To study a very simple case we take a matrix of 2 random forms in a polynomial ring in 3 variables.

```
i58 : R = KK[x,y,z];

i59 : F = random(R^1, R^{-2,-3})

o59 = | -8514x2+5827xy-9613y2+7886xz+5702yz+11835z2  -15791x3+12712 ...

               1      2
o59 : Matrix R  <--- R
```

The `random` command makes a 1×2 matrix F whose elements have degrees $\{2,3\}$ (that is, F is a random map to the free module $R^1 = R(0)$, which has its one generator in the default degree, namely 0, from the free module $R(-2) \oplus R(-3)$ with generators in degrees, $\{2,3\}$). We can compute:

```
i60 : GB = generators gb F

o60 = | x2-14140xy+5100y2+12053xz+14443yz+7878z2  xy2-6589y3-2097xy ...

               1      3
o60 : Matrix R  <--- R

i61 : LT = leadTerm generators gb F

o61 = | x2 xy2 y4 |
```

```
                  1     3
o61 : Matrix R  <--- R

i62 : betti LT
      total: 1 3
         0: 1 .
         1: . 1
         2: . 1
         3: . 1
```

The betti command shows that there are Gröbner basis elements of degrees 2, 3, and 4. This result is dependent on the monomial order in the ring *R*; for example we could take the lexicographic order

```
i63 : R = KK[x,y,z, MonomialOrder => Lex];
```

For more possibilities, use the *help* facility:

```
i64 : help MonomialOrder

MonomialOrder -- a key used with monoids to indicate a
     monomial order other than the default (graded reverse lexicog ...

Values:
  "GRevLex" -- graded reverse lexicographic order (the default)
  "GLex" -- graded lexicographic order
  "Lex" -- lexicographic order
  "RevLex" -- reverse lexicographic order
  "Eliminate" n -- elimination order, eliminating first n variab ...
  "ProductOrder"{n1, n2, ..., nv} -- product order
Eventually, more general monomial orders will be allowed.
```

We get

```
i65 : F = random(R^1, R^{-2,-3})

o65 = | -9394x2+8312xy+12075xz-3147y2-11542yz+3907z2 -13293x3-1763 ...

                  1     2
o65 : Matrix R  <--- R

i66 : GB = generators gb F

o66 = | x2+2417xy+2168xz+13360y2-639yz+14449z2 xy2+198xyz-4686xz2+ ...

                  1     5
o66 : Matrix R  <--- R

i67 : LT = leadTerm generators gb F

o67 = | x2 xy2 xyz2 xz4 y6 |

                  1     5
o67 : Matrix R  <--- R

i68 : betti LT
      total: 1 5
         0: 1 .
         1: . 1
         2: . 1
         3: . 1
         4: . 1
         5: . 1
```

and there are Gröbner basis elements of degrees 2, 3, 4, 5, and 6.

Division with Remainder

A major application of Gröbner bases is to decide whether an element is in a give ideal, and whether two elements reduce to the same thing modulo an ideal. For example, everyone knows that the trace of a nilpotent matrix over a field is 0. We can produce an ideal *I* that defines the variety X of nilpotent 3 × 3 matrices by taking the ideal generated by the entries of the cube of a generic matrix. Here's how:

```
i69 : R = KK[a..i];

i70 : M = genericMatrix(R,a,3,3)

o70 = | a d g |
      | b e h |
      | c f i |

             3       3
o70 : Matrix R  <--- R

i71 : N = M^3

o71 = | a3+2abd+bde+2acg+bfg+cdh+cgi       a2d+bd2+ade+de2+cdg+af ...
      | a2b+b2d+abe+be2+bcg+ach+ceh+bfh+chi abd+2bde+e3+bfg+cdh+2e ...
      | a2c+bcd+abf+bef+c2g+cfh+aci+bfi+ci2 acd+cde+bdf+e2f+cfg+f2 ...

             3       3
o71 : Matrix R  <--- R

i72 : I = ideal N

o72 = ideal | a3+2abd+bde+2acg+bfg+cdh+cgi a2b+b2d+abe+be2+bcg+ach ...

o72 : Ideal of R
```

But the trace is not in *I*! This is obvious from the fact that the trace has degree 1, but the polynomials in *I* are of degree 3. However, even the 6th power of the trace is not in *I* (the seventh is; Bernard Mourrain has worked out a general formula telling which power is necessary). We could also check by division with remainder:

```
i73 : Tr = trace M

o73 = a + e + i

o73 : R

i74 : Tr^6 % I

           2 2 2          2 2        3 2       4 2         2 ...
o74 = 90 a e i + 90 b d e i + 90 a e i + 90 e i + 90 c e ...

o74 : R

i75 : Tr^7 % I

o75 = 0

o75 : R
```

If we actually want to see the quotient, we need to specify the order of the generators of the ideal, that is, we have to give the matrix. The quotient is

```
i76 : Q = Tr^6 // generators I
```

```
o76 = | a3+6a2e+3bde+15ae2+22e3+3bfg+3cdh-12efh+6a2i+30aei+60e2i+3 ...
      | 18de2+18efg+18dei+18fgi                                    ...
      | 18deh+18egi+18dhi+18gi2                                    ...
      | -27abe-63be2-36ach+9ceh+18bfh-72abi-162bei+90chi           ...
      | -2a3+15a2e+21bde+6ae2+e3-6bfg-6cdh-36afh+15efh+60a2i+72bdi ...
      | 63beg+36a2h+18bdh-18aeh+18e2h+9cgh+18fh2+117bgi-162ahi-27e ...
      | -18a2c+18ace+18abf-18bef+36cfh+45aci+18cei-81bfi+153ci2    ...
      | -18cde-36a2f-36bdf+72aef+9e2f-45cfg-18f2h-108cdi+162afi+9e ...
      | 16a3+18abd-30a2e-42bde-30ae2-44e3+18acg-18ceg+3bfg+3cdh+18 ...

              9      1
o76 : Matrix R  <--- R
```
and then
```
i77 : (generators I)*Q + (Tr^6 % I) == Tr^6

o77 = true
```
gives "true".

The trace is nonzero mod *I* simply because the generators of *I* (the entries of the cube of the matrix) do not generate the ideal of all forms vanishing on *X* — this we could find (with a little time!) using the `radical` command as follows.
```
i78 : time rI = radical I
      -- used 234.73 seconds

o78 = ideal | a+e+i bd+e2+cg+fh+ei+i2 ceg-bfg-cdh-efh-cgi-2fhi-i3  ...

o78 : Ideal of R

i79 : J = ideal (trace M,
               trace exteriorPower(2,M),
               trace exteriorPower(3,M))

o79 = ideal | a+e+i -bd+ae-cg-fh+ai+ei -ceg+bfg+cdh-afh-bdi+aei |

o79 : Ideal of R

i80 : rI == J

o80 = true
```
Thus we see that the radical is generated by the trace, the determinant, and the sum of the principal 2×2 minors, that is, by the coefficients of the characteristic polynomial.

Elimination Theory

Consider the problem of projecting the "twisted cubic", a curve in $\mathbb{P}^3$ defined by the three 2×2 minors of a certain 2×3 matrix. Such problems can be solved in a simple direct way using Gröbner bases. The technique lends itself to many extensions and in its developed form can be used to find the closure of the image of any map of affine varieties.

In this section we shall give first a direct treatment of the problem above, and then show how to use *Macaulay 2* as a general tool to solve the problem.

We first clear the earlier meaning of x to make it into a subscripted variable,
```
i81 : erase quote x;
```
and then set
```
i82 : R = KK[x_0..x_3];
```
the homogeneous coordinate ring of $\mathbb{P}^3$ and
```
i83 : M = matrix(table(2,3, (i,j)->x_(i+j)))
```

```
o83 = | x_0 x_1 x_2 |
       | x_1 x_2 x_3 |

                     2       3
o83 : Matrix R  <--- R

i84 : I = minors(2,M)

o84 = ideal | -x_1^2+x_0x_2 -x_1x_2+x_0x_3 -x_2^2+x_1x_3 |

o84 : Ideal of R
```

the ideal of the twisted cubic.

As projection center we take the point defined by

```
i85 : pideal = ideal(x_0+x_3, x_1, x_2)

o85 = ideal | x_0+x_3 x_1 x_2 |

o85 : Ideal of R
```

To find the image we must intersect the ideal *I* with the subring generated by the generators of pideal. We make a change of variable so that these generators become the last three variables in the ring; that is, we write the ring as $KK[y_0, \ldots, y_3]$ where

$$y_0 = x_0, \quad y_1 = x_1, \quad y_2 = x_2, \quad y_3 = x_0 + x_3$$

and thus $x_3 = y_3 - y_0$, etc. We want the new ring to have an "elimination order" for the first variable.

```
i86 : erase quote y;

i87 : S = KK[y_0..y_3,MonomialOrder => Eliminate 1];
```

Here is one way to make the substitution

```
i88 : I1 = substitute(I, matrix{{y_0,y_1,y_2,y_3-y_0}})

o88 = ideal | y_0y_2-y_1^2 -y_0^2+y_0y_3-y_1y_2 -y_0y_1-y_2^2+y_1y ...

o88 : Ideal of S
```

The elimination of one variable from the matrix of Gröbner basis elements proceeds as follows:

```
i89 : J = selectInSubring(1,generators gb I1)

o89 = | y_1^3+y_2^3-y_1y_2y_3 |

                     1       1
o89 : Matrix S  <--- S
```

and gives (a matrix whose single entry is) the cubic equation of a rational curve with one double point in the plane. However, we are still in a ring with 4 variables, so if we really want a plane curve (and not the cone over one) we must move to yet another ring:

```
i90 : S1 = KK[y_1..y_3];

i91 : J1 = substitute(J, S1)

o91 = | y_1^3+y_2^3-y_1y_2y_3 |
```

```
               1         1
     o91 : Matrix S1  <--- S1
```

This time we didn't have to give so much detail to the `substitute` command because of the coincidence of the names of the variables.

Having shown the primitive method, we now show a much more flexible and transparent one: we set up a ring map from the polynomial ring in 3 variables (representing the plane) to R/I, taking the variables y to the three linear forms that define the projection center. Then we just take the kernel of this map! ("Under the hood", *Macaulay 2* is doing a more refined version of the same computation as before.)

Here is the ring map

```
     i92 : Rbar = R/I;

     i93 : f = map(Rbar, S1,
                       matrix(Rbar,{{x_0+x_3, x_1,x_2}})
                  )

     o93 = map(Rbar,S1,| x_0+x_3 x_1 x_2 |)

     o93 : RingMap Rbar <--- S1
```

and the desired ideal

```
     i94 : J1 = ker f

     o94 = ideal | y_2^3-y_1y_2y_3+y_3^3 |

     o94 : Ideal of S1
```

Quotients and Saturation

Another typical application of Gröbner bases and syzygies is to the computation of ideal quotients and saturations. Again we give an easy example that we can treat directly, and then introduce the tool used in *Macaulay 2* to treat the general case.

If I and J are ideals in a ring R, we define $(I:J)$, the ideal quotient, by

$$(I:J) = \{f \in R \mid fJ \subset I\}$$

In our first examples we consider the case where J is generated by a single element g. This arises in practice, for example, in the problem of homogenizing an ideal. Suppose we consider the affine space curve parametrized by $t \mapsto (t, t^2, t^3)$. The ideal of polynomials vanishing on the curve is easily seen to be $(b - a^2, c - a^3)$ (where we have taken a, b, c as the coordinates of affine space). To find the projective closure of the curve in $\mathbb{P}^3$, we must homogenize these equations with respect to a new variable d, getting $(db - a^2, d^2c - a^3)$. But these forms do not define the projective closure! In general, homogenizing the generators of the ideal I of an affine variety one gets an ideal $I1$ that defines the projective closure up to a component supported on the hyperplane at infinity (the hyperplane $d = 0$). To see the ideal of the closure we must remove any such components, for example by replacing $I1$ by the union $I2$ of all the ideals $(I1 : d^n)$, where n ranges over positive integers. This is not so hard as it seems: First of all, we can successively compute the increasing sequence of ideals $(I1 : d)$, $(I1 : d^2), \ldots$, until we get two that are the same; all succeeding ones will be

equal, so we have found the union. A second method involves a special property of
the reverse lex order, and is much more efficient in this case. We shall illustrate both.
First we set up the example above:

```
i95 : R = KK[a,b,c,d];

i96 : I1 = ideal(d*b-a^2, d^2*c-a^3)

o96 = ideal | -a2+bd -a3+cd2 |

o96 : Ideal of R
```

We discuss now a method for computing the ideal quotient. If I is generated by
$f_1, \ldots, f_n$. We see that $s \in (I : J)$ if and only if there are ring elements r_i such that

$$\sum_{i=1}^{n} r_i f_i + sg = 0.$$

Thus it suffices to compute the kernel (syzygies) of the $1 \times (n+1)$ matrix

$$(f_1, \ldots, f_n, g)$$

and collect the coefficients of g, that is, the entries of the last row of a matrix representing the kernel. Thus in our case we may compute $(I1 : d)$ by concatenating the
matrix for $I1$ with the variable d.

```
i97 : I1aug = generators I1 | d

o97 = | -a2+bd -a3+cd2 d |

              1       3
o97 : Matrix R  <--- R

i98 : augrelations = generators ker I1aug

o98 = | -a      d    |
      | 1       0    |
      | ab-cd   a2-bd |

              3       2
o98 : Matrix R  <--- R
```

There are 3 rows (numbered 0, 1, 2) and 2 columns (numbered 0, 1), so to extract the
last row we may do this.

```
i99 : I21 = submatrix(augrelations, {2}, {0,1})

o99 = | ab-cd a2-bd |

              1       2
o99 : Matrix R  <--- R
```

But this is not an "ideal", properly speaking: first of all, it is a matrix, rather than a
submodule of R^1, and second of all its target is not R^1 but $R(-1)$, the free module of
rank 1 with generator in degree 1. We can fix both of these problems with

```
i100 : I21 = ideal I21

o100 = ideal | ab-cd a2-bd |

o100 : Ideal of R
```

This is larger than the original ideal, having two quadratic generators instead of a
quadric and a cubic, so we continue. Instead of doing the same computation again,
we introduce the built-in command for computing ideal quotients.

```
i101 : I22 = I21 : d
```

```
o101 = ideal | b2-ac ab-cd a2-bd |
```

```
o101 : Ideal of R
```

The result is again larger than $I21$, having three quadratic generators. We repeat.

```
i102 : I23 = I22 : d
```

```
o102 = ideal | b2-ac ab-cd a2-bd |
```

```
o102 : Ideal of R
```

We get an ideal which is the same as $I22$. Thus the homogeneous ideal $I2$ of the projective closure is equal to $I23$ (this is the homogeneous ideal of the twisted cubic, already encountered above).

A more perspicuous way of approaching the computation of the union of the ideals $(I : d^n)$, which is called the saturation of I with respect to d, and written $(I : d^\infty)$, is first to compute a reverse lex Gröbner basis.

```
i103 : generators gb I1
```

```
o103 = | a2-bd abd-cd2 b2d2-acd2 |
```

```
                 1       3
o103 : Matrix R  <--- R
```

We see that the second generator is divisible by d, and the third is divisible by d^2. General theory says that we get the right answer simply by making these divisions, that is, the saturation is

$$(a^2 - cd, ac - cd, c^2 - cd),$$

as previously computed. The same thing can be accomplished in one line by

```
i104 : I2 = divideByVariable(generators gb I1,d)
```

```
o104 = | a2-bd ab-cd b2-ac |
```

```
                 1       3
o104 : Matrix R  <--- R
```

This ideal $I2$ is the ideal of the projective closure that we wanted to compute.

C.2 Local Cohomology of Graded Modules

Mathematical Background

If R is a ring, m is an ideal of R, and M is an R-module, we define the zero-th local cohomology (with supports in m) to be the submodule

$$H^0_m(M) \subset M$$

which is the set of those elements of M that are annihilated by some power of m. It is easy to see that H^0_m is a left-exact functor on the category of R-modules, and we define the i-th local cohomology functor H^i_m to be the i-th right derived functor of H^0_m. The local cohomology functors appear frequently in commutative algebra. They generalize the (Zariski) cohomology of coherent sheaves (on a projective or toric variety) and the computation of local cohomology in the corresponding cases gives part of the cohomology of sheaves. In this tutorial we learn how to handle these computations. We shall state without proof the mathematical results on which these computations are based. We refer the reader to the original book [G67] by A. Grothendieck or to the somewhat more accessible summaries of special cases given in [Ei95, Appendix 4], [BH93], and Chapter 8 and Appendix A of this book for more information.

One useful property of the local cohomology is its independence of the ring over which it is computed. We may codify this property as follows.

Proposition C.1. *Suppose that* $R' \to R$ *is a homomorphism of rings, and* $\mathfrak{m}' \subset R'$ *is an ideal. Let* $\mathfrak{m} = \mathfrak{m}'R$. *If* M *is an* R-*module, and* M' *is the same set as* M, *regarded as an* R'-*module, then the local cohomology of* M *with respect to* $\mathfrak{m}$ *is the same as the local cohomology of* M' *with respect to* $\mathfrak{m}'$.

Because of Proposition C.1, we may, in computing the local cohomology of a module over a finitely generated algebra, always suppose that the algebra is simply a polynomial ring.

The modules $H^i_{\mathfrak{m}}(M)$ may not be finitely generated, but in the situation where R and M are positively graded, and $\mathfrak{m}$ is the homogeneous maximal ideal of R, the module $H^i_{\mathfrak{m}}(M)$ is graded. Moreover, for any integer e, the submodule generated by the homogeneous elements of degrees at least e is finitely generated (in fact, $H^i_{\mathfrak{m}}(M)$ is nonzero only in finitely many degrees).

A more refined version of this assertion is the key to the most general method of computing local cohomology. To express it we write $a_i(M)$ for the maximum of the degrees of the minimal generators of the i-th syzygy module of M.

Theorem C.2. *Let* R_0 *be a Noetherian ring, and let* $R = R_0[x_1, \ldots, x_n]$ *be a graded polynomial ring over* R. *Let* $\mathfrak{m} = (x_1, \ldots, x_n) \subset R$ *and let* M *be a finitely generated graded* R-*module. If* J *is any homogeneous ideal of* R *containing a power of* $\mathfrak{m}$ *then for* $i \geq 1$ *there is a natural map* $\mathrm{Ext}^i_R(R/J, M) \to H^i_{\mathfrak{m}}(M)$ *which is an isomorphism in all large degrees. The map is an isomorphism in all degrees* $\geq e$ *if* J *is contained in* $\mathfrak{m}^d$, *where* $d = \max(a_{n-i}(M), a_{n-i+1}(M)) - n + 1 - e$.

Proof sketch. The i-th local cohomology is the homology of the subcomplex

$$\ldots \to E_{i-1} \to E_i \to E_{i+1} \to \ldots$$

of the graded injective resolution of M consisting of all elements annihilated by some power of $\mathfrak{m}$; the Ext is the homology of the subcomplex of elements annihilated by J. Thus they agree in degrees $\geq e$ if J annihilates the parts of E_{i-1} and E_i of degrees $\geq e$, and this will be true as soon as J is contained in the d-th power of $\mathfrak{m}$ for $d = a - e + 1$, where a is the maximum degree of a socle element of E_{i-1} or E_i.

The degrees of the socle elements of E_i are (by local duality – see below) minus the degrees of the generators of

$$\mathrm{Ext}^{n-i}(M, R(-n)) = \mathrm{Ext}^{n-i}(M, R)(-n).$$

Since $\text{Ext}^{n-i}(M,R)$ is a subquotient of the dual of the free resolution of M, we see that

$$a \leq \max(a_{n-i}(M), a_{n-i+1}(M)) - n,$$

which yields the desired estimate.   □

It happens that in many important cases the local cohomology is nonzero in only finitely many degrees, and then of course we can compute it all.

Another way to get at the whole local cohomology module is through the local duality theorem, which also allows quick computation of the local cohomology in some simple cases. Here is the version we use:

Theorem C.3. *Let k be a field, and let $R = k[x_1, \ldots, x_n]$ be a graded polynomial ring over R. Let $\mathfrak{m} = (x_1, \ldots, x_n) \subset R$ and let M be a finitely generated graded R-module. Writing $W = R(-n)$ (the "relative canonical module") we have*

$$H_{\mathfrak{m}}^i(M) = \text{Hom}(\text{Ext}_R^{n-i}(M,W), k).$$

Finally, to know where to look for interesting invariants, it helps to know the following:

Theorem C.4. *Let k be a field, and let $R = k[x_1, \ldots, x_n]$ be a graded polynomial ring over R. Let $\mathfrak{m} = (x_1, \ldots, x_n) \subset R$ and let M be a finitely generated graded R-module. The local cohomology $H_{\mathfrak{m}}^i(M)$ is zero for $i < \text{depth } M$ and for $i > \dim M$. It is nonzero for $i = \text{depth } M$ or $i = \dim M$ (and may be zero or not for values in between).*

Macaulay 2 **Routines**

With these techniques in hand, we can explain the routines that *Macaulay 2* uses to compute local cohomology. Here is the routine using Theorem C.2.

```
i1 : localCohomology = method();

i2 : localCohomology(ZZ,Module,ZZ) := (i,M,e) -> (
            -- Use the formula
            -- H^i(M) = Ext^i(R/J,M) in degrees >= e,
            -- valid whenever J is an ideal contained in the
            -- p th power of the max ideal (of finite
            -- colength),
            -- where
            -- p >= max(a(r-i)-r+1, a(r-i+1)-r+1)-e,
            -- and
            -- a(i) = max degree of an i th syzygy
            -- of M over a polynomial ring R in r variables.
            -- We need to place M into a polynomial ring,
            -- and compute the resolution there, at least
            -- r-i+2 steps back.
            -- To find the degree bound we need, we must
            -- work over a polynomial ring, so we move
            -- to the presenting ring, if necessary:
            A := ring M;
            F := presentation A;
            R := ring F;
            -- Make a copy N of M over the polynomial ring
            N := coker lift(presentation M,R) ** coker F;
            -- now compute the resolution. We need to get
            -- the r-i+1 differential right.
            r := numgens R;
            C := res (N,LengthLimit => r-i+2);
```

```
        -- at this point it costs nothing to check
        -- whether depth M <= i <= dim M, and
        -- return 0 otherwise
        if i > dim N or length C < r-i
        then A^0
        else (
                -- Find a degree bound p for the cohomology
                -- computation to come:
                deg1 := first max degrees C_(r-i);
                p := deg1-r+1-e;
                if length C >= r-i+1 then (
                        deg2 := first max degrees C_(r-i+1);
                        p = max(p, deg2-r+1-e);
                );
                p = max(p,0);
                -- Find J. Since we are going to do Koszul
                -- homology, we can work directly in A = ring M.
                -- We compute Ext^i(R/J,M)
                -- as the cohomology of
                -- the Koszul complex on J,
                -- or equivalently as the homology,
                -- in complementary degree, twisted by rp:
                J := map(A^1,A^{r:-p},
                        {apply(numgens A, j -> A_j^p)});
                HM := homology(M**koszul(r-i,J),
                                M**koszul(r-i+1,J));
                -- since a nonminimal presentation is not so
                -- useful, we use prune to get a minimal one.
                prune (A^{r*p} ** HM)));
```

To implement the routine using Theorem C.3, we need a method to ascertain the largest and smallest degrees in which a module of finite length has a nonzero component. The routine degreeRange below provides, for a module M, a list of those degrees in which M has a nonzero component.

```
i3 : degreeList = (M) -> (
        if dim M > 0
        then error "expected module of finite length";
        H := poincare M;
        T := (ring H)_0;
        H = H // (1-T)^(numgens ring M);
        exponents H / first);
```

Here are the main routines. In the first, localCohomology1, we produce a module M that is, in a certain range of degrees, the dual of the module we want. We begin with a routine to find the dual of the interesting piece of M. More precisely, this routine assumes that the underlying ring of M is a polynomial ring. It returns the dual of the finite length module $M_{<e}$ that agrees with M in degrees $< e$ and is zero in larger degrees.

```
i4 : truncatedDual = method();

i5 : truncatedDual(Module,ZZ) := (M,e) -> (
        -- find (k-dual M), truncated in degrees >= e.
        R := ring M;
        n := numgens R;
        ww := R^{-n};
        M1 := prune (M / (truncate(-e+1,M)));
        Ext^n(M1,ww));
```

Now the routine that computes local cohomology via local duality:

```
i6 : localCohomology1= method();

i7 : localCohomology1(ZZ,Module,ZZ) := (i,M,e) -> (
        -- compute the degree >= e part of the
        -- local cohomology H^i(M).
        -- The method used here is local duality:
        -- H^i(M) is k-dual to Ext^(n-i)(M,R(-n)), where
```

```
-- the ring R of M is a polynomial ring in n
-- variables.
--
-- First step: bring M back to a polynomial ring,
-- if necessary.
A := ring M;
F := presentation A;
R := ring F;
M = coker lift(presentation M,R) ** coker F;
-- Second step: compute the Ext
n := numgens R;
ww := R^{-n};
E := prune Ext^(n-i)(M,ww);
-- Third step: dualize.  If E is finite length,
-- this may be done as Ext^n(E,ww).
-- Otherwise, we must truncate degrees (or else the
-- result will not be finitely generated).
result := (
    if dim E <= 0
    then Ext^n(E,ww)
    else truncatedDual(E,e)
    );
-- Finally, we put the module back into
-- the original ring, if A is not R.
prune (result ** A)
);
```

Example: The Rational Quartic Curve in $\mathbb{P}^3$

As an example we compute the local cohomology of the ring A which is the homogeneous coordinate ring of the smooth rational quartic curve in $\mathbb{P}^3$.

First we set up the homogeneous coordinate ring of projective space.

```
i8 : kk = ZZ/32003;

i9 : R = kk[x_0..x_3];
```

Then we construct the ideal of the quartic.

```
i10 : I = monomialCurve(R,{1,3,4})

o10 = ideal | x_1x_2-x_0x_3 x_2^3-x_1x_3^2 x_0x_2^2-x_1^2x_3 x_1^3 ...

o10 : Ideal of R

i11 : S = R/I;
```

And now we compute the local cohomology in degrees ≥ -5.

```
i12 : X0 = localCohomology(0, S^1, -5)

o12 = 0

o12 : S - module
```

Nothing there; we're below depth *S*.

```
i13 : X1 = localCohomology(1, S^1, -5)

o13 = cokernel | x_3 x_2 x_1 x_0 |

                             1
o13 : S - module, quotient of S
```

One dimensional, concentrated in degree 1; this is the "Hartshorne-Rao module" of the quartic.

```
i14 : X2 = localCohomology(2, S^1, -5)
```

```
o14 = cokernel | 0    0    0    0    0    0    0   -x_3 0    0    ...
               | 0    0    0    0    0    0   -x_3 0    0    0    ...
               | 0    0    0    0    0    0    0    x_2 0   -x_3  ...
               | 0    0    0    0    0    0    x_2 0   -x_3 0    ...
               | 0    0    0    0    0    0    0    0    0    x_2  ...
               | 0    0    0    0    0    0    0    0    x_2 0    ...
               | 0    0    0    0    0    x_3 0    0    0    0    ...
               | 0    0    0    0    0    0    0    0    0    0    ...
               | -x_3 0   -x_2 -x_1 0    0    0    0    0    0    ...
               | x_2 -x_3 0    x_0 -x_1 0    0    0    0    0    ...
               | 0    x_2  x_1  0    x_0  0    0    0    0    0    ...
                                              11
o14 : S - module, quotient of S
```

This module is more complicated! The second local cohomology is the dual of the canonical module, and is in particular infinite-dimensional.

```
i15 : X3 = localCohomology(3, S^1, -5)

o15 = 0

o15 : S - module
```

Again 0; we're above dim S.

C.3 Cohomology of a Coherent Sheaf

Mathematical Background

Next we turn to the problem of computing the cohomology of a sheaf F on a projective variety X. We suppose that X is already embedded in a projective space P^r. We may thus represent the variety by giving its homogeneous coordinate ring R. Given a graded R-module M there is an associated sheaf $F = \tilde{M}$, and every quasicoherent sheaf F can be represented in this way. If M is finitely generated, then F is coherent, and every coherent sheaf can be represented by a finitely generated graded module. However, these representations are not unique, and sometimes one may even represent a coherent sheaf by modules that are not finitely generated.

A result corresponding to Proposition C.1 above shows that for the computation of cohomology it does not really matter whether we regard F as a sheaf on X or on. We will therefore write $H^i(F)$, not mentioning X, for the i-th cohomology of F.

Given a graded module M we can produce another one by "shifting" degrees: we define $M(n)$ to be the module whose degree d part is M_{n+d}. The corresponding sheaf is

$$\widetilde{M(n)} = F(n) = F \otimes O_{P^r}(n),$$

called the n-th "twist" of F, where $O_{P^r}(n)$ is the n-th tensor power of the tautological bundle on P^r. There is a also a canonical graded module representing F, given by

$$M_\infty = H^0_*(F) = \bigoplus_{n \in Z} H^0(F(n))$$

but this may fail to be finitely generated even when F is coherent (this happens when F has associated varieties of dimension 0 in the projective space).

We now assume that F is coherent. For any integer e, the module obtained as the sum of the global sections of all the twists $F(n)$ for $n \geq e$, that is

$$M = H^0_{\geq e}(F) = \bigoplus_{n \geq e} H^0(F(n))$$

has $\tilde{M} = F$, and our first task is to compute M. We call it a *truncation* of M_∞. Of course we will also want to compute each of the corresponding truncations of the higher cohomology sums,

$$H^i_{\geq e}(F) = \bigoplus_{n \geq e} H^i(F(n)).$$

The computations for $i > 0$ reduce to the computations of local cohomology because of the isomorphism

$$H^i_*(F) = H^{i+1}_{\mathfrak{m}}(M),$$

valid for any module M representing a sheaf F and any integer $i > 0$, where $\mathfrak{m}$ is the maximal homogeneous ideal of the homogeneous coordinate ring of the projective space. Thus it would suffice to do the case $i = 0$. However, there is a formulation which works for $i = 0$ and is the same for all i, and we shall present this version. Again we let $a_i(M)$ be the maximal degree of a minimal generator of the i-th syzygy module for M. The following result corresponds to the computation of local cohomology made in Theorem C.2.

Theorem C.5. *Let k be a field, and let $R = k[x_0, \ldots, x_r]$ be the graded polynomial ring over R. Let $\mathfrak{m} = (x_0, \ldots, x_r) \subset R$, and set $n = r + 1$, the number of variables. Let M be a finitely generated graded module over R containing no nonzero submodule of finite length, and let $\tilde{M}$ be the corresponding sheaf on P^r. If J is any homogeneous ideal of R containing some power of $\mathfrak{m}$ then for $i \geq 0$ there is a natural map*

$$\text{Ext}^i_R(J, M) \to H^i_*(\tilde{M})$$

which is an isomorphism in all large degrees. The map is an isomorphism in all degrees $\geq e$ if J is contained in $\mathfrak{m}^d$, where

$$d \geq \max(a_{n-i}(M), a_{n-i+1}(M)) - n + 1 - e.$$

Proof sketch. If $i > 0$ then

$$\text{Ext}^i_R(J, M) = \text{Ext}^{i+1}_R(R/J, M)$$

and

$$H^i_*(M) = H^{i+1}_{\mathfrak{m}}(M)$$

and the result reduces to Theorem C.2 above (note that we have not used the hypothesis that M has no nonzero submodule of finite length in this part).

Now suppose that $i = 0$. Since $J = R$ locally on the punctured spectrum of R, any map $J \to M$ of degree d induces a global section of $\widetilde{M(d)}$, and this defines a natural map

$$\text{Hom}(J,M) \to H^0_*(\tilde{M}).$$

If M contains no nonzero submodule of finite length we use the short exact sequence

$$0 \to J \to R \to R/J \to 0$$

to obtain the exact sequence

$$0 \to \text{Hom}(R,M) \to \text{Hom}(J,M) \to \text{Ext}^1(R/J,M) \to 0$$

and with the exact sequence

$$0 \to M \to H^0_*(\tilde{M}) \to H^1_{\mathfrak{m}}(M) \to 0$$

we get the following commutative diagram.

$$
\begin{array}{ccccccc}
0 & \longrightarrow & M & \longrightarrow & \text{Hom}(J,M) & \longrightarrow & \text{Ext}^1(R/J,M) & \longrightarrow & 0 \\
 & & \| & & \downarrow & & \downarrow & & \\
0 & \longrightarrow & M & \longrightarrow & H^0_*(\tilde{M}) & \longrightarrow & H^1_{\mathfrak{m}}(M) & \longrightarrow & 0.
\end{array}
$$

From the snake Lemma we see at once that the map $\text{Hom}(J,M) \to H^0_*(\tilde{M})$ is an isomorphism in degree d if and only if the map

$$\text{Ext}^1(R/J,M) \to H^1_{\mathfrak{m}}(M)$$

is an isomorphism in degree d. Thus the result reduces to Theorem C.2 above. $\qquad\square$

Macaulay 2 Routines for Sheaf Cohomology

The following routines call one version or another of the localCohomology routines for the higher cohomology, and the globalSections routine for H^0.

```
i16 : sheafCohomology = method();

i17 : sheafCohomology(ZZ, Module, ZZ) := (i,M,e) -> (
          if i === 0
          then globalSections(M,e)
          else localCohomology(i+1, M, e)
          );

i18 : sheafCohomology1 = method();

i19 : sheafCohomology1(ZZ, Module, ZZ) := (i,M,e) -> (
          if i === 0
          then globalSections(M,e)
          else localCohomology1(i+1, M, e)
          );

i20 : globalSections = method();

i21 : globalSections(Module,ZZ) := (M,e) -> (
          -- if M has a submodule of finite length,
          -- kill it:
          M = M / saturate 0_M;
          -- To compute degree bounds we need to work over
          -- a polynomial ring, so we make a copy of
          -- M over a polynomial ring if we're not already
          -- there:
```

```
A := ring M;
F := presentation A;
R := ring F;
N := coker lift(presentation M,R) ** coker F;
r := numgens R;
wR := R^{-r};
-- Find the degree bound for the ideal J below.
if pdim N < r-1
then M
else (
        -- We must find an ideal J that
        -- annihilates H^1(local,M) or,
        --   equivalently, annihilates the dual ext
        E1 := Ext^(r-1)(N,wR);
        -- If E1 has finite length, then we don't
        -- need to truncate the answer below, and  we
        -- may use the spread of degrees in which
        --   E1 is nonzero to estimate the annihilator
        -- (1) If E1 is finite length,
        --      let p := max deg E1 - min deg E1 + 1;
        -- (2) If E1 is not finite length,
        -- but the part of
        --      degree >= e is desired, choose
        --      p := min deg E1 + e + 1
        p := max(0,
                if dim E1 <= 0
                then (
                        max degreeList E1
                        - min degreeList E1 + 1
                        )
                else min degrees E1 + e + 1
                );
        -- We move back to A and M:
        J := ideal apply(numgens A, j -> A_j^p);
        Hom(module J,M))
);
```

Example: The Hodge Diamond of a K3 Surface

The Hodge diamond of a projective variety X of dimension n is the set of numbers $h^{p,q} = H^q(\Omega_X^p)$, where Ω_X^p denotes the p-th exterior power of the cotangent bundle of X. If X is smooth there are equalities $h^{p,q} = h^{q,p}$ and $h^{p,q} = h^{n-q,n-p}$. For a K3 surface the numbers are

$$
\begin{array}{ccccccccc}
 & & & & h^{2,2} & & & & \\
 & & & h^{3,0} & & h^{0,3} & & & \\
 & & h^{2,0} & & h^{1,1} & & h^{0,2} & & \\
 & & & h^{1,0} & & h^{0,1} & & & \\
 & & & & h^{0,0} & & & &
\end{array}
\qquad = \qquad
\begin{array}{ccccc}
 & & 1 & & \\
 & 0 & & 0 & \\
1 & & 20 & & 1 \\
 & 0 & & 0 & \\
 & & 1 & &
\end{array}
$$

Any nonsingular quartic surface in $\mathbb{P}^3$, for example the "Fermat" quartic below, is a K3 surface.

```
i22 : kk = ZZ/32003;

i23 : R = kk[x_0..x_3];

i24 : F = sum(4,i->x_i^4)

       4    4    4    4
o24 = x  + x  + x  + x
       0    1    2    3
```

```
o24 : R

i25 : S = R/F;
```

We compute the cotangent bundle as the homology of the complex

$$S^g \xrightarrow{\;J\;} S^{r+1} \xrightarrow{\;xx\;} S$$

where xx is the row of variables and J is the Jacobian matrix of the ideal presenting S.

```
i26 : cotangentBundle = (S) -> (
         F := presentation S;
         J := jacobian F ** S;
         xx:= vars ring F ** S;
         prune homology(xx,J));
```

Here is the computation of the Hodge diamond for the Fermat quartic in $\mathbb{P}^3$ using the routine sheafCohomology, presented in the format

$$\begin{array}{ccc} h^{0,0} & h^{0,1} & h^{0,2} \\ h^{1,0} & h^{1,1} & h^{1,2} \\ h^{2,0} & h^{2,1} & h^{2,2}. \end{array}$$

```
i27 : Omega = cotangentBundle S

o27 = cokernel | 0     0    -x_1 -x_2 x_3^3 0     0     x_0^3 |
                | 0    -x_1 0    x_3  x_2^3 0     x_0^3 0     |
                | 0    x_2  x_3  0    x_1^3 x_0^3 0     0     |
                | -x_2 0    x_0  0    0     -x_3^3 0    x_1^3 |
                | x_3  x_0  0    0    0     -x_2^3 x_1^3 0    |
                | x_1  0    0    x_0  0     0     -x_3^3 x_2^3 |
                                        6
o27 : S - module, quotient of S

i28 : time matrix table(3,3,(p,q)->
         hilbertFunction(0,
           sheafCohomology(q,exteriorPower(p,Omega),0)
         ))
     -- used 106.24 seconds

o28 = | 1 0  1 |
      | 0 20 0 |
      | 1 0  1 |
                3       3
o28 : Matrix ZZ  <--- ZZ
```

Here is the corresponding computation using sheafCohomology1.

```
i29 : time matrix table(3,3,(p,q)->
         hilbertFunction(0,
           sheafCohomology1(q,exteriorPower(p,Omega),0)
         ))
     -- used 66.06 seconds

o29 = | 1 0  1 |
      | 0 20 0 |
      | 1 0  1 |
                3       3
o29 : Matrix ZZ  <--- ZZ
```

References

[Ab90]   Abhyankar, S.: *Algebraic Geometry for Scientists and Engineers*. Mathematical Surveys and Monographs **95**, Amer. Math. Soc., Providence, 1990.

[AL90]   Adams, W. W., Loustaunau, P.: *An Introduction to Gröbner Bases*. Graduate Studies in Mathematics **3**, Amer. Math. Soc., Providence, 1994.

[AG90]   Allgower, E. L., Georg, K., Eds.: *Computational Solution of Nonlinear Systems of Equations*. Lectures in Applied Mathematics **26**, Amer. Math. Soc., Providence, 1990.

[Al78]   Almkvist, G.: K–theory of endomorphisms. *J. Algebra* **55** (1978), 308–340; *Erratum, J. Algebra* **68** (1981), 520–521.

[AS95]   Armendáriz, I., Solernó, P.: On the computation of the radical of polynomial complete intersection ideals. In: Proceedings 11th AAEEC, Lecture Notes in Computer Science **948**, Springer, Berlin Heidelberg New York, 1995, 106–119.

[AM69]   Atiyah, M. F., Macdonald, I. G.: *Introduction to Commutative Algebra*. Addison–Wesley, Reading, 1969.

[Au61]   Auslander, M.: Modules over unramified regular local rings. *Illinois J. Math.* **5** (1961), 631–647.

[AB59]   Auslander, M., Buchsbaum, D.: On ramification theory in Noetherian rings. *American J. Math.* **81** (1959), 749–765.

[ASt88]   Auzinger, W., Stetter, H. J.: An elimination algorithm for the computation of all zeros of a system of multivariate polynomial equations. In: *Numerical Mathematics*, Singapore 1988 (R. P. Agarval, Y. M. Chow and S. J. Wilson, Eds.), International Series of Numerical Mathematics **86**, Birkhäuser, Boston Basel Berlin, 1988, 11–30.

[BM93]   Bayer, D., Mumford, D.: What can be computed in Algebraic Geometry? In: *Computational Algebraic Geometry and Commutative Algebra*, Proceedings, Cortona 1991 (D. Eisenbud and L. Robbiano, Eds.), Cambridge University Press, 1993, 1–48.

[BS92]   Bayer, D., Stillman, M.: *Macaulay*: A system for computation in algebraic geometry and commutative algebra. 1992. Available via anonymous ftp from `math.harvard.edu`.

[BSa87]   Bayer, D., Stillman, M.: A criterion for detecting m–regularity. *Invent. Math.* **87** (1987), 1–11.

[BSb92]   Bayer, D., Stillman, M.: Computation of Hilbert functions. *J. Symbolic Computation* **14** (1992), 31–50.

[Be77]   Becker, J.: On the boundedness and unboundedness of the number of generators of ideals and multiplicity. *J. Algebra* **48** (1977), 447–453.

[BKW93]   Becker, T., Kredel, H., Weispfenning, V.: *Gröbner Bases*. Springer, Berlin Heidelberg New York, 1993.

[Ber70]   Berlekamp, E. R.: Factoring polynomials over large finite fields. *Math. Comp.* **24** (1970), 713–715.

[Bi97]   Bigatti, A.: Computation of Hilbert–Poincaré series. *J. Pure & Applied Algebra* **119** (1997), 237–253.

[BCR93]   Bigatti, A., Caboara, M., Robbiano, L.: On the computation of Hilbert-Poincaré series. *Applicable Algebra in Eng., Commun. and Comput.* **2** (1993), 21–33.

[BV93]   Brennan, J., Vasconcelos, W. V.: Effective computation of the integral closure of a morphism. *J. Pure & Applied Algebra* **86** (1993), 125–134.

[BV01]   Brennan, J., Vasconcelos, W. V.: On the structure of closed ideals. *Math. Scand.* **88** (2001), 3–16.

[Bro83]   Brodmann, M.: Einige Ergebnisse der lokalen Kohomologie Theorie und ihre Anwendung. Osnabrücker Schriften zur Mathematik, Reihe M, Heft 5, (1983).

[Br87]   Bronwnawell, D.: Bounds for the degrees in the Nullstellensatz. *Ann. Math.* **126** (1987), 577–591.

[BSV88]   Brumatti, P., Simis, A., Vasconcelos, W. V.: Normal Rees algebras. *J. Algebra* **112** (1988), 26–48.

[Bru86]   Bruns, W.: Length formulas for the local cohomology of exterior powers. *Math. Z.* **191** (1986), 145–158.

[BH93]   Bruns, W., Herzog, J.: *Cohen–Macaulay Rings*. Cambridge University Press, Cambridge, 1993.

[BVV97]   Bruns, W., Vasconcelos, W. V., Villarreal, R.: Degree bounds in monomial subrings. *Illinois J. Math.* **41** (1997), 341–353.

[BVe88]   Bruns, W., Vetter, U.: *Determinantal Rings*. Lecture Notes in Mathematics **1327**, Springer, Berlin Heidelberg New York, 1988.

[Buc85]   Buchberger, B.: Gröbner bases: An algorithmic method in polynomial ideal theory. *Recent Trends in Mathematical Systems Theory* (N.K. Bose, Ed.), D. Reidel, Dordrecht, 1985, 184–232.

[Can89]   Caniglia, L.: *Complejidad de Algoritmos en Geometria Algebraica Computacional*. Ph.D. Thesis, Universidad de Buenos Aires, 1989.

[CNR95]   Capani, A., Niesi, G., Robbiano, L.: CoCoA: A system for doing computations in commutative algebra. 1995. Available via anonymous ftp from `lancelot.dima.unige.it`.

[CGW91]   Char, B., Geddes, K., Watt, S.: *Maple V Library Reference Manual*. Springer, Berlin Heidelberg New York, 1991.

[Ch83]   Chvátal, V.: *Linear Programming*. W. H. Freeman & Company, New York, 1983.

[CHV96]   Conca, A., Herzog, J., Valla, G.: Sagbi bases with applications to blow–up algebras. *J. reine angew. Math.* **474** (1996), 113–138.

[CT91]   Conti, P., Traverso, C.: Computing the conductor of an integral extension. *Disc. App. Math.* **33** (1991), 61–72.

[CHV98]   Corso, A., Huneke, C., Vasconcelos, W. V.: On the integral closure of ideals. *Manuscripta Math.* **95** (1998), 331–347.

[CVV98]   Corso, A., Vasconcelos, W. V., Villarreal, R.: Generic gaussian ideals. *J. Pure & Applied Algebra* **125** (1998), 117–127.

[CLO92]   Cox, D., Little, J., O'Shea, D.: *Ideals, Varieties, and Algorithms*. Springer, Berlin Heidelberg New York, 1992.

[DST88] Davenport, J. H., Siret, Y., Tournier, E.: *Computer Algebra*. Academic Press, San Diego, 1988.

[DEP82] DeConcini, C., Eisenbud, D., Procesi, C.: *Hodge Algebras*. *Astérisque* **91**, Soc. Math. France, 1982.

[D1882] Dedekind, R., Weber, H.: Theorie der algebraischen Funktionen einer Veränderlichen. *J. reine angew. Math.* **92** (1882), 181–290.

[De95] De Dominicis, G.: *Algoritmi di decomposizione primaria in anelli polinomiali.* Tesi di Laurea, Università di Genova, 1995.

[DS91] Dickenstein, A. M., Sessa, C.: Duality methods for the membership problem. In: *Effective Methods in Algebraic Geometry*, Progress in Mathematics **94**, Birkhäuser, Boston Basel Berlin, 1991, 89–103.

[Der99] Derksen, H.: Computation of invariants of reductive groups. *Adv. Math.* **141** (1999), 366–384.

[DK97] Derksen, H., Kraft, H.: Constructive invariant theory. In: *Algèbre non commutative, groupes quantiques et invariants* (Reims, 1995), 221–224, Sémin. Congr. bf 2, Soc. Math. France, Paris, 1997.

[DG96] Doering, L. R., Gunston, T.: Algebras arising from planar bipartite graphs. *Comm. in Algebra* **24** (1996), 3589–3598.

[DGV98] Doering, L. R., Gunston, T., Vasconcelos, W. V.: Cohomological degrees and Hilbert functions of graded modules. *American J. Math.* **120** (1998), 493–504.

[EN62] Eagon, J., Northcott, D.G.: Ideals defined by matrices and a certain complex associated with them. *Proc. Royal Soc.* **269** (1962), 188–204.

[ES76] Eakin, P., Sathaye, A.: Prestable ideals. *J. Algebra* **41** (1976), 439–454.

[Ei80] Eisenbud, D.: Homological algebra on a complete intersection with an application to group representations. *Trans. Amer. Math. Soc.* **260** (1980), 35–64.

[Ei93] Eisenbud, D.: Open problems in computational Algebraic Geometry. In: *Computational Algebraic Geometry and Commutative Algebra*, Proceedings, Cortona 1991 (D. Eisenbud and L. Robbiano, Eds.), Cambridge University Press, 1993, 49–70.

[Ei95] Eisenbud, D.: *Commutative Algebra with a view toward Algebraic Geometry.* Springer, Berlin Heidelberg New York, 1995.

[EE73] Eisenbud, D., Evans, E. G.: Every algebraic set in *n*–space is the intersection of *n* hypersurfaces. *Invent. Math.* **19** (1973), 107–112.

[EiG84] Eisenbud, D., Goto, S.: Linear free resolutions and minimal multiplicities. *J. Algebra* **88** (1984), 89–133.

[EHV92] Eisenbud, D., Huneke, C., Vasconcelos, W.V.: Direct methods for primary decomposition. *Invent. Math.* **110** (1992), 207–235.

[EiS94] Eisenbud, D., Sturmfels, B.: Finding sparse systems of parameters. *J. Pure & Applied Algebra* **94** (1994), 143–157.

[EiS96] Eisenbud D., Sturmfels, B.: Binomial ideals. *Duke Math. J.* **84** (1996), 1–45.

[FGLM] Faugère, J. C., Gianni, P., Lazard, D., Mora, T.: Efficient computation of zero–dimensional Gröbner bases by change of ordering. *J. Symbolic Computation* **16** (1993), 329–344.

[Fer69] Ferrand, D.: Descent de la platitude par un homomorphisme fini. *C. R. Acad. Sc. Paris* **269** (1969), 946–949.

[GT89] Galligo, A., Traverso, C.: Practical determination of the dimension of an algebraic variety. In: *Computers in Mathematics* (E. Kaltofen and S. M. Watt, Eds.), Springer, Berlin Heidelberg New York, 1989, 46–52.

[GM91]   Gallo, G., Mishra, B.: Efficient algorithms and bounds for Wu–Ritt characteristic sets. In: *Effective Methods in Algebraic Geometry*, Progress in Mathematics **94**, Birkhäuser, Boston Basel Berlin, 1991, 119–142.

[GMT89]   Gianni, P., Miller, V., Trager, B.: Decomposition of algebras. Proceedings IS-SAC'88, Lecture Notes in Computer Science **358**, Springer, Berlin Heidelberg New York, 1989, 300–308.

[GM89]   Gianni, P., Mora, T.: Algebraic solution of systems of polynomial equations using Gröbner bases. Proceedings 5th AAEEC, Lecture Notes in Computer Science **356**, Springer, Berlin Heidelberg New York, 1989, 247–257.

[GTZ88]   Gianni, P., Trager, B., Zacharias, G.: Gröbner bases and primary decomposition of polynomial ideals. *J. Symbolic Computation* **6** (1988), 149–167.

[Gil72]   Gilmer, R.: *Multiplicative Ideal Theory*. Marcel Dekker, New York, 1972.

[Gil84]   Gilmer, R.: *Commutative Semigroup Rings*. University of Chicago Press, Chicago, 1984.

[GH91]   Giusti, M., Heintz, J.: Algorithmes - disons rapides - pour la décomposition d'une variété algébrique en composantes irréductibles et équidimensionelles. In: *Effective Methods in Algebraic Geometry*, Progress in Mathematics **94**, Birkhäuser, Boston Basel Berlin, 1991, 169–194.

[GTr94]   González-Vega, L., Trujillo, G.: Topological degree methods determining the existence of a real solution for a polynomial system of equations. Preprint, 1994.

[Got87]   Goto, S.: Integral closedness of complete intersection ideals. *J. Algebra* **108** (1987), 151–160.

[Go78]   Gotzmann, G.: Eine Bedingung für die Flachheit und das Hilbertpolynom eines graduierten Ringes. *Math. Z.* **158** (1978), 61–70.

[Gr89]   Gräbe, H.-G.: Moduln Über Streckungsringen. *Results in Mathematics* **15** (1989), 202–220.

[GS96]   Grayson, D., Stillman, M.: *Macaulay2*. 1996. Available via anonymous ftp from `math.uiuc.edu`.

[Gre89]   Green, M.: Restrictions of linear series to hyperplanes, and some results of Macaulay and Gotzmann. In: E. Ballico and C. Ciliberto, Eds., *Algebraic curves and projective geometry*, Lecture Notes in Mathematics **1389**, Springer, Berlin Heidelberg New York, 1989, 76–86.

[GPS95]   Greuel, G.-M., Pfister, G., Schoenemann, H.: *Singular*: A system for computation in algebraic geometry and singularity theory. 1995. Available for download from `http://www.singular.uni-kl.de`.

[Gro49]   Gröbner, W.: *Moderne Algebraische Geometrie*. Springer, Wien Innsbruck, 1949.

[Gro70]   Gröbner, W.: *Algebraische Geometrie*. Bibliographisches Institut, Mannheim, 1970.

[G67]   Grothendieck, A.: *Local Cohomology*. Lecture Notes in Mathematics **41**, Springer, Berlin Heidelberg New York, 1967.

[Gu72]   Gulliksen, T. H.: On the length of faithful modules over Artinian local rings. *Math. Scand.* **31** (1972), 78–82.

[Har66]   Hartshorne, R.: Connectedness of the Hilbert scheme. Publications Math. I.H.E.S. **29** (1966), 261–304.

[Har77]   Hartshorne, R.: *Algebraic Geometry*. Springer, Berlin Heidelberg New York, 1977.

[Hea87]   Hearn, A.: *Reduce User's Manual*. The RAND Corporation, Santa Monica, California, 1987. See www.dib.de/pub/reduce.

[Her26]   Hermann, G.: Die Frage der endlich vielen Schritte in der Theorie der Polynomideale. *Math. Annalen* **95** (1926), 736–788.

[He70]   Herzog, J.: Generators and relations of abelian semigroups and semigroup rings. *Manuscripta Math.* **3** (1970), 153–193.

[He78]   Herzog, J.: Ein Cohen–Macaulay Kriterium mit Anwendungen auf den Konormalenmodul und Differentialmodul. *Math. Z.* **163** (1978), 149–162.

[HK71]   Herzog, J., Kunz, E.: *Der kanonische Modul eines Cohen–Macaulay Rings. Lecture Notes in Mathematics* **238**, *Springer, Berlin Heidelberg New York, 1971.*

[HSV87]   Herzog, J., Simis, A., Vasconcelos, W. V.: On the canonical module of the Rees algebra and the associated graded ring of an ideal. *J. Algebra* **105** (1987), 285–302.

[HO98]   Hibi, T., Ohsugi, H.: Normal polytopes arising from finite graphs. *J. Algebra* **207** (1998), 409–426.

[Hil90]   Hilbert, D.: Über die Theorie der algebraischen Formen. *Math. Annalen* **36** (1890), 473–531.

[Hil93]   Hilbert, D.: Über die vollen Invariantensysteme. *Math. Annalen* **42** (1893), 313–370.

[HR74]   Hochster, M., Roberts, J. L.: Rings of invariants of reductive groups acting on regular rings are Cohen–Macaulay. *Adv. in Math.* **13** (1974), 115–175.

[Ho72]   Hochster, M.: Rings of invariants of tori, Cohen–Macaulay rings generated by monomials, and polytopes. *Annals of Math.* **96** (1972), 318–337.

[Hre94]   Hreinsdóttir, F.: A case where choosing a product order makes the calculation of a Gröbner basis much faster. *J. Symbolic Computation* **18** (1994), 373–378.

[Hu82]   Huneke, C.: On the finite generation of symbolic blow–ups. *Math. Z.* **179** (1982), 465–472.

[Hu87]   Huneke, C.: The primary components of and integral closures of ideals in 3-dimensional regular local rings. *Math. Annalen* **275** (1987), 617–635.

[Hu92]   Huneke, C.: Uniform bounds in Noetherian rings. *Invent. Math.* **107** (1992), 203–223.

[HSV89]   Huneke, C., Simis, A., Vasconcelos, W. V.: Reduced normal cones are domains. *Amer. Math. Soc., Contemporary Mathematics* **88** (1989), 95–101.

[HU93]   Huneke, C., Ulrich, B.: General hyperplane sections of algebraic varieties. *J. Algebraic Geometry* **2** (1993), 487–505.

[Jac74]   Jacobson, N.: *Basic Algebra* I. W. H. Freeman, San Francisco, 1974.

[JS92]   Jenks, R. D., Sutor, R. S.:*AXIOM: The Scientific Computation System.* Springer, Berlin Heidelberg New York, 1992.

[KS95]   Kalkbrener, M., Sturmfels, B.: Initial complexes of prime ideals. *Adv. in Math.* **116** (1995), 365–376.

[Kal92]   Kaltofen, E.: Polynomial factorization 1987–1991. Proceedings Latin'92, Lecture Notes in Computer Science **583**, Springer, Berlin Heidelberg New York, 1992, 294–313.

[KK88]   Kandri-Rody, A., Kapur, D.: Computing a Gröbner basis of a polynomial ideal over an Euclidean domain. *J. Symbolic Computation* **6** (1988), 37–57.

[Kap66]   Kaplansky, I.: *Fields and Rings.* University of Chicago Press, Chicago, 1966.

[Kap74]   Kaplansky, I.: *Commutative Rings.* University of Chicago Press, Chicago, 1974.

[KL92]   Kapur, D., Lakshman, Y. N.: Elimination methods: an introduction. In: *Symbolic and Numerical Computation for Artificial Intelligence* (B.R. Donald, D. Kapur and J.L.Mundy, Eds.) Academic Press, San Diego, 1992, 45–87.

[KM89]   Kapur, D., Madlener, K.: A completion procedure for computing a canonical basis for a *k*-subalgebra. In: *Computers and Mathematics*, Springer, Berlin Heidelberg New York, 1989, 1–11.

[Ke79]   Kempf, G. R.: The Hochster–Roberts theorem of invariant theory. *Michigan Math. J.* **26** (1979), 19–32.

[Ke91]   Kempf, G. R.: More on computing invariants, Algebraic Geometry. Proceedings of the US–USSR Symposium (S. Block, I. V. Dolgachev and W. Fulton, Eds.), Lecture Notes in Mathematics **1479**, Springer, Berlin Heidelberg New York, 1991, 87–89.

[Ko88]   Kollár, J.: Sharp effective Nullstellensatz. *J. Amer. Math. Soc.* **1** (1988), 963–975.

[Kr84]   Kraft, H.: *Geometrische Methoden in der Invariententheorie.* Vieweg, Wiesbaden, 1984.

[KLo91]   Krick, T., Logar, A.: An algorithm for the computation of the radical of an ideal in the ring of polynomials. In: Proceedings 9th AAEEC, Lecture Notes in Computer Science **539**, Springer, Berlin Heidelberg New York, 1991, 195–205.

[Kro82]   Kronecker, L.: Grundzüge einer arithmetischen Theorie der algebraischen Größen. *J. reine angew. Math.* **92** (1882), 1–122.

[Kun85]   Kunz, E.: *Introduction to Commutative Algebra and Algebraic Geometry.* Birkhäuser, Boston Basel Berlin, 1985.

[Kun86]   Kunz, E.: *Kähler Differentials.* Vieweg, Wiesbaden, 1986.

[LaL91]   Lakshman, Y. N., Lazard, D.: On the complexity of zero–dimensional algebraic systems. In: *Effective Methods in Algebraic Geometry*, Progress in Mathematics **94**, Birkhäuser, Boston Basel Berlin, 1991, 217–225.

[Laz77]   Lazard, D.: Algèbre linéaire sur $K[x_1, \ldots, x_n]$ et élimination. *Bull. Soc. Math. France* **105** (1977), 165–190.

[Laz81]   Lazard, D.: Résolution des systèmes d'equations algébriques. *Theoretical Comp. Sci.* **15** (1981), 77–110.

[Len92]   Lenstra, Jr., H. W.: Algorithms in algebraic number theory. *Bull. Amer. Math. Soc.* **26** (1992), 211–244.

[Lev90]   Levin, G.: Personal communication. 1990.

[Lic66]   Lichtenbaum, S.: On the vanishing of Tor in regular local rings. *Illinois J. Math.* **10** (1966), 220–226.

[Lip69]   Lipman, J.: On the Jacobian ideal of the module of differentials. *Proc. Amer. Math. Soc.* **21** (1969), 422–426.

[LS81]   Lipman, J., Sathaye, A.: Jacobian ideals and a theorem of Briançon–Skoda. *Michigan Math. J.* **28** (1981), 199–222.

[Log89]   Logar, A.: A computational proof of the Noether normalization lemma. In: Proceedings 6th AAEEC, Lecture Notes in Computer Science **357**, Springer, Berlin Heidelberg New York, 1989, 259–273.

[Mac16]   Macaulay, F. S.: *The Algebraic Theory of Modular Systems.* Cambridge University Press, Cambridge, 1916.

[Mac27]   Macaulay, F. S.: Some properties of enumeration in the theory of modular systems. *Proc. London Math. Soc.* **26** (1927), 531–555.

[Mat80]   Matsumura, H.: *Commutative Algebra.* Benjamin/Cummings, Reading, 1980.

[MMe82]   Mayr, E., Meyer, A.: The complexity of the word problem for commutative semi-groups and polynomial ideals. *Adv. in Math.* **46** (1982), 305–329.

[Mil92]   Milne, P. S.: On the solutions of a set of polynomial equations. In: *Symbolic and Numerical Computation for Artificial Intelligence* (B.R. Donald, D. Kapur and J.L.Mundy, Eds.) Academic Press, San Diego, 1992, 89–101.

[Mis93]   Mishra, B.: *Algorithmic Algebra.* Springer, Berlin Heidelberg New York, 1993.

[Mol91]   Möller, H.M.: Computing syzygies à la Gauss–Jordan. In: *Effective Methods in Algebraic Geometry*, Progress in Mathematics **94**, Birkhäuser, Boston Basel Berlin, 1991, 335–345.

[MM86] Möller, H. M., Mora, T.: New constructive methods in classical ideal theory. *J. Algebra* **100** (1986), 138–178.

[MRo88] Mora, T., Robbiano, L.: The Gröbner fan of an ideal. *J. Symbolic Computation* **6** (1988), 183–208.

[Mor92] Morgan, A. P.: Polynomial continuation and its relationship to the symbolic reduction of polynomial systems. In: *Symbolic and Numerical Computation for Artificial Intelligence* (B.R. Donald, D. Kapur and J.L.Mundy, Eds.) Academic Press, San Diego, 1992, 23–44.

[Nag62] Nagata, M.: *Local Rings*. Interscience, New York, 1962.

[Nag64] Nagata, M.: *Lectures on the Fourteenth Problem of Hilbert*. Tata Institute, Bombay, 1964.

[Noe21] Noether, E.: Idealtheorie in Ringbereichen. *Math. Annalen* **83** (1921), 24–66.

[NVa93] Noh, S., Vasconcelos, W. V.: The S_2–closure of a Rees algebra. *Results in Mathematics* **23** (1993), 149–162.

[Nor63] Northcott, D. G.: A homological investigation of a certain residual ideal. *Math. Annalen* **150** (1963), 99–110.

[NR54] Northcott, D. G., Rees, D.: Reductions of ideals in local rings. *Proc. Camb. Phil. Soc.* **50** (1954), 145–158.

[Ooi82] Ooishi, A.: Castelnuovo's regularity of graded rings and modules. *Hiroshima Math. J.* **12** (1982), 627–644.

[Ost66] Ostrowski, A. M.: *Solutions of Equations and Systems of Equations*. Prentice Hall, New York, 1966.

[Pes66] Peskine, C.: Une généralisation du main theorem de Zariski. *Bull. Sc. Math.* (2) **90** (1966), 119–127.

[PSz74] Peskine, C., Szpiro, L.: Liaison des variétés algébriques. *Invent. Math.* **26** (1974), 271–302.

[Ra079] Rao, A. P.: Liaison among curves in $\mathbb{P}^3$. *Invent. Math.* **50** (1979), 205–217.

[RR78] Ratliff, Jr., L. J., Rush, D. E.: Two notes on reductions of ideals. *Indiana Univ. Math. J.* **27** (1978), 929–934.

[Ray70] Raynaud, M.: *Anneaux Locaux Henséliens*. Lecture Notes in Mathematics **169**, Springer, Berlin Heidelberg New York, 1970.

[Ree58] Rees, D.: On a problem of Zariski. *Illinois J. Math.* **2** (1958), 145–149.

[Rit50] Ritt, J. F.: *Differential Algebra*. AMS Colloquium Publ. **33**, New York, 1950.

[Ro86] Robbiano, L.: On the theory of graded structures. *J. Symbolic Computation* **2** (1986), 139–170.

[Ro88] Robbiano, L.: Introduction to the theory of Gröbner bases. Queen's Papers in Pure and Applied Mathematics, 1988.

[Ro85] Robbiano, L.: Term orderings on the polynomial ring. In: Proceedings EUROCAL 85, Lecture Notes in Computer Science **204**, Springer, Berlin Heidelberg New York, 1985, 513–517.

[RoS90] Robbiano, L., Sweedler, M.: Subalgebra bases. In: *Commutative Algebra*, Proceedings, Salvador 1988 (W. Bruns and A. Simis, Eds.), Lecture Notes in Mathematics **1430**, Springer, Berlin Heidelberg New York, 1990, 61–87.

[PRo87] Roberts, P.: Le théorème d'intersection. *C. R. Acad. Sc. Paris* **304** (1987), 177–180.

[Sal76] Sally, J. D.: Bounds for numbers of generators for Cohen–Macaulay ideals. *Pacific J. Math.* **63** (1976), 517–520.

[Sal78] Sally, J. D.: *Number of Generators of Ideals in Local Rings*. Lecture Notes in Pure & Applied Math. **35**, Marcel Dekker, New York, 1978.

[Sam64]   Samuel, P.: Anneaux gradués factoriels et modules réflexifs. *Bull. Soc. Math. France* **92** (1964), 237–249.

[ST95]   Scavo, T. R., Thoo, J. B.: On the geometry of Halley's method. *Amer. Math. Monthly* **102** (1995), 417–426.

[ScS75]   Scheja, G., Storch, U.: Über Spurfunktionen bei vollständigen Durchschnitten. *J. reine angew. Math.* **278** (1975), 174–190.

[Sc80]   Schreyer, F.-O.: *Die Berechnung von Syzygien mit dem verallgemeinerten Weierstrass'schen Divisionssatz.* Diplom Thesis, University of Hamburg, 1980.

[Sch05]   Schur, I.: Zur Theorie der Vertauschbären Matrizen. *J. reine angew. Math.* **130** (1905), 66–76.

[Sei66]   Seidenberg, A.: Derivations and integral closure. *Pacific J. Math.* **16** (1966), 167–173.

[Sei74]   Seidenberg, A.: Constructions in algebra. *Trans. Amer. Math. Soc.* **197** (1974), 273–313.

[Sei75]   Seidenberg, A.: Construction of the integral closure of a finite integral domain II. *Proc. Amer. Math. Soc.* **52** (1975), 368–372.

[Sei84]   Seidenberg, A.: On the Lasker–Noether decomposition theorem. *American J. Math.* **106** (1984), 611–638.

[Ser65]   Serre, J.-P.: *Algèbre Locale–Multiplicités.* Lecture Notes in Mathematics **11**, Springer, Berlin Heidelberg New York, 1965.

[ShS86]   Shannon, D., Sweedler, M.: Using Gröbner bases to determine algebra membership, split surjective algebra homomorphisms determine birational equivalence. *J. Symbolic Computation* **6** (1986), 267–273.

[ShY96]   Shimoyama, T., Yokoyama, K.: Localization and primary decomposition of polynomial ideals. *J. Symbolic Computation* **22** (1996), 247–277.

[Sim96]   Simis, A.: Effective computation of symbolic powers by jacobian matrices. *Comm. in Algebra* **24** (1996), 3561–3565.

[ST88]   Simis, A., Trung, N. V.: Divisor class group of ordinary and symbolic blow–ups. *Math. Z.* **198** (1988), 479–491.

[SVV94]   Simis, A., Vasconcelos, W. V., Villarreal, R.: On the ideal theory of graphs. *J. Algebra* **167** (1994), 389–416.

[SVV98]   Simis, A., Vasconcelos, W. V., Villarreal, R.: The integral closure of subrings associated to graphs. *J. Algebra* **199** (1998), 281–289.

[Sjo73]   Sjödin, G.: On filtered modules and their associated graded modules. *Math. Scand.* **33** (1973), 229–249.

[Somb97] Sombra, M.: Bounds for the Hilbert function of polynomial ideals and for the degrees in the Nullstellensatz. *J. Pure & Applied Algebra* **117-118** (1997), 565–599.

[Spe97]   Spear, D.: A constructive approach to commutative ring theory. In: *1977 MACSYMA Users Conference*, Proceedings, NASA CP-2012, 369–376.

[Spr77]   Springer, T. A.: *Invariant Theory.* Lecture Notes in Mathematics **585**, Springer, Berlin Heidelberg New York, 1977.

[Sta78]   Stanley, R. P.: Hilbert functions of graded algebras. *Adv. in Math.* **28** (1978), 57–83.

[Sta91]   Stanley, R. P.: On the Hilbert function of a graded Cohen–Macaulay domain. *J. Pure & Applied Algebra* **73** (1991), 307–314.

[Sto68]   Stolzenberg, G.: Constructive normalization of an algebraic variety. *Bull. Amer. Math. Soc.* **74** (1968), 595–599.

[Stc72]   Storch, U.: Bemerkung zu einem Satz von M. Kneser. *Arch. Math.* **23** (1972), 403–404.

[SVo86]   Stückrad, J., Vogel, W.: *Buchsbaum Rings and Applications*. Springer, Vienna New York, 1986.

[Stu93]   Sturmfels, B.: Sparse elimination theory. In: *Computational Algebraic Geometry and Commutative Algebra*, Proceedings, Cortona 1991 (D. Eisenbud and L. Robbiano, Eds.), Cambridge University Press, 1993, 264–298.

[Stu91]   Sturmfels, B.: Gröbner bases of toric varieties. *Tôhoku Math. J.* **43** (1991) 249–261.

[Stu93]   Sturmfels, B.: *Algorithms in Invariant Theory*. Springer, Vienna New York, 1993.

[Stu95]   Sturmfels, B.: *Gröbner Bases and Convex Polytopes*. AMS University Lecture Series **8**, Providence, 1995.

[SWh91]   Sturmfels, B., White, N.: Computing combinatorial decompositions of rings. *Combinatorica* **11** (1991), 275-293.

[STV95]   Sturmfels, B., Trung, N. V., W. Vogel, W.: Bounds on degrees of projective schemes. *Math. Annalen* **302** (1995), 417–432.

[Swe95]   Sweedler, M.: Sturm, winding number and degree of maps. Preprint, 1995.

[Tat70]   Tate, J.: The different and the discriminant. *Appendix* to: B. Mazur and L. Roberts, Local Euler characteristics, *Invent. Math.* **9** (1970), 201-234.

[Tay66]   Taylor, D.: *Ideals generated by monomials in an R–sequence*. Ph.D. Thesis, University of Chicago, 1966.

[Tra86]   Traverso, C.: A study on algebraic algorithms: the normalization. Rend. Semin. Mat. Univ. Politec. Torino (fasc. spec.) *Algebraic Varieties of Small Dimension*, 1986, 111–130.

[Tru87]   Trung, N. V.: Reduction exponent and degree bound for the defining equations of graded rings. *Proc. Amer. Math. Soc.* **101** (1987), 229–236.

[Ulr92]   Ulrich, B.: *Linkage Theory and the Homology of Noetherian Rings*. Lecture notes, 1992.

[UVp]   Ulrich, B., Vasconcelos, W. V.: On the complexity of the integral closure. *Trans. Amer. Math. Soc.*, to appear.

[Val79]   Valla, G.: On form rings which are Cohen–Macaulay. *J. Algebra* **58** (1979), 247–250.

[Val81]   Valla, G.: Generators of ideals and multiplicities. *Comm. in Algebra* **9** (1981), 1541–1549.

[Vas86]   Vasconcelos, W. V.: What is a prime ideal? Atas IX Escola de Algebra, IMPA, Rio de Janeiro, 1986, 141–149.

[Vas89]   Vasconcelos, W. V.: Symmetric algebras and factoriality. In: *Commutative Algebra* (M. Hochster, C. Huneke and J. D. Sally, Eds.), MSRI Publications **15**, Springer, Berlin Heidelberg New York, 1989, 467–496.

[Vas93]   Vasconcelos, W. V.: Constructions in commutative algebra. In: *Computational Algebraic Geometry and Commutative Algebra*, Proceedings, Cortona 1991 (D. Eisenbud and L. Robbiano, Eds.), Cambridge University Press, 1993, 151–197.

[Vas91]   Vasconcelos, W. V.: Jacobian matrices and constructions in algebra. In: Proceedings 9th AAEEC, Lecture Notes in Computer Science **539**, Springer, Berlin Heidelberg New York, 1991, 48–64.

[Va91a]   Vasconcelos, W. V.: Computing the integral closure of an affine domain. *Proc. Amer. Math. Soc.* **113** (1991), 633–638.

[Vas92]   Vasconcelos, W. V.: The top of a system of equations. *Bol. Soc. Mat. Mexicana* (issue dedicated to José Adem) **37** (1992), 549–556.

[Vas94]   Vasconcelos, W. V.: *Arithmetic of Blowup Algebras*. London Math. Soc., Lecture Note Series **195**, Cambridge University Press, Cambridge, 1994.

[Va94a]   Vasconcelos, W. V.: The integral closure. In: *Commutative Algebra*, Proceedings, Trieste 1992 (G. Valla, V.T. Ngo and A. Simis, Eds.), World Scientific, Singapore, 1994, 263–290.

[Vas97]   Vasconcelos, W. V.: Testing flatness and torsionfree morphisms. *J. Pure & Applied Algebra* **122** (1997), 313–321.

[Vas96]   Vasconcelos, W. V.: The reduction number of an algebra. *Compositio Math.* **104** (1996), 189–197.

[Vas98]   Vasconcelos, W. V.: The homological degree of a module. *Trans. Amer. Math. Soc.* **350** (1998), 1167–1179.

[Vaz91]   Vazquez, A. T.: Rational desingularization of a curve defined over a finite field. In: *Number Theory, New York Seminar 1989–1990* (D. V. Chudnovsky *et al.* Eds.), Springer, Berlin Heidelberg New York, 1991, 229–250.

[Vil90]   Villarreal, R.: Cohen–Macaulay graphs. *Manuscripta Math.* **66** (1990), 277–293.

[Vil95]   Villarreal, R.: Rees algebras of edge ideals. *Comm. in Algebra*, **23** (1995), 3513–3520.

[Vil01]   Villarreal, R.: *Monomial Algebras*. Monographs and Textbooks in Pure and Applied Mathematics **238**, Marcel Dekker, New York, 2001.

[War70]   van der Waerden, B. L.: *Algebra*. Vols. I,II, Ungar, New York, 1970.

[Wan92]   Wang, D.: Irreducible decomposition of algebraic varieties via characteristic sets and Gröbner bases. *Computer Aided Geometric Design* **9** (1992), 471–484.

[Wat87]   Watanabe, J.: m–full ideals. *Nagoya Math. J.* **106** (1987), 101–111.

[Wa92]   Watkins, A.: *Hilbert functions of face rings arising from graphs*. M.S. Thesis, Rutgers University, 1992.

[Wei94]   Weibel, C.: *An Introduction to Homological Algebra*. Cambridge University Press, Cambridge, 1994.

[Wol96]   Wolfram, S.: *The Mathematica Book*. Wolfram Media and Cambridge University Press, 1996.

[Wu78]   Wu, W.-T.: On the decision problem and the mechanization of theorem proving in elementary geometry. *Scientia Sinica* **21** (1978), 157–179.

[Zar65]   Zariski, O.: Studies in equisingularities, I. *American J. Math.* **87** (1965), 507–536.

[ZS60]   Zariski, O., Samuel, P.: *Commutative Algebra*. Vol. I, Van Nostrand, Princeton, 1960.

[An74]   Anderson, B. D. O.: Orthogonal decomposition defined by a pair of skew-symmetric forms. *Linear Algebra and its Applications* **8** (1974), 91–93.

[BK98]   Bruns, W., Koch, R.: *Normaliz*: A program to compute normalizations of semigroups. Available by anonymous ftp from ftp.mathematik.Uni-Osnabrueck.DE/pub/osm/kommalg/software/

[Con03]   Conca, A.: Reduction numbers and initial ideals. *Proc. Amer. Math. Soc.* **131** (2003), 1015–1020.

[CLO98]   Cox, D., Little, J., O'Shea, D.: *Using Algebraic Geometry*. Springer, Berlin Heidelberg New York, 1998.

[EGSS0]   Eisenbud, D., Grayson, D., Stillman, M., Sturmfels, B.: *Computations in Algebraic Geometry with Macaulay 2*. Springer, Berlin, 2002. Geometry — Achievements and Perspectives. *J. Symbolic Computation* **30** (2000), 253–289.

[GKi0]   Galligo, A., Kwieciński, M.: Flatness and fibred powers over smooth varieties. *J. Algebra* **232** (2000), 48-63.

[Giu84]   Giusti, M.: Some effectivity problems in polynomial ideal theory. In: *EUROSAM* 1984, Lecture Notes in Computer Science **174**, Springer, Berlin Heidelberg New York, 1984, 159–171.

[GrP02]   Greuel, G.-M., Pfister, G.: A **Singular** Introduction to Commutative Algebra. Springer, Belin, 2002.

[Kno89]   Knop, F.: Der kanonische Modul eines Invariantenringes. *J. Algebra* **127** (1989), 40–54.

[KrR00]   Kreuzer, M., Robbiano, L.: *Computational Commutative Algebra.* 1. Springer, Berlin, 2000.

[Kwi98]   Kwieciński, M.: Flatness and fibred powers. *Manuscripta Math.* **97** (1998), 163–173.

[Mts98]   Matsumoto, R.: Simple computation of the radical of an ideal in positive characteristic. Preprint, 1998.

[Mts00]   Matsumoto, R.: On computing the integral closure. *Comm. in Algebra* **28** (2000), 401–405.

[Noe50]   Noether, E.: Idealdifferentiation und Differente. *J. reine angew. Math.* **188** (1950), 1–21.

[MMa84]   Möller, H. M., Mora, T.: Upper and lower bounds for the degree of Gröbner bases. In: *EUROSAM* 1984, Lecture Notes in Computer Science **174**, Springer, Berlin Heidelberg New York, 1984, 172–183.

[Sti90]   Stillman, M.: Methods for computing in algebraic geometry and commutative algebra. *Acta Applicandae Mathematicae* **21** (1990), 77–103.

[STa99]   Stillman, M., Tsai, H.: Using SAGBI bases to compute invariants. *J. Pure & Applied Algebra* **139** (1999), 285–304.

[Stu01]   Sturmfels, B: Gröbner bases of abelian matrix groups. *Contemp. Math.* **286** (2001), 141–143.

[Tru03]   Trung, N. V.: Constructive characterization of the reduction numbers. *Compositio Math.* **137** (2003), 99–113.

[Vas00]   Vasconcelos, W. V.: Divisorial extensions and the computation of integral closures. *J. Symbolic Computation* **30** (2000), 595–604.

[Yao02]   Yao, Y.: Primary decomposition: compatibility, independence and linear growth. *Proceedings Amer. Math. Soc.* **130** (2002), 1629–1637.

Index

Printing: Strauss GmbH, Mörlenbach
Binding: Schäffer, Grünstadt